Satellite Communication Systems Engineering

Satellite Communication Systems Engineering

Wilbur L. Pritchard

Joseph A. Sciulli

Prentice-Hall, Inc., Englewood Cliffs, N.J. 07632

Library of Congress Cataloging-in-Publication Data

Pritchard, Wilbur L.
  Satellite communication systems engineering.

  Bibliography: p.
  Includes index.
  1. Artificial satellites in telecommunication—
Systems engineering. I. Sciulli, Joseph A. II. Title.
TK5104.P74 1986   621.38′0422   85-19434
ISBN 0-13-791245-5

Editorial/production supervision and interior design: Jane Zalenski
Cover design: 20/20 Services, Inc.
Manufacturing buyer: Gordon Osbourne

© 1986 by Prentice-Hall, Inc., Englewood Cliffs, New Jersey 07632

All rights reserved. No part of this book may be
reproduced, in any form or by any means,
without permission in writing from the publisher.

Printed in the United States of America

10 9 8 7 6 5 4 3

ISBN   0-13-791245-5   025

Prentice-Hall International (UK) Limited, *London*
Prentice-Hall of Australia Pty. Limited, *Sydney*
Prentice-Hall Canada Inc., *Toronto*
Prentice-Hall Hispanoamericana, S.A., *Mexico*
Prentice-Hall of India Private Limited, *New Delhi*
Prentice-Hall of Japan, Inc., *Tokyo*
Prentice-Hall of Southeast Asia Pte. Ltd., *Singapore*
Editora Prentice-Hall do Brasil, Ltda., *Rio de Janeiro*
Whitehall Books Limited, *Wellington, New Zealand*

The authors dedicate this book to the working satellite communications systems engineer.

Contents

PREFACE xi

1 INTRODUCTION TO SATELLITE COMMUNICATIONS 1

    1.0   Historical Background 1
    1.1   Basic Concepts of Satellite Communications 9
    1.2   Communications Networks and Services 23
    1.3   Comparison of Transmission Technologies 24
    1.4   Growth of Satellite Communications 27

2 ORBITS 29

    2.0   General Considerations 29
    2.1   Foundation of the Theory 30
    2.2   Conservation of Energy and Momentum 31
    2.3   Kepler's Laws 32
    2.4   The Satellite Orbit in Space 35
    2.5   Relations Among Orbit Parameters 36
    2.6   The Geostationary Orbit 37
    2.7   Orbital Transfers 39
    2.8   Transfer Between Coplanar Elliptic Orbits 43
    2.9   Change in Longitude 44
    2.10   Orbital Perturbations 45
    2.11   Other Orbits for Satellite Communication 55

3 THE GEOMETRY OF THE GEOSTATIONARY ORBIT 57

    3.0   General Considerations 57
    3.1   Basic Geometry 57
    3.2   Satellite Coordinates 61
    3.3   Eclipse Geometry 62
    3.4   Sun Interference 68
    3.5   Satellite Topocentric Coordinates and Ground Traces 73
    3.6   The Nonspherical Earth 77
    3.7   The Apparent Position of an Almost Geostationary Satellite 79
    3.8   The Track of an Almost Synchronous Satellite 80

4 LAUNCH VEHICLES AND PROPULSION 82

    4.0   Introduction 82
    4.1   Launch Missions 83
    4.2   The Rocket Equation 83
    4.3   Powered Flight 90
    4.4   Injection into Final Orbit and Orbital Maneuvers 96
    4.5   Mission Possibilities 98
    4.6   Low-Thrust Variations 99

5 SPACECRAFT 109

    5.0   Introduction 109
    5.1   Subsystems 110
    5.2   Structural Subsystem 110
    5.3   Attitude Control 112
    5.4   Primary Power 119
    5.5   Thermal Subsystems 125
    5.6   Propulsion Subsystem (RCS) 127
    5.7   Telemetry, Tracking and Command 131
    5.8   Estimating the Mass of Communications Satellites 134
    5.9   System Reliability and Design Lifetime 139

6 THE RF LINK 145

    6.0   General Considerations 145
    6.1   Noise 146

Contents ix

    6.2   The Basic RF Link 147
    6.3   Three Special Types of Limits on Link Performance 150
    6.4   Satellite Links (Up and Down) 152
    6.5   Composite Performance 154
    6.6   Optimization of the RF Link 158
    6.7   Noise Temperature 162
    6.8   Antenna Temperatures 166
    6.9   Overall System Temperature 171
    6.10   Rain Attenuation Model 173

7 MODULATION AND MULTIPLEXING 182

    7.0   Introduction 182
    7.1   Source Signals: Voice, Data, Video 184
    7.2   Analog Transmission Systems 187
    7.3   Digital Transmission Systems 203
    7.4   Television Transmission 231

8 MULTIPLE ACCESS SYSTEMS 241

    8.0   Introduction 241
    8.1   Systems Engineering Considerations 244
    8.2   Definitions 245
    8.3   FDMA Systems 246
    8.4   TDMA Systems 260
    8.5   Beam Switching and Satellite-Switched TDMA 269
    8.6   Code-Division Multiple Access 270
    8.7   Comparison of Multiple Access Techniques 274

9 TRANSPONDERS 279

    9.0   Introduction 279
    9.1   Function of the Transponder 279
    9.2   Transponder Implementations 282
    9.3   Transmission Impairments 291
    9.4   Using Manufacturers Data and Experimental Results 300
    9.5   Other Aspects of Transponders 305

10 EARTH STATIONS 308

    10.0   Introduction 308
    10.1   Transmitters 310

	10.2	Receivers 314
	10.3	Antennas 320
	10.4	Tracking Systems 337
	10.5	Terrestrial Interface 340
	10.6	Primary Power 341
	10.7	Test Equipment 342

11 INTERFERENCE 348

	11.0	Introduction 348
	11.1	Calculation of C/I for a Single Interfering Satellite 349
	11.2	Calculation of C/I for Multiple Interfering Satellites 352
	11.3	Two Types of Satellites 353
	11.4	Homogeneous Satellites 356
	11.5	Protection Ratio 358

12 SPECIAL PROBLEMS IN SATELLITE COMMUNICATIONS 366

	12.0	Background 366
	12.1	Echo Control 367
	12.2	Satellite Data Communications 372
	12.3	Orbital Variations and Digital Network Synchronization 382

INDEX   **387**

Preface

This book is the outgrowth of a series of seminars and short courses taught by the authors in the years between 1974 and 1984. As the notes for these courses improved and matured, it became clear that they could be expanded into a textbook that would fill a conspicuous need. We saw that there was no single book that addressed the problems of the systems engineer. Such a book should cover concepts and calculations used across the entire field of satellite communications at a level sufficiently sophisticated and profound to help make major systems planning decisions. This book is the result. We have purposely avoided the highly theoretical analyses found in the literature on special disciplines. We hope we have met the needs of the working engineer.

The objective of our book is to provide the tools necessary for the calculation of such basic parameters in a satellite communication system as channel capacities, picture quality, signal to noise ratios, bit error rate, earth station antenna sizes, spacecraft mass, spacecraft primary power, and orbital velocity requirements; in short, all those elements necessary for system planning. These procedures are inherently quantitative and use mathematics, but for these purposes nothing beyond the normal undergraduate level. Our book is written for systems engineers with an academic background of an undergraduate degree in engineering, physics, or applied sciences. We have tried to include the practical methods of analysis used by working engineers to make numerical computations, but we have avoided abstruse generalizations and all advanced mathematics.

One of the charming aspects of satellite communication for technical people is the wide variety of scientific and engineering disciplines employed.

This book is comprehensive in that it attempts to cover all these disciplines to a sufficient depth to aid in system planning, project management, and similar work. It deals with almost all the aspects of satellite communication likely to be significant in system planning. Many of the practical approaches and calculations are not to be found elsewhere, and we hope they will enhance the book's utility.

The authors would like particularly to thank Harley Radin for his contribution of Chapter 11. They also thank Kathleen Pritchard for preparing the index and glossary and supervising the tedious task of proofreading, and Lee Sciulli for help in preparation and editing. In addition, they would like to thank the following for constructive comments and assistance in reading and criticizing the manuscript: Kenneth Seidelmann of the U.S. Naval Observatory, Robert Nelson of the University of Maryland, and Douglas A. Kerr, telecommunications consultant to Satellite Systems Engineering.

Naturally the authors themselves accept full responsibility for any errors.

WILBUR L. PRITCHARD
Satellite Systems Engineering, Inc.

JOSEPH A. SCIULLI
Telecommunications Techniques Corporation

Organization of the Book

Communication satellite systems engineering can be divided into several widely disparate fields, that is, the design of the communications transponder, the space platform on which to carry it, a launch system for placing it into orbit, the earth stations for communicating and the interconnection lines with terrestrial systems. It is impossible for a systems engineer to be skilled in all these difficult subjects but, at the same time, it is necessary for him to appreciate the compromises among conflicting elements. Each of the chapters of this book tries to give the systems engineer a broad understanding of a different element of the system and its relation to the remainder of the system.

Following the overview in Chapter 1, Chapter 2 addresses the physics of orbits. In dealing with satellites, approximately three-quarters of the space system cost is absorbed in placing the satellite in orbit and providing a congenial atmosphere in which the communications equipment can function. Orbital physics or, as it is usually called, celestial mechanics, is classical Newtonian physics at its best. Although the far reaches of the subject can be extremely abstruse and mathematical, a satisfactory understanding of major system trade-offs is not difficult to achieve. In the same chapter, the closely related problems of orbital geometry are also addressed. The geometric interrelations amongst the satellite, earth, sun and moon can be, although straightforward, surprisingly complicated. The results are not widely available in communications literature, and Chapter 2 provides a good collection of working mathematics to deal with this class of problem. Not only is the orbital physics of two-body problems discussed, but also the perturbations, both gravitational and other, that prevent a satellite from remaining in exactly the prescribed orbit are explained. They are important because they create substantial requirements for propulsion and concomitant complications and costs.

Chapter 3 focuses more specifically on the geometry and characteristics of the geostationary orbit. Subjects such as satellite positioning, sun outage and eclipse, and the effects of the nonspherical earth and the moon are discussed.

Chapter 4 deals with the related problems of launching satellites into orbit. Indeed, the launch mission itself is a succession of powered flight and orbital maneuvers, the understanding of which depends on the material in Chapter 2. A brief introduction to the physics and engineering of rocket engines is included in this chapter and several simple and useful rocket equations are presented. The characteristics of some of the most important launch vehicles are tabulated and discussed briefly, more by way of example and illustration than attempting to keep the reader up to date with launch vehicle technology.

Chapter 5 is a discussion of integrated spacecraft design. This subject is about as complicated and varied a discipline as one can find in engineering and it is possible only to highlight the major problems and subsystems. Particular attention is given to the choice of attitude control methods and the basic design of the primary power subsystem because they characterize the whole spacecraft and determine its cost and utility. Methods for estimating primary power and spacecraft mass in orbit are given. These methods are useful both in predicting cost and in making the complicated and expensive arrangements for launch vehicles.

Chapter 6 is the first of the communications chapters. As already mentioned, the communications problem itself can be divided neatly into two parts, that is, the radio frequency link between the spacecraft and the earth terminals and the connections from the earth terminals, through the terrestrial network, to the user. We have chosen in the presentation of the system theory to divide the problem in half at the computation of carrier-to-noise ratio. That is, given the characteristics of the transmitters, receivers, antennas, geometry, propagation, media, etc., what overall carrier-to-noise ratio is available for communications purposes? How does one go about calculating it from the point of view of the communications systems engineer?

Chapter 7 is entitled "Modulation and Multiplexing." It covers the principal characteristics of frequency modulation, digital phase shift keying, pulse code modulation and conventional terrestrial multiplexing systems; notably, frequency division multiplexing and time division multiplexing. The purpose of this chapter is two-fold: first, to keep the book complete so that the reader can get this material without reference to other books; and more importantly, to identify and highlight those elements of conventional modulation theory that are so important to the satellite communication problem. Much of the

Organization of the Book

material in Chapter 7 should be familiar, at least in other contexts, to the experienced telecommunications engineer.

Chapter 8, entitled "Multiple Access Systems," is a very important chapter. As previously mentioned, the outstanding advantage of a satellite is geometric, but it must be exploited. We focus on a description of two true multiple access techniques in which transponders are used in common by many earth stations. In frequency division multiple access (FDMA), separate carrier frequencies are used. In time division multiple access (TDMA), separate time slots are used. A third system, one in which separate pseudorandom decoding systems are used, is called code division multiple access (CDMA). FDMA and TDMA are by far the most important commercially and most of this chapter is devoted to them. CDMA is a special technique used mostly for operating military systems in the presence of high interference and is discussed briefly in this book. Military systems principally operate using TDMA or FDMA in the "clear" mode. We consider space division multiple access as a pseudo method in which different earth stations use the same satellite by using different antenna beams. It is not true multiple access since earth stations in the *same* beam cannot use it. It is an important idea, nonetheless, with wide applicability. Chapters 7 and 8 combine to formulate the second part of the satellite communications problem dealing with the link between the earth station and the user.

Chapter 9 is devoted to a description of satellite transponders. We have only highlighted the main ideas, above all, those that affect critical system choices and compromises. Block diagram organization of various kinds of transponders (single conversion, double conversion and remodulating or regenerative) are presented. Transmission impairments are described which affect the RF signal as it passes through the transponder.

Antenna theory, on the other hand, to the extent necessary for satellite system planning, is discussed in Chapter 10, Earth Stations. Chapter 10 does for earth stations what the previous chapter does for transponders. It deals with the various block diagram organizations of earth stations from the most elaborate fixed service INTELSAT-type earth station to the simplest receive-only or transmit-only station used in broadcast and data gathering services. Earth station designs tend to be dominated by the antenna. We hope Chapter 6 has demonstrated convincingly that the performance of the satellite communication system on both up and downlinks is determined by transmitter power used on that link and by the physical size of the antenna on the ground. This is a point of considerable confusion in the satellite world. One often comes across gross misunderstandings based on the idea that, since the antenna gain increases as the frequency increases, somehow improved performance is achieved at higher frequencies. As shown in Chapter 6, performance depends only on the physical size of the antenna and thus considerable attention is devoted to the antenna problems in this chapter. All the earth station subsystems are discussed to some extent, especially high power amplifiers, low noise amplifiers and tracking systems. We are more concerned here with the radio terminal part of the earth station than with the terrestrial interface.

Chapter 11 deals with an increasingly important problem, one which will, in the long run, dominate the field of satellite communication system planning—interference. In the early days of satellite communication, performance was strictly limited by the amount of power that could be carried on the spacecraft, in other words, by the carrier-to-noise level on the downlink. As the technology for using directed antennas on the spacecraft developed, it became common for channel capacity to be limited by the available bandwidth. In today's world of closely-spaced satellites in geostationary orbit, another element must be considered, that is, interference. Transmission from adjacent satellites and earth stations intending to work with them are, essentially, noise in the desired channel. In addition to the thermal and intermodulation carrier-to noise components to be considered in the overall performance, we now have a carrier-to-noise interference ratio which, in some systems, can dominate the design.

Interference calculations are divided neatly into two halves: first, the composite level of all the interfering signals and second, comparing this interference to ordinary thermal noise. Both problems are discussed in Chapter 11, and methods are given for calculating carrier-to-interference ratios on a system basis and for assessing the relative effect of different kinds of interference. Some ITU standards and procedures are also included, because interference planning is inherently international in nature.

Chapter 12 is an especially interesting chapter dealing with some problems unique to satellite communications. Attention is given to the problems inherent in echo control and coding that are aggravated by the long time delay. This earth-station-to-earth-station delay is an undesirable feature of satellites at geostationary altitude. It gives rise to interesting and important difficulties, notably in data transmission and echo control. Both are discussed at length in Chapter 12. Another important requirement of a satellite system is connected with the exploitation of the geometric advantage in cases where there are many nodes in a traffic network, each with small amounts of traffic. This has led to a concept in satellite communication called "demand assignment", in which channels are made available from a pool and assigned when needed. Both the demand assignment system that goes with frequency division multiple access (SCPC) and that which goes with time division multiple access (DSI, or digital speech interpolation) are explained.

1
Introduction to Satellite Communications

1.0 HISTORICAL BACKGROUND

The first operational communications satellite was the moon, used as a passive reflector by the U.S. Navy in the late 1950s for low-data-rate communications between Washington, D.C., and Hawaii. The first communication from an artificial earth satellite took place in October 1957 when the Soviet satellite, *Sputnik I*, transmitted telemetry information for 21 days. This achievement was followed by a flurry of space activity by the United States beginning with *Explorer I*. That satellite, launched in January 1958, transmitted telemetry for nearly five months. The first artificial satellite used for voice communication was *Score*, launched in December 1958, and used to broadcast President Eisenhower's Christmas message of that year.

In those early years, serious limitations were imposed on payload size by the capacity of launch vehicles and the reliability of space-born electronics. In one attempt to solve some of those problems, an experimental passive repeater, *Echo I*, was placed in medium-altitude orbit in 1960. Signals were reflected from the metallized surface of this satellite, which was simply a large balloon. The approach was simple and reliable, but huge transmitters were needed on the ground to transmit even very low rate data. During the same year, *Courier*, a "store-and-forward" satellite that put messages on magnetic tape for retransmission later during the orbit, became part of the early history of satellite communications.

The first nongovernment ventures into space communications occurred in July 1962, when the Bell System designed and built *Telstar I*, an active real-time repeater. Telstar was placed in a medium-altitude elliptical orbit by NASA and it demonstrated the feasibility of using broadband microwave repeaters for commercial telecommunications. The government experiments continued, with NASA launching *Relay I* in December 1962. This satellite, built by RCA, was used for early experiments with the transmission of voice, video, and data.

Perhaps the most important questions considered in the early 1960s centered around the best orbit to use for a communications satellite. Medium-altitude systems have the advantage of low launch costs, higher payloads, and relatively short radio-frequency propagation times. Their disadvantage is the need to track the satellite in orbit with tracking earth stations and to transfer operations from one satellite to another. Therefore, no single satellite link is available at all times for all stations in the network. The synchronous orbit was first suggested by Arthur C. Clarke in the mid-1940s. (This orbit is in the equatorial plane and the orbital period is synchronized to the rotation of the earth.) Despite its convenience, it was thought by many to have serious limitations because of the long propagation delay and the cost and complexity of the launch. The conspicuous advantage of this orbit is that nearly the whole earth can be covered with three satellites, each maintaining a stationary position and able to "see" one-third of the earth's surface. No "hand-over" is needed and earth station tracking is used only for the correction of minor orbital perturbations.

The first attempt at a synchronous orbit was made by NASA, launching *SYNCOM I* in February 1963. Although *SYNCOM I* was lost at the point of orbit injection, *SYNCOM II* and *SYNCOM III*, launched in July 1963, and August 1964, respectively, were able to accomplish successful synchronous orbit placement and to demonstrate communications by means of such a link. Almost a quarter of a century after the loss of *SYNCOM I*, the injection into final circular equatorial orbit continues to be the riskiest of the orbital maneuvers.

The Communications Satellite Act, signed by President Kennedy in 1962, was at that time the most important piece of American communications legislation since the Communications Act of 1934. It has had profound international consequences. The Act allowed for the formation of the Communications Satellite Corporation (Comsat) and provided the environment to spawn one of the most successful multinational ventures ever undertaken— *Intelsat*. An organization that numbers well over 100 nations, Intelsat was formed in July 1964, in accordance with Resolution 1721 of the United Nations General Assembly. The interim agreements were signed on August 20, 1964, and Intelsat has been a thriving entity ever since. Prompted by the political and operational desirability of fixed satellite assignments, Intelsat made the courageous decision to "go synchronous" and launched *Early Bird* (*Intelsat I*) in April 1965 into that orbit. It was a milestone in the development of satellite communications for commercial use. The evolution of the Intelsat system from that time until the present has been a dramatic succession of increasing satellite capacities and enlarging earth station networks. Table 1-1 summarizes the Intelsat series of satellites from *Early Bird* through *Intelsat VI*. *Intelsat I* through *Intelsat IVA* were successively larger spin-stabilized spacecraft. A major departure into three-axis stabilization was made with *Intelsat V*. Besides the typical increase in capacity, it carried the first commercial experiments with frequencies higher than C-band. *Intelsat VI*, interestingly, will again be a large spinning satellite. The differences and compromises between spinning and body-stabilized spacecraft are discussed in Chapter 5.

From the mid-1960s through the 1970s, Intelsat continued to expand and flourish. By the mid-1970s, a new aspect of the satellite communications industry, domestic satellites, began to form. The cost associated with satellite transmission had dropped dramatically from those early years and it was practical to consider domestic and regional satellites to create telecommunications networks over areas much smaller than the visible earth. Virtually every major country in the world has or is planning a satellite communications system of its own. In the United States alone there are eight major domestic satellite communications carriers and many others involved in partnerships, joint ventures, and other organizational structures that are typical for this burgeoning industry. The Soviet Union, which launched its first communications satellite into a highly elliptical, high-altitude orbit in April 1965, uses a satellite called *Molniya*, which provides television and voice communications in the Soviet Union. The special attributes of this orbit are discussed in Chapter 4.

Tables 1-2 through 1-7 summarize some representative technical characteristics of various types of communication satellites. Although satellite communication began with "fixed" service, it also has great potential applications in mobile and broadcast services. For those services, the inherent geometric advantages of satellites are overwhelming. Submarine cables, fiber optics, and microwave radio provide effective competition to satellites for geographically fixed, wideband service. However, there seems to be no alternative to satellites for the provision of wideband transmissions to mobile terminals. Broadcast transmissions to, and data collection from, many small terminals whose locations are not known a priori are other services in which satellites seem to have a substantial inherent advantage over terrestrial means. The latter services are just beginning to be exploited. Their slow development is the result of the differing political and institutional viewpoints on the broadcast and mobile services—both domestically and internationally.

In the brief period of a quarter of a century, satellite communications has evolved from a notion into a multibillion-dollar industry with the whole world as its market. The evolution is continuing.

TABLE 1-1 The Intelsat Series

	Intelsat I	Intelsat II	Intelsat III	Intelsat IV
System				
Spacecraft manufacturer	Hughes	Hughes	TRW	Hughes
Number of satellites	2	5	8	8
Communications capacity	240 voice circuits or one TV channel	240 voice circuits or one TV channel	1500 voice circuits or four TV channels	4000 voice circuits or two TV channels
Number of transponders	2 (no multiple access)	2 (multiple access)	2	12 (36 MHz)
Stabilization	Spin	Spin	Spin (despun antenna)	Spin (despun antenna)
Mass in orbit[a]	38.6 kg BOL	86.4 kg BOL	151.8 kg BOL	731.8 kg BOL
Dimensions	59 cm long, 72.1 cm diameter	67.3 cm long, 142.2 cm diameter	104.1 cm long, 142.2 cm diameter	2.81 m long (5.26 m with antenna), 2.38 m diameter
Description	Cylindrical structure	Cylindrical structure	Cylindrical structure	Cylindrical structure
Design life	18 months	3 years	5 years	7 years
Operational life	3.5 years	3–5 years	5–7 years	7–10 years
Launch vehicle	Three-stage thrust-augmented Delta	Three-stage thrust-augmented Delta	Three-stage long-tank Delta	Atlas Centaur
Status	Retired from service	Retired from service	Retired from service	Two spacecraft still used for domestic services

TABLE 1-1 (continued)

Intelsat IV-A	Intelsat V	Intelsat V-A	Intelsat VI
Hughes	Ford Aerospace	Ford Aerospace	Hughes
6	9[b]	6[c]	5 under contract[d]
6000 voice circuits and two TV channels	12,000 voice circuits and two TV channels	14,000 voice circuits and two TV channels	Approx. 40,000 voice circuits and two TV channels
20	21 C-band and 6 Ku-band	32 C-band and 6 Ku-band	36 C-band and 10 Ku-band
Spin (despun antenna)	Three-axis	Three-axis	Spin (despun antenna)
862.6 kg BOL	1012 kg BOL[a]	1160 kg BOL	2004 kg BOL
2.81 m long (6.78 m with antenna), 2.38 m diameter	6.44 m high at launch, 15.85 m long with deployed arrays	6.44 m high at launch, 15.85 m long with deployed arrays	11.8 m high (fully deployed); solar drums are about 6 m long, have 3.6 m diameter
Cylindrical structure	Boxlike structure with one solar panel on each side and antenna structure	Boxlike structure with one solar panel on each side and antenna structure	Telescoping dual-cylindrical structure with deployable antennas
7 years 7–10 years (prob.)	7 years	9 years	10 years
Atlas Centaur	Atlas Centaur, Ariane	Atlas Centaur, Ariane	Shuttle or Ariane
Five spacecraft still operational	Seven launched	Planned for 1985 launch	Planned for 1986 launch

[a] BOL, Beginning of Life, after firing of apogee kick motor.
[b] Last four satellites of nine carry maritime mobile payload (seven additional transponders).
[c] Last three spacecraft of six are being modified for specialized business services.
[d] Intelsat has options to buy 11 more spacecraft.

TABLE 1-2 U.S. Fixed-Service Communications Satellites

System	Comstar	Galaxy	Satcom	SBS	Telstar	Westar
Operator	Comsat General	Hughes Comm.	RCA Americom	SBS	AT&T	Western Union
Prime contractor	Hughes	Hughes	RCA Astro-Electronics	Hughes	Hughes	Hughes
Number of satellites	4[a]	2[a]	6[a]	3[a]	1[a]	5[a]
Design life (years)	7	10	10	7	10	10
Satellite mass in orbit (kg)	811	643	598	546	659	584
Solar array power EOL (W)	610	741	1100	900	800	684
Stabilization	Spin	Spin	Three-axis	Spin	Spin	Spin
Number of transponders	24	24	24	10	24	24
Frequencies	C-band	C-band	C-band	Ku-band	C-band	C-band
Transponder bandwidth (MHz)	34	36	34	43	36	36
Satellite EIRP (dBW)	33	34	33	44	34	34
Receive G/T (dB/K)	−8.8	−7.0	−5.0	+2.0	−5.0	−6.0
Launch vehicle	Atlas Centaur	Delta 3920	Delta 3924	Delta 3910 and Shuttle	Delta 3920	Delta 3910
Launch year for first spacecraft	1976	1983	1975	1980	1983	1974

TABLE 1-2 (continued)

	ASC	Fordsat	GSTAR	Hughes-Ku	RCA-Ku	Spacenet
	American Satellite	FASSC	GTE	Hughes Comm.	RCA Americom	GTE Spacenet
	RCA Astro-Electronics	Ford Aerospace	RCA Astro-Electronics	Hughes	RCA Astro-Electronics	RCA Astro-Electronics
	3	3[b]	3	3[b]	3	3
	10	10	10	10	10	10
	665	1528	670	1150 est.	600	670
	1215	1300 est.	1350	1200 est.	1395	1150
	Three-axis	Three-axis	Three-axis	Spin	Three-axis	Three-axis
	24	54	16	16	16	24
	C-band and Ku-band	C-band and Ku-band	Ku-band	Ku-band	Ku-band	C-band and Ku-band
	36 and 72	36	54	54	54	36 and 72
	34 and 39	35 and 48	42	48	41	33 and 40
	−4 and −5	−3 and +4	−1.5	0	−1.5	−5 and −3
	Shuttle	Shuttle	Ariane	Ariane or Shuttle	Ariane or Shuttle	Ariane
	1985[b]	1988[b]	1985	1987[b]	1985[b]	1984

[a] In orbit at end of 1983.
[b] Planned.

TABLE 1-3 Canadian Fixed-Service Communications Satellites

	Anik A	Anik B	Anik C	Anik D
System				
Operator	Telesat Canada	Telesat Canada	Telesat Canada	Telesat Canada
Prime contractor	Hughes	RCA Astro-Electronics	Hughes	Spar Aerospace
Number of satellites[a]	4	1	2	1
Design life (years)	7	7	10	10
Satellite mass in orbit (kg)	272	440	567	635
Solar array power EOL (W)	260	650	800 est.	830
Stabilization	Spin	Three-axis	Spin	Spin
Number of transponders	24	18	16	24
Frequencies	C-band	C-band and Ku-band	Ku-band	C-band
Transponder bandwidth (MHz)	36	36 and 72	54	36
Satellite EIRP (dBW)	33	36 and 47	47	36
Receive G/T (dB/K)	−7	−6 and −1	+2	−3
Launch vehicle	Delta 2914	Delta 3914	Shuttle	Delta 3910

[a] In orbit at end of 1984.

1.1 BASIC CONCEPTS OF SATELLITE COMMUNICATIONS

This book focuses on communications using a satellite placed in geostationary orbit at an altitude of 35,786 km above the surface of the earth. Over the past 20 years, the use of that orbit for satellite communications has been prolific. Its special characteristics and features are discussed in Chapter 3. Despite the convenience and utility of the geostationary orbit, other orbits are utilized for communications purposes and are discussed to some extent in Chapter 2. Most of the communications theory described here applies equally well to any orbit.

Figure 1-1 illustrates the end-to-end communications required in establishing a satellite link. The link is shown in its most general form with transmit and receive facilities at both ends. Such facilities are characteristic of the fixed and mobile services, but broadcast and data collection applications are transmit only at one end and receive only at the other end of the link. The overall problem can be conveniently divided into two parts. The first deals with the satellite radio-frequency (RF) link which establishes communications between a transmitter and a receiver using the satellite as a repeater. In describing the satellite radio link, we quantify its capability in terms of the *overall available carrier-to-noise ratio* $(C/N)_A$. This figure of merit, representing the ratio of the carrier power to the noise power measured in a bandwidth B, is directly related to the channel-carrying capability of the satellite link. The value of $(C/N)_A$ depends on a number of factors, which in turn depend on the available power and bandwidth.

The second part of the problem concentrates on the link between the earth terminal and the user environment. In the user environment customers are typically concerned with establishing voice, data, or video communications with either simplex or duplex connections. The quality of these "baseband" links is characterized by various figures of merit such as transmission rates, error rate, signal-to-noise ratio, and other performance measures. For example, a data communications link used to transmit financial account balances must exhibit an extremely low rate of error to be effective. The error-rate specification for such a data communications service is directly translated into a required carrier-to-noise ratio per channel. The two parts of the problem can then be linked together when the available carrier-to-noise ratio of the satellite link is compared to the required carrier-to-noise ratio dictated by the user application. In the paragraphs that follow, a brief overview of these notions is provided. In later chapters, more detailed discussions of each of the contributing design factors is addressed.

1.1.1 The Radio-Frequency Satellite Link

As illustrated in Figure 1-2, a communications satellite operates as a distant line-of-sight microwave repeater providing communications services

TABLE 1-4 Other Non-U.S. Fixed-Service Communications Satellites

System	CS-2	ECS	Insat	Palapa A
Operator	NASDA	Eutelsat	ISRO	Perumtel
Prime contractor	Mitsubishi	MESH	Ford	Hughes
Number of satellites[a]	2	1	1	2
Design life (years)	5	7	7	7
Satellite mass in orbit (kg)	350	610	600	300
Solar array power EOL (W)	375	900	1090	259
Stabilization	Spin	Three-axis	Three-axis	Spin
Number of transponders	8	9	14	12
Frequencies	C-band and K-band	Ku-band	C-band	C-band
Transponder bandwidth (MHz)	180 and 130	80	36	34
Satellite EIRP (dBW)	30 and 37	35	33 and 43	32
Receive G/T (dB/K)	-8 and -5	-5	-4 and -3	-7
Launch vehicle	N-2	Ariane	Shuttle	Delta 2914
Launch year for first spacecraft	1983	1983	1982	1976

[a] In orbit at end of 1983.

among multiple earth stations in various geographic locations. The performance of a satellite link is typically specified in terms of its channel capacity, and there are several definitions that are relevant.

A *Channel* is a one-way link from a transmitting earth station through the satellite to the receiving earth station.

A *Circuit* is a full-duplex link between two earth stations.

A *Half-circuit* is a two-way link between an earth station and the satellite only.

The capacity of a link is specified by the types and numbers of channels and

TABLE 1-4 (continued)

Palapa B	Arabsat	Australia	Brazilsat	Telecom
Perumtel	Arab League	Aussat	Embratel	French PTT
Hughes	Aerospatiale	Hughes	Spar Aerospace	Matra Espace
1	2	3	2	2
8	7	7	8	7
628	680	650 est.	650 est.	653
831	1300	860	800	1045
Spin	Three-axis	Spin	Spin	Three-axis
24	26	15	24	12
C-band	C-band and S-band	Ku-band	C-band	C-band, X-band, and Ku-band
36	33	45	36	36, 40, and 120
34	32 and 41	36 and 42	35 est.	29, 35, and 47.5
−5	−7.5	−4	−4.5 est.	−13.6 and +6.5
Shuttle	Ariane	Ariane or Shuttle	Ariane	Ariane
1983	1st-Feb. 1985 2nd planned 1985	3 planned: 2-1985 1-1986	1st-Feb. 1985 2nd planned 1985	1st 1984 2nd planned 1985 3rd planned 1986

the performance requirements of each channel. In practical terms, a communications common carrier providing voice service must provide circuits to its customers. The term "channel," however, may also apply to television and data circuits as well. In the case of an international system, a link from a transmitting station to the satellite may originate in one country and the link from the satellite to the receiving earth station may terminate in a second country. In this case, the concept of a half-circuit is used for accounting purposes. For broadcast and data collection applications, one-way channels are typical.

The channel-carrying capacity of a satellite RF link (typically expressed in terms of voice channels per transponder) is directly related to the overall

TABLE 1-5 U.S. Direct Broadcast Satellites

	CBS	DBSC	RCA	STC	USSB (HUBBARD)
System operator					
Prime contractor	Not yet chosen	Ford Aerospace	RCA Astro-Electronics	RCA Astro-Electronics	RCA Astro-Electronics
Design life (years)	7	7	7	7	7
Satellite mass in orbit (kg)	1000 est.	1358	1096	670	1044
Solar array power (W)	4000	4600	3660	1700	3600 est.
Stabilization	Three-axis or spin	Three-axis	Three-axis	Three-axis	Three-axis or spin
Number of transponders	3	6 + 8 spots	6	3	6
Frequencies	K-band	K-band	K-band	K-band	K-band
Transponder bandwidth (MHz)	24	24	24	24	24
Satellite EIRP (dBW)	60	54	59	57	57
Receive G/T (dB/K)	8.2	+0 to +10	−3.6	7.7	7.7
Launch vehicle	Shuttle or Ariane	Ariane	Shuttle	Shuttle	Shuttle or Ariane
Launch year for first spacecraft	Temporary abandoned plans	1988[a]	1989[a]	Temporary abandoned plans	1988[a]

[a] Planned.

TABLE 1-6 Non-U.S. Direct Broadcast Satellites

System	BS-2	L-Sat	TDF/TV-Sat	Tele-X	Unisat
Operator	NASDA	ESA	France/Germany	Sweden	United Satellites
Prime contractor	Toshiba	British Aerospace	Eurosatellite	Aerospatiale	British Aerospace
Number of satellites[a]	2	1	2	1	2
Design life (years)	5	7	7	7	10
Satellite mass in orbit (kg)	350	1380	1060	1300	850 est.
Solar array power EOL (W)	690	3000 est.	2500	3000 est.	2000 est.
Stabilization	Three-axis	Three-axis	Three-axis	Three-axis	Three-axis
Number of transponders	2	9	3	4	8
Frequencies	K-band	Ku-band and K-band	K-band	K-band	K-band
Transponder bandwidth (MHz)	70	27, 36, and 150	27	27, 40, and 86	36
Satellite EIRP (dBW)	55	41, 51, and 61	65	58 to 60	45 and 65
Receive G/T (dB/K)	−8.2	+2 to +5	+10 est.	n.a.[b]	3.5 and 12
Launch vehicle	N-2	Ariane	Ariane	Ariane	Ariane or Shuttle
Launch year for first spacecraft	1st 1984 2nd planned 1985	1987[a]	1986[a]	1987[a]	1986[a]

[a] Planned.
[b] n.a., not available.

TABLE 1-7 Military and Mobile Satellites

System	DSCS-2	DSCS-3	Fleetsatcom	Marecs	Marisat
Operator	DoD	DoD	DoD	ESA	Comsat General
Spacecraft manufacturer	TRW	G.E. Space Division	TRW	British Aerospace	Hughes
Number of satellites[a]	8	4	5	2	3
Design life (years)	5	10	5	7	5
Satellite mass in orbit (kg)	500	816	1005	572	326
Solar array power EOL (W)	357	837	1425	748	300
Stabilization	Spin	Three-axis	Three-axis	Three-axis	Spin
Number of transponders	2	7	12	2	2
Frequencies	X-band	X-band and UHF	X-band and UHF	C-band and L-band	C-band and L-band
Transponder bandwidth (MHz)	50 to 185	50 to 85	0.005 to 0.5	4.75 and 5.9	4
Satellite EIRP (dBW)	28 to 40	23 to 40	16.5 to 28	18.8 to 29.5	18.8 to 29.5
Receive G/T (dB/K)	8.5 to 20	−16 to −1	−16.6	−12.8 to −16.5	−17 to −25.4
Launch vehicle	Titan III-C or 34-D	Titan 34-D	Atlas Centaur	Ariane	Delta 2914

[a] In orbit at end of 1984.

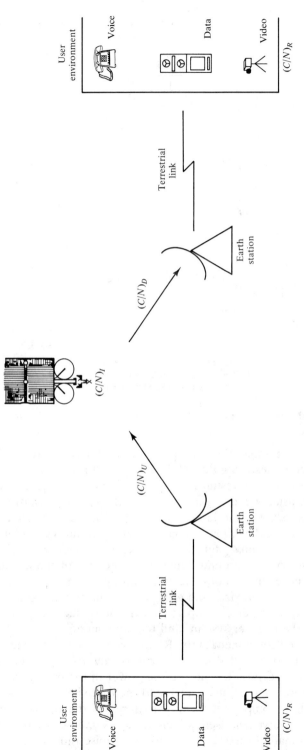

Figure 1-1 End-to-end satellite communications link.

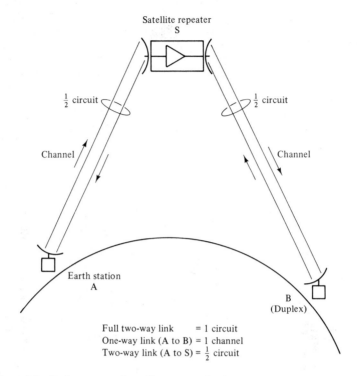

Figure 1-2 Basic concept of satellite communications.

available carrier-to-noise ratio. Exclusive of interference, three basic elements are considered in designing this RF link. The first is the *uplink*, representing the channel from the transmitting earth station to the satellite. The quality of this link is usually expressed in terms of the *uplink carrier-to-noise ratio*, $(C/N)_U$. As we shall see in later chapters, $(C/N)_U$ depends on the power of the transmitting earth station, the gain of the transmitting antenna, the gain of the receiving antenna, and the system noise temperature. The power of the transmitter on the ground depends on the size of the power amplifier employed. The gains of both the transmitting and receiving antennas are directly related to their sizes and efficiencies. The system noise temperature is composed of the receiver temperature, the noise due to losses between the antenna and the receiver system, and the antenna noise.

The second component in the RF link is the *downlink*. The corresponding figure of merit is called the *downlink carrier-to-noise ratio*, $(C/N)_D$. As with the uplink, $(C/N)_D$ depends on the power of the transmitter, the transmitting and receiving gains, and the receiving system noise temperature. The third component to be considered in the RF link design is the *satellite electronics system* itself, which produces undesirable noiselike signals which are normally expressed in a carrier-to-noise ratio which we shall call

$(C/N)_I$. Several impairments, primarily intermodulation effects caused by the nonlinear operation of the satellite amplifier, can be included in the $(C/N)_I$ component. Interference from other satellites and terrestrial systems can also be collectively characterized by a carrier-to-interference ratio. If one assumes that the thermal noise and all the noise-like impairments are low level and additive, they can easily be combined to yield a composite carrier-to-noise ratio, $(C/N)_A$.

The overall carrier-to-noise ratio available in the link is a combination of the three elements $[(C/N)_U, (C/N)_I, (C/N)_D]$, and they combine in the same way as do resistors in parallel. That is, the overall carrier-to-noise ratio can never be any larger than the smallest of the three individual components. Chapter 6 deals in detail with link analysis and with the calculation of the overall carrier-to-noise ratio available in an RF link.

Two basic elements are required to establish a satellite link. The first is the satellite repeater, usually called a *transponder*, and the second is a satellite earth station. The following paragraphs briefly introduce the characteristics of each of these elements.

1.1.2 Satellite Transponders

From a communications standpoint, a satellite may be considered as a distant microwave repeater which receives uplink transmissions and provides filtering, amplification, processing, and frequency translation to the downlink band for retransmission. This kind of transponder is a quasi-linear repeater amplifier, a block diagram of which is shown in Figure 1-3. The uplink and downlink bands are separated in frequency to prevent oscillation within the satellite amplifier while permitting simultaneous transmission and reception at different frequencies through a device called the *diplexer*. Moreover, the lower-frequency band is normally used on the downlink to exploit the lower-atmospheric losses (sometimes called *path loss*), thereby minimizing satellite power amplifier requirements.

Through the late 1970s, frequencies between 2 and 8 GHz bands were

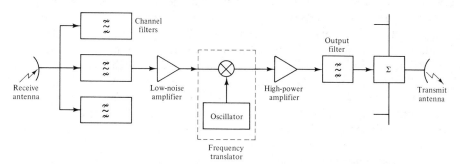

Figure 1-3 Basic satellite repeater.

predominantly employed in satellite communications. These C- and X-band frequencies have the advantages of adequate bandwidth, negligible fading, low rain loss (for earth station elevation angles above 5°), and the availability of affordable and reliable microwave devices. Unfortunately, the available orbital locations and thus the number of times the available spectrum can be reused are being consumed quickly. The frequency coordination and spectral utilization problems have become very severe. Therefore, the major trend toward using higher frequencies, which began in the 1920s, continues. The Ku-band at 14 and 12 GHz has become popular in recent satellite design as the device technology has developed. Although Ku-band has the advantage of ameliorating the interference problem, a substantial fading margin must be provided to accommodate the severe rain loss. Despite this, Ku-band is rapidly taking its place along with C-band for satellite communications.

Satellite transponder amplifiers must provide relatively large gains (in the range 80 to 100 dB) while maintaining relatively low-noise operation. The high-gain requirements typically require multiple-stage low-noise amplification. The first stages in modern transponder amplifier chains are provided by solid-state FET amplifiers. These devices require careful design to minimize noise and intermodulation effects. Channelizing filters must also employ careful design practices to minimize interference from adjacent channels, as well as other factors, including intersymbol interference and group-delay distortion. Frequency translation from the uplink to the downlink band causes the uplink noise to be amplified along the multiple-stage amplifier. Final stages of amplification in the transponder are typically provided by traveling-wave-tube amplifiers (TWTAs), which operate well for constant envelope signals. It is in this high-power output amplifier stage that most of the impairments that affect $(C/N)_I$ are generated. In multiple-carrier operation, intermodulation is usually the dominant impairment. Other factors, such as AM-to-PM conversion, must also be considered. AM-to-PM results from amplitude variations that produce unintentional phase modulation at the output of the TWT. The AM-to-PM is also affected by variations in the passband of the channelizing filters. Adjacent channel and adjacent satellite interference must also be included in an overall consideration of impairments. Therefore, the designer, in addition to considering thermal noise effects, must consider a wide range of typically nonlinear impairments. These impairments are related to both the design of the satellite hardware components and the design operating points in the RF link. It is interesting to note that such impairments are typically less serious in terrestrial systems, including earth stations, where primary power and a wide range of hardware are readily available. In a satellite, however, the limitations of spacecraft mass and power in orbit force systems engineers to balance the available power against the acceptable distortions due to nonlinear impairments.

Virtually all satellite transponders employed to this time have been the quasi-linear repeater type. In the future, more sophisticated transponder designs employing regenerating-type repeaters will surely be used for digital transmission. In a regenerating transponder the digital signal is demodulated and remodulated within the transponder itself. This approach has the distinct advantage of separating the uplink and the downlink into two independent paths.

These subjects and their effects are described in much more detail in Chapter 6, covering the RF link, and in Chapter 9, on transponders.

1.1.3 Earth Stations

The second basic element in a satellite link is an earth terminal. A block diagram of a typical earth station configuration is shown in Figure 1-4. Earth stations are available in a wide variety of size, function, sophistication, and cost. They are categorized by function, by the size of the antenna, and by the level of the radiated power. Antenna diameters range in size from as small as 0.7 m for direct broadcast receive-only applications to as large as 30 m in diameter for large international gateway stations. Larger stations may require tracking systems to maintain the pointing of the antenna at the satellite.

An earth station consists of an antenna subsystem, a power amplifier subsystem, a low-noise receiver subsystem, and a ground communications equipment (GCE) subsystem. Most stations are equipped with separate power supply systems plus control, test and monitoring facilities, sometimes called *telemetry, tracking, and command systems* (TT&C). Smaller stations usually do not require tracking systems because of the large beamwidth of the antenna compared to larger aperture stations.

The performance of an earth station is specified by its equivalent isotropic radiated power (EIRP) and its gain-to-system noise temperature ratio (G/T). EIRP is the product of the power output of the high-power amplifier at the antenna, and the gain of the transmitting antenna. The receiving system sensitivity is specified by G/T, which is the ratio of the receive gain of the antenna to the system noise temperature. The antenna gain is proportional to the square of the diameter and is dependent on the efficiency of the feed/reflector system. The system noise temperature is composed of three components: the noise of the receiver, the noise due to losses between the antenna feed system and the receiver, and the antenna noise. Although the performance of an earth station is typically limited by thermal noise, it can also be plagued with some of the same difficulties caused by nonlinear impairments in a satellite transponder. In general, the larger the station, the more affordable power levels and equipment becomes, and fewer system design problems are encountered. The system designer must account for impairments such as intermodulation distortion in the high-power amplifier located in earth stations as well as in the satellite transponder. However, the level of difficulty presented the designer in the earth station is less restrictive than that in the satellite transponder. Chapter 10 deals more specifically with earth station technology and design considerations. This chapter, coupled with the material

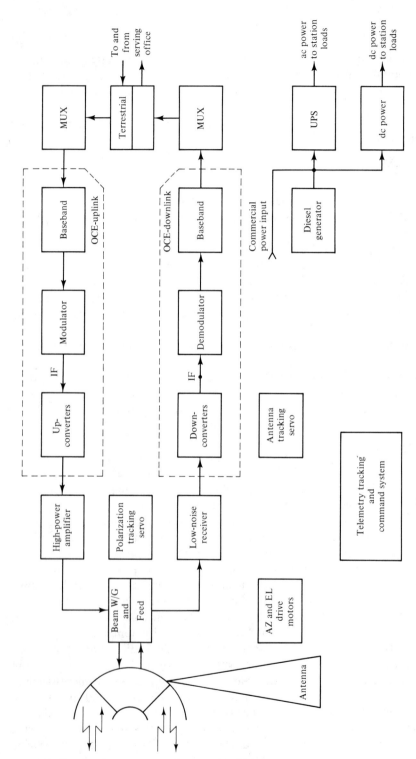

Figure 1-4 Typical earth station configuration.

in Chapter 6 on the RF link design, will tie together the systems engineering problems.

1.1.4 The Terrestrial Link

Referring again to Figure 1-1, the second part of the end-to-end satellite communications problem is embedded in the link between the satellite earth station and the user environment. This part of the problem deals more specifically with the baseband signal (i.e., the signal after demodulation). To provide adequate satellite service to a user, the service requirements must be well defined in terms of quality. Quality of service, specified in terms of parameters such as link availability (grade of service), bit error rate, and signal-to-noise ratio, may then be translated into a required carrier-to-noise ratio $(C/N)_R$ in the RF link. The required carrier-to-noise ratio $(C/N)_R$ is then compared with the available carrier-to-noise ratio $(C/N)_A$ to determine the overall capacity of the link. In designing systems to meet required quality and grades of service, several fundamental baseband processing elements must be considered.

The first level of processing is source coding and/or modulation wherein a source signal (voice, data, or imagery) is coded in digital form or processed in analog form to prepare it for transmission. In this book we cover only the most common analog and digital techniques used in satellite telecommunications. In analog transmission they are amplitude modulation (AM) and frequency modulation (FM). For digital transmission, pulse-code modulation (PCM), delta modulation (DM), and variations of these techniques are used most often.

Following individual channel coding, the next processing level is multiplexing. For analog transmission, channels are combined using *frequency-division multiplex* (FDM), which employs frequency-separated carriers, each of which accommodates one channel. In digital transmission, multiple channels are combined into higher-level digital signals using time-division multiplexing (TDM), which employs separate time slots, each carrying the information from one channel.

The next level of processing is channel coding, which may be used to improve the quality of digital transmission by adding redundancy prior to transmission to reduce the overall error rate. Finally, the process of RF modulation is used to modulate either single- or multiple-channel signals onto radio-frequency carriers high enough for transmission on the satellite link. Analog transmission typically uses FM (although AM is also used occasionally), and digital transmission usually employs some form of phase-shift keying (PSK).

The final level in signal processing is *multiple access*. To exploit the satellite's geometric advantage, we must design systems that permit more than one earth station pair to utilize a transponder simultaneously. Multiple access techniques have been used extensively in satellite communications. For commercial applications, two types have been employed most often. The first is *frequency-division multiple access* (FDMA), which employs multiple carriers within the same transponder, as illustrated in Figure 1-5a. It is the most common method in use because it was a natural extension of the FDM systems that were in use for many years in terrestrial carrier systems. In FDMA, each uplink RF carrier occupies a defined frequency slot and is assigned a specific bandwidth in a multicarrier repeater. The receiving station selects the desired carrier by RF filtering. In FDMA, multiple access is achieved through frequency planning.

The second basic type used in commercial satellite communications is *time-division multiple access* (TDMA). TDMA employs a single carrier which is time-shared among many users, as illustrated in Figure 1-5b. It operates in *burst mode*, such that the transmission bursts from all stations arrive at the satellite transponder consecutively. The bursts are contiguously interleaved without overlapping in time, as illustrated in Figure 1-5b. Each earth station receives all bursts from all stations (including its own) and selects those destined for its users. Utilizing time-shared single-carrier TDMA provides some distinct advantages over FDMA. In particular, a single-carrier TDMA approach avoids the intermodulation distortion that must be accommodated in multiple-carrier FDMA systems.

A third kind of multiple access system, known as *code-division multiple access* (CDMA), is one in which all uplink signals occupy the same frequency band at the same time. Each has its own pseudorandom code which is chosen

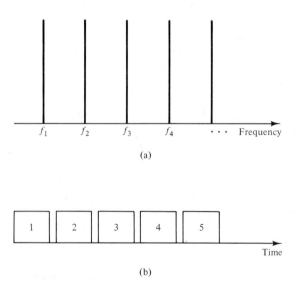

Figure 1-5 (a) Frequency-division multiple access (FDMA), multiple carriers; (b) time-division multiple access (TDMA), single carrier.

from an orthogonal set, which is used to separate the desired signal from the other signals. This system has been used primarily in military applications for security purposes. Although it will not be discussed extensively in this text, commercial applications of CDMA (or *spread spectrum*, as it is sometimes called) are beginning to emerge for lower-speed data communications networking.

1.2 COMMUNICATIONS NETWORKS AND SERVICES

A telecommunications system must service several types of networks to achieve full interconnectivity among users. The shortest networks consist of buses used to distribute information from PBXs or distributed computer systems operating within a building or a small campus environment. Such networks are typically less than 0.1 km in maximum communications distance and are normally serviced through wire pairs, or other short-distance techniques, such as line-of-sight optical or infrared systems. In the next level of the network hierarchy are local networks extending over distances from 0.1 km to about 10 km. Examples of such networks are those serving large campuses or small cities. These networks are potentially implemented using one of several technologies, such as fiber optics, coaxial cable systems, and multipoint digital radio distribution (sometimes referred to as *digital termination service*—DTS). Still larger networks, such as regional networks, extend from 1 km to several hundred kilometers. These include systems serving large cities or compact multistate areas. Long-haul networks greater than 100 km are the most cost-effective applications for fixed-service satellite transmission. Although most satellite networking applications cover distances greater than 1000 km, shorter distances are becoming economically feasible.

Created and encouraged by the continued development of digital technology, many new telecommunication services have emerged in recent years. Still heading the list is traditional telephone service, accounting for approximately 80 percent of the value of telecommunication services currently in place. Although telephone service has had a long evolutionary history, spanning more than 100 years, the last decade has produced a revolution because of the deregulatory environment, the divestiture of the Bell System, and the emergence of digital technology.

New video services have also emerged from the digital revolution. Once these services were limited to full-motion broadcast-quality video. New services have been developed to create revolutionary communication systems, such as video teleconferencing for business applications. New transmission systems have evolved for communicating image signals at data rates and bandwidths far less than normally required for broadcast-quality video. Facsimile and graphics systems are also developing rapidly as new tools to communicate image information while saving on the cost of travel.

Perhaps the fastest-growing area is that of data communication services. Seeded by the financial community and large corporations seeking to manage their businesses better, new data communication services have emerged. Database access and transfer through widely dispersed processing systems, transactional service for retail operations, inventory management systems, network management, remote access, and electronic mail are just a few examples of emerging new services. A wide range of data rates is required to support these various services. For discussion purposes, we classify data services in terms of speed, roughly in accordance with the bandwidth capabilities of the transmission media.

Speed Range	Data Rate
Low speed (narrowband)	≤ 300 b/s
Medium speed (voiceband)	300–16 kb/s
High speed (wideband)	>16 kb/s

Chapter 7 contains a description of some typical systems together with methods of performance analysis used to evaluate systems at these data rates.

1.3 COMPARISON OF TRANSMISSION TECHNOLOGIES

Many new technologies are emerging during the 1980s. Table 1-8 summarizes the five most popular: satellites, digital microwave radio, fiber and coaxial cables, local distribution radio, and wire pairs. Each of them will expand greatly in the coming years. Historically, as each new technique was introduced, those associated with its creation were often so zealous in promoting it that the inexperienced were misled into thinking that it would dominate the telecommunications field. This has not often been the case. Each technique tends to apply best under a particular set of circumstances, and as telecommunications expand, each finds its proper place in the overall communications framework.

Table 1-8 lists each technology and illustrates those applications for which it performs best. Also noted are the primary issues or problems associated with that particular technology. *Satellite transmission*, for example, applies best in long-haul applications for medium-band to wideband transmission. Because of its geometric advantage, it is suited for multipoint-to-multipoint (multiple access), multipoint-to-point (data gathering), or point-to-multipoint (broadcast) applications. In these services its inherent advantage over terrestrial techniques is formidable.

One of the early uncertainties in satellite technology was the effect of the long time delay to geosynchronous orbit. Although it certainly must be considered in any satellite communications systems design, proven solutions

TABLE 1-8 Comparison of Technologies

Technology	Network Distance	Data Rate	Connectivity	Primary Consideration
Satellite	Long haul	Medium to high speed	Multipoint-to-multipoint Point-to-multipoint Multipoint-to-point	Propagation time delay
Digital microwave radio	Local/regional	High speed	Point-to-point	Transmission-path geometry
Fiber optics/Coaxial cable	Local/regional	High speed	Point-to-point	Construction cost (cable duct availability)
Local distribution radio	Local/regional	Medium to high speed	Point-to-point Point-to-multipoint	Transmission-path geometry
Twisted pairs	Local	Low to medium speed	Point-to-point	Construction cost (cable duct availability)

to the problems exist. Chapter 12 deals with those problems and the solutions used to accommodate long time delay. As illustrated in Figure 1-6, the unique geometric advantage of a satellite in geosynchronous orbit makes it possible to service complex, fully interconnected mesh networks with individual nodes widely separated on the earth's surface. Note that with n links, the geometric advantage of the satellite position allows $[N(N - 1)/2]$ links of interconnection without introducing switching. Note also that communications between any two nodes in such a network is independent of the distance between them. This distance insensitivity is a distinct characteristic of satellite systems which applies within the coverage area of a particular RF beam. One should be careful not to interpret distance insensitivity within a particular beam to mean that the cost of a satellite link is independent of the size of the total geographic area covered by the beam. As discussed in Chapter 6, as coverage area increases, more satellite power is needed and thus higher costs are incurred.

Digital microwave radio, evolving at the same time, fits well in local or regional network applications and is applicable to wideband transmission in point-to-point applications. Typically, the issue for the systems engineer is to accommodate the transmission path geometry and coordinate frequency and power levels to share spectrum with other systems.

Fiber optics and *coaxial cable systems*, now rapidly developing, are also applicable to wideband transmission. Point-to-point and in some cases point-

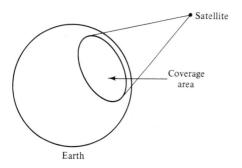

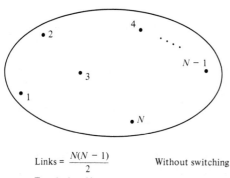

Links = $\dfrac{N(N-1)}{2}$ Without switching
Terminals = N
Inherent geometric advantage $\sim N$

Figure 1-6 Satellite networking.

to-multipoint network applications are possible. The salient issue is the cost of construction of the network. Regardless of the cost of the optical fiber itself, the cost of the cable ducts (or just the cost of installing the fiber in existing ducts) can make a network too expensive compared to other methods. However, the installed cost of fiberlinks is dropping rapidly, thus providing strong competition for satellite communications in many applications.

Local distribution radio, the technology for which has been developing for many years, is now emerging rapidly because it allows instantaneous setup of local distribution of digital services between a central site and a network node. Data rates and services from low speed (1200 b/s) to high speed (1.544 Mb/s) are easily set up covering network distances of approximately 5 to 15 miles. Transmission-path geometry, as in the case of any radio system, is the critical item.

Twisted pairs are one of the greatest natural communications resources for local distribution existing in the world today. Narrowband and medium-band data rates are possible on the wire pairs that have been installed by telephone companies over the past 100 years. Use of wire pairs is limited to point-to-point applications and the critical issue is construction cost, which is often related to the availability of pathways for cable installation. We can expect that the pathways for existing local telephone systems will be used in the future to accommodate coaxial cables and optical fibers where twisted pair is now used.

1.4 GROWTH OF SATELLITE COMMUNICATIONS

In fewer than 25 years, satellite communication has grown from a few experiments by major governments into a multibillion-dollar industry developed with private funding and support from large corporations and governments of all sizes. The growth of the industry can be appreciated by considering the number of satellites and earth stations in operation and planned in the coming years. For example, more than forty satellites are spaced along less than 80° of the geostationary orbital arc over North America for use in domestic satellite communications alone. In fact, the density of satellites occupying this precious orbital arc has become a major concern in the satellite communications field in recent years.

Earth station growth, in sheer numbers, is even more dramatic. The estimated value of the annual earth station marketplace by the end of the decade is well over $200,000,000. There are multiple market segments employing large numbers of earth stations. These markets include common carriers, large businesses, government and television distribution. The continued long-term growth is concentrated in large business communications, government and common carriers. It should be noted that small home terminals for direct broadcast and mobile satellite applications may number in the hundreds of millions.

The growth rates predicted for the satellite communications industry are spectacular. Some forecasters are predicting annual growth rates during the 1980s in excess of 100% per year. This will provide many opportunities for new entrepreneurial ventures and keep systems engineers rather busy.

REFERENCES

Communications Act of 1934, as amended, latest edition, Sup't of Pub. Docs. USGPO Washington, D.C. 20402 (Stock Number 0400-0264).

"Determination of Coordination Area," CCIR Report 382-3, CCIR *Green Books*, Volume IX, ITU, Geneva, Switzerland.

EDELSON, BURTON I.: "Global Satellite Communications," *Scientific American*, Vol. 236, No. 2, Feb. 1977.

IEEE Transactions on Communications: Special Issue on Satellite Communication, Oct. 1979.

Manual of Regulations and Procedures for Radio Frequency Management, OTP (now NTIA), Washington, D.C. (Available to government agencies, and perhaps to their contractors.)

MIYA, K.: *Satellite Communications Engineering*, Lattice Co., Japan, 1975.

MORGAN, L. W.: "Communications Satellites—1973 to 1983," IEEE, ICC-78, Vol. 1, Toronto, June 4–7, 1978.

PELTON, J.: "An Overview of Satellite Communications," Satellite Commun. Users Conf., Denver, CO, Aug. 1980, and *Satellite Communications*, Oct. 1980.

PRITCHARD, W. L.: "Satellite Communication—An Overview of the Problems and Programs," *Proc. IEEE*, Special Issue on Satellite Communications, March 1977.

Radio Regulations, Edition of 1976. ITU, Geneva, Switzerland (Price given in the magazine, *Telecommunications Journal*, a monthly publication of the ITU.)

Rules and Regulations of the FCC, Parts 21 and 25, FCC or GPO, Washington, D.C.

SMITH, E. K.: "The History of the ITU with Particular Attention to the CCITT and the CCIR . . .," *Radio Science*, Vol. 11, No. 6, pp. 497–507, June 1976.

SPILKER, J. J.: *Digital Communications by Satellite*, Prentice-Hall, Engelwood Cliffs, NJ, 1977.

VAN TREES, H. L., ed.: *Satellite Communications*, IEEE Press Selected Reprint Series, IEEE Press, New York, 1979a.

2

Orbits

2.0 GENERAL CONSIDERATIONS

The communication satellite is, to begin with, a satellite. Something like three-fourths of its cost is associated with launching and maintaining it in its operational orbit so that the communications package or payload can function satisfactorily. It is therefore essential that the systems engineer have a working knowledge of orbital mechanics and be able to deal, at the system level, with the problems of transfer from parking orbit into operational orbit, station-keeping maneuvers, and the geometry of the orbit itself.

Orbital mechanics, as applied to artificial earth satellites, is based on celestial mechanics, an extraordinarily successful branch of classical physics, which started with Kepler and Newton and was expanded and elaborated by most of the giants of theoretical physics during the eighteenth and nineteenth centuries. Lagrange, Laplace, Gauss, Hamilton, and many others made substantial contributions to the mathematical refinement of the theory, starting with the basic notions of universal gravitation, Newton's laws of motion, and the principles of conservation of energy and momentum. The nineteenth century saw the triumphant prediction of the existence and orbital parameters of a new planet, Neptune, based on the observed perturbations in the orbit of Uranus and the use of Newtonian mechanics. The theory finally became so refined in its ability to predict the motions of celestial bodies that when discrepancies appeared between theoretical predictions and observations, astronomers were confident enough of the theory to suspect that the difficulty was with the timekeeping scale, that is, that the rotation of the earth itself was not constant. The astronomers then changed time scale from one based on the earth's rotation (universal time) to one based on the orbital motions of the planets (ephemeris time) with several orders of magnitude increase in accuracy. Quite recently, atomic timekeeping has made it possible to eliminate astronomical time scales completely from the fundamental definition of the second.

2.1 FOUNDATION OF THE THEORY

Fortunately, the communication satellite engineer does not have to deal with celestial mechanics in its full glory but only with a restricted class of problems. We intend here to show the framework of the theory in its simplest form, indicate how the results are derived, and then elaborate on those results that are of utility to the problem at hand.

The most accessible way of developing orbital mechanical theory, at least as applied to earth satellites, starts with Newton's law of universal gravitation and the second law of motion, as seen in Eqs. (2-1) and (2-2):

$$\mathbf{F} = -\frac{GMm}{r^2}\frac{\mathbf{r}}{r} \tag{2-1}$$

$$\mathbf{F} = m\frac{d\mathbf{v}}{dt} \tag{2-2}$$

The vector $\mathbf{r}$ is from M to m, and the force is on m, which accounts for the minus sign in Eq. (2-1).

Equations (2-1) and (2-2) can be written for n bodies and made to include the effects of nongravitational disturbances, such as atmospheric drag, and of gravitational anomalies, such as those due to the nonspherical earth. Closed-form solutions of such complicated systems are impossible and numerical computer solutions are commonly used. For artificial earth satellites many important results can be found from the simple two-body problem.

We define an inertial frame of reference (X', Y', Z') as one in which Newton's laws apply and we consider two bodies, M and m, in this frame as seen in Figure 2-1 and assume that each acts on the other in accordance with Eqs. (2-1) and (2-2). We have

$$m\ddot{\mathbf{r}}_m = -\frac{GMm}{r^2}\frac{\mathbf{r}}{r}$$

$$M\ddot{\mathbf{r}}_M = +\frac{GMm}{r^2}\frac{\mathbf{r}}{r}$$

and

$$\mathbf{r} = \mathbf{r}_m - \mathbf{r}_M$$

Sec. 2.2   Conservation of Energy and Momentum   31

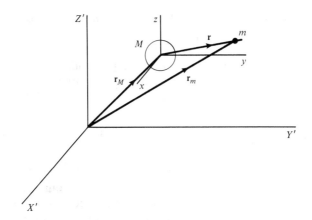

Figure 2-1   Gravitational interaction of two particles.

Subtracting yields

$$\ddot{\mathbf{r}} = -\frac{G(M + m)}{r^3}\mathbf{r} \qquad (2\text{-}3)$$

This is the basic vector differential equation for the two-body problem. It can be expanded in any convenient coordinate system. As long as XYZ is a nonrotating frame, the magnitude and direction of $\mathbf{r}$ and $\ddot{\mathbf{r}}$ will be the same as in $X'Y'Z'$, and the problem can be dealt with in the noninertial frame XYZ centered on the large mass M.

For artificial earth satellites (but not the moon, since it is too large for the approximation)

$$M \gg m \quad \text{and} \quad G(M + m) \cong GM = \mu$$

The value of μ is 398601.8 km³/s² (Soop, 1983). Equation (2-3) then becomes

$$\ddot{\mathbf{r}} + \frac{\mu}{r^3}\mathbf{r} = 0 \qquad (2\text{-}4)$$

This is the two-body differential equation used in the study of artificial earth satellites.

2.2 CONSERVATION OF ENERGY AND MOMENTUM

Either by the judicious use of vector algebra (*Handbook of Chemistry and Physics*, Mathematics Supplement) or by assuming the validity of the laws of conservation of energy and momentum, the following two simple and basic results are derivable:

$$E = \frac{v^2}{2} - \frac{\mu}{r} \qquad (2\text{-}5)$$

where E is the total energy (called the *vis-viva equation*) and

$$\mathbf{h} = \mathbf{r} \times \mathbf{v} \qquad (2\text{-}6)$$

where $\mathbf{h}$ is the angular momentum per unit mass. Both E and $\mathbf{h}$ are constant. Since $\mathbf{h}$ is a vector, $\mathbf{r}$ and $\mathbf{v}$ must always be in the same plane—an important result.

2.3 KEPLER'S LAWS

Before Newton put celestial mechanics on a firm theoretical basis, Kepler had stated his three laws based on inferences drawn from the extensive observations of Mars by Tycho Brahe. They were stated for planetary motion about the sun, but are equally applicable to earth satellites and are a good starting point:

1. The orbit of each planet (satellite) is an ellipse with the sun at a focus.
2. The line joining the planet (satellite) to the sun sweeps out equal areas in equal times.
3. The square of the period is proportional to the cube of the mean distance from the focus.

The second and third laws can be inferred from the conservation of momentum and energy, as just developed, but proving the first law requires a solution of differential equation (2-4). It can be written in polar coordinates[1] as

$$\ddot{r} - r\dot{v}^2 = -\frac{\mu}{r^2} \qquad \text{radial component}$$
$$r\ddot{v} + 2\dot{r}\dot{v} = 0 \qquad \text{transverse component} \qquad (2\text{-}7)$$

These equations can be integrated in a complicated but straightforward procedure. The result is that any conic section (i.e., an ellipse, parabola, or hyperbola) will satisfy the differential equations. The choice depends on the initial conditions. In fields such as ballistic missiles and interplanetary flight, all the solutions are of interest, but for artificial earth satellites we need only be concerned with the ellipse. Equation (2-8) is the polar coordinate equation of an ellipse with the origin at a focus, and is given as both a function of the angle v and the angle E[2]. In astronomical literature, these angles are called, respectively, the *true* and *eccentric anomalies*. The curious terminology is

---

[1] v is used as the angular measure rather than the more common θ, which is reserved for later use as an elevation angle. Some celestial mechanics texts use v, which is liable to confusion with velocity.

Sterne, Theodore E., *An Introduction to Celestial Mechanics, Interscience Tracts on Physics and Astronomy No. 9.* Interscience Publishers, Inc. New York, NY, 1960.

Sec. 2.3   Kepler's Laws

traceable to the Greek astronomers, who used the term "anomaly" to mean "a departure." It was the rotational angle of the epicycle—a geometrical device necessary to account for the departure between the observed planetary motions and those predicted assuming a simple circular orbit.

$$r = \frac{a(1 - e^2)}{1 + e \cos v} \quad \text{or} \quad r = a(1 - e \cos E) \tag{2-8}$$

The relation between v and E can be written in several convenient ways:

$$\cos E = \frac{e + \cos v}{1 + e \cos v}$$

$$\cos v = \frac{e - \cos E}{e \cos E - 1} \tag{2-9}$$

$$\tan \frac{v}{2} = \left(\frac{1 + e}{1 - e}\right)^{\frac{1}{2}} \tan \frac{E}{2} \quad \text{(Gauss)}$$

These equations are derivable by routine trigonometry from the diagram in Figure 2-2.

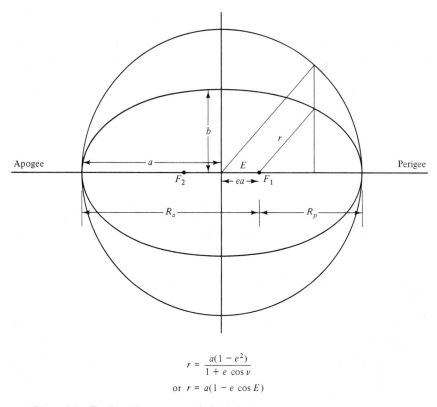

$$r = \frac{a(1 - e^2)}{1 + e \cos v}$$

$$\text{or } r = a(1 - e \cos E)$$

Figure 2-2   Earth orbit parameters in orbital plane.

The mean anomaly M is defined by

$$M = n(t - \tau) \tag{2-10}$$

where n = satellite mean angular motion = $2\pi/P$

P = period of satellite revolution

τ = time at reference epoch (perigee)

$$P = 2\pi \sqrt{\frac{a^3}{\mu}} \tag{2-11}$$

M is the mean anomaly of a planet (or satellite), the position the satellite would have if it moved at a constant angular velocity in a circle in its own plane.

The relation between M and E is famous, and was first derived empirically by Kepler:

$$M = E - e \sin E \tag{2-12}$$

To find the position of a satellite at any time, given the orbital parameters, is a classical problem. The mean anomaly M is easily calculated from Eq. (2-10), but regrettably Eq. (2-12) is not solvable in closed form for E. Once E is determined, Eq. (2-9) can be used to find v, the true anomaly, but Eq. (2-12) must be solved numerically. A solution for Eq. (2-12) in the form of a trigonometric series was developed by Lagrange:

$$E = M + 2 \sum_{n=1}^{\infty} \frac{1}{n} J_n(ne) \sin nM \tag{2-13}$$

where J_n is a Bessel function of the first kind of order n.

Even more usefully, a series expansion for v, the true anomaly, can be found directly in terms of M. It is called the *equation of the center*[2] and is

$$v = M + \left(2e - \frac{e^3}{4}\right) \sin M + \frac{5}{4} e^2 \sin 2M + \frac{13}{12} e^3 \sin 3M + \cdots \tag{2-14}$$

---

[2]The terminology derives from medieval astronomy when off-center circles (equants) were commonly used as models of "geocentric" planetary orbits. The difference between the observed position of a planet and its mean longitude could be approximated reasonably well if the angles to the planet were measured from a point off the center of the circle. The anomaly was thus related to this difference, hence the term "equation of the center."

Sec. 2.4 The Satellite Orbit in Space

If Eq. (2-14) is not sufficiently accurate, numerical methods (e.g., Newton's method) should be used to solve Kepler's equation, and ν then calculated from Eq. (2-9).

2.4 THE SATELLITE ORBIT IN SPACE

To specify completely the size and shape of an elliptical orbit and its orientation relative to the earth, five parameters are needed. Further, to specify the position of a satellite in its orbit, a sixth or time reference is needed. These six orbital parameters are

	a	semimajor axis
	e	eccentricity
	i	inclination
	Ω	right ascension of ascending node
	ω	argument of perigee
	τ	time of perigee

The semimajor axis a and eccentricity e are as shown in Figure 2-2. The remaining parameters, other than τ, time of perigee, are illustrated on Figure 2-3.

The satellite orbit is projected on the geocentric celestial sphere. This is a concept from ancient astronomy that is still extremely convenient. This sphere is conceived as being of infinite radius and its equator is the great-

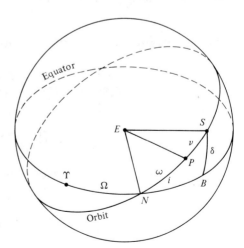

Figure 2-3 Earth orbit parameters in inertial coordinates.

circle intersection with the plane of the earth's equator. The reference points on Figure 2-3 are

E	center of Earth
S	satellite
N	ascending node of orbit
P	perigee
♈	vernal equinox (first point of Aries)

The inclination i is the angle between the orbit and the equator as each is projected on the celestial sphere; it is thus the angle between the plane of the orbit and the plane of the equator. The right ascension of the ascending node Ω is measured eastward from the vernal equinox ♈. It defines the orientation of the orbital plane. The argument of perigee ω defines the orientation of the elliptical orbit within the orbital plane. Finally, the time of perigee τ defines the "timetable" with which the satellite follows the orbit.

Occasionally, ϖ, the "longitude of perigee" is stated $\varpi = \Omega + \omega$ and is measured along a path lying in two different planes. When it is specified, either Ω or ω will usually be omitted.

The location of the satellite at a particular instant can be described by giving the distance to the satellite r and its true anomaly v. They constitute a set of polar coordinates within the plane of the orbit, with the semimajor axis of the orbital ellipse (to perigee) as the reference axis.

The location of the satellite can also be described in *geocentric inertial coordinates*, comprising the satellite's *right ascension* α, equal to the arc $\widehat{♈\beta}$, and its declination, δ, *equal to the arc* $\widehat{BS}$.

2.5 RELATIONS AMONG ORBIT PARAMETERS

It is not difficult to show that the total specific mechanical energy E of Eq. (2-5) is equal to $-\mu/2a$ (Smart, 1977). From that relation, another useful vis-viva equation follows easily:

$$v^2 = \mu\left(\frac{2}{r} - \frac{1}{a}\right) \qquad (2\text{-}15)$$

In the same reference it is shown that the product of the perigee and apogee velocities is constant and given by $v_p v_a = \mu/a$. Also,

$$v_p r_p = v_a r_a \qquad (2\text{-}16)$$

It is possible using the equations already given (and much tedious manipulation) to relate apogee and perigee velocities and radii to the ellipse parameters, energy, momentum, and orbital period in a wide variety of ways.

These relations are summarized in Table 2-1. This is a particularly useful collection.

2.6 THE GEOSTATIONARY ORBIT

Although a number of different orbits have been used for special purposes in satellite communication—and probably will continue to be used—the overwhelming interest for this application is in the *geostationary* orbit. This designation is preferable to the often-heard terms "synchronous" or "geosynchronous," since any orbit whose period of revolution is some multiple of the earth's can be considered as synchronous. Only a circular orbit whose period of rotation is equal to that of the earth's and whose plane is in the plane of the equator is geostationary. Such a satellite remains fixed in apparent position relative to the earth. The satellite's period of revolution must be synchronized with that of the earth in inertial space. To a very close approximation, this period is equal to that of the sidereal day. The negligible error is the result of measuring the sidereal day with reference to the vernal equinox, a point that is itself in slow motion. This universal reference point for all observational astronomy is in the direction of the line of intersection of the plane of the earth's equator and the plane of its orbit about the sun. Its extension to the fixed stars is at the point called the *vernal equinox*. Its symbol ♈ is that of the constellation Aries, in which the point was during the period of ancient astronomy. Note that the plane of the earth's orbit, the *ecliptic*, is also the plane of the sun's orbit about the earth, when looked at geocentrically. This is a convenient picture for artificial earth satellite work, and is perfectly correct geometrically. From that point of view, the equinoxes are those two points where the apparent path of the sun on the celestial sphere crosses the earth's equator. The vernal equinox is the crossing where the sun is going north and the autumnal equinox where it is going south. With either picture, because of the gravitational torque exerted by the sun and the moon on the nonspherical rotating earth, the plane of the earth's equator is in slow rotation with a precessional period of about 26,000 years. In addition, the plane of the ecliptic is itself in slow rotation because of the gravitational effects of other planets, particularly Venus and Jupiter, on the plane of the earth's orbit. As a result of this combined motion, there is a mean motion of the vernal equinox of $50''.25$ seconds per year; this produces a discrepancy of $0^s.008$ second between the earth's sidereal period of rotation and the period of rotation in inertial space.*

Figure 2-4 shows the characteristics of the geostationary orbit together with the transfer maneuver into this orbit, starting with a launch from Cape

---

*In this chapter and the next, we will use astronomical notation (*e.g.* $12^h\ 15^m\ 6^s.05$) for all time and angle values.

TABLE 2-1 Elliptical Orbit Parameter Relationships

Given Parameters	Semimajor Axis, a	Semiminor Axis, b	Apogee Radius, r_a	Perigee Radius, r_p	Eccentricity, e	Apogee Velocity, V_a	Perigee Velocity, V_p	Angular Momentum, h	Total Specific Energy, E	Period, P
a,e	a	$a\sqrt{1-e^2}$	$a(1+e)$	$a(1-e)$	e	$\sqrt{\dfrac{\mu}{a}\dfrac{1-e}{1+e}}$	$\sqrt{\dfrac{\mu}{a}\dfrac{1+e}{1-e}}$	$\sqrt{\mu a(1-e^2)}$	$-\dfrac{\mu}{2a}$	$2\pi\sqrt{\dfrac{a^3}{\mu}}$
r_a,r_p	$\tfrac{1}{2}(r_a+r_p)$	$\sqrt{r_a r_p}$	r_a	r_p	$\dfrac{r_a-r_p}{r_a+r_p}$	$\sqrt{\dfrac{2\mu}{r_a+r_p}\dfrac{r_p}{r_a}}$	$\sqrt{\dfrac{2\mu}{r_a+r_p}\dfrac{r_a}{r_p}}$	$\sqrt{\dfrac{2\mu r_a r_p}{r_a+r_p}}$	$-\dfrac{\mu}{r_a+r_p}$	$\pi\sqrt{\dfrac{(r_a+r_p)^3}{2\mu}}$
a,r_a	a	$\sqrt{r_a(2a-r_a)}$	r_a	$2a-r_a$	$\dfrac{r_a-a}{a}$	$\sqrt{\dfrac{\mu}{a}\dfrac{2a-r_a}{r_a}}$	$\sqrt{\dfrac{\mu}{a}\dfrac{r_a}{2a-r_a}}$	$\sqrt{\dfrac{\mu}{a}r_a(2a-r_a)}$	$-\dfrac{\mu}{2a}$	$2\pi\sqrt{\dfrac{a^3}{\mu}}$
a,r_p	a	$\sqrt{r_p(2a-r_p)}$	$2a-r_p$	r_p	$\dfrac{a-r_p}{a}$	$\sqrt{\dfrac{\mu}{a}\dfrac{r_p}{2a-r_p}}$	$\sqrt{\dfrac{\mu}{a}\dfrac{2a-r_p}{r_p}}$	$\sqrt{\dfrac{\mu}{a}r_p(2a-r_p)}$	$-\dfrac{\mu}{2a}$	$2\pi\sqrt{\dfrac{a^3}{\mu}}$
e,r_a	$\dfrac{r_a}{1+e}$	$r_a\sqrt{\dfrac{1-e}{1+e}}$	r_a	$r_a\dfrac{1-e}{1+e}$	e	$\sqrt{\dfrac{\mu}{r_a}(1-e)}$	$\sqrt{\dfrac{\mu}{r_a}\dfrac{(1+e)^2}{1-e}}$	$\sqrt{\mu r_a(1-e)}$	$-\dfrac{\mu}{2r_a}(1+e)$	$2\pi\sqrt{\dfrac{r_a^3}{\mu(1+e)^3}}$
e,r_p	$\dfrac{r_p}{1-e}$	$r_p\sqrt{\dfrac{1+e}{1-e}}$	$r_p\dfrac{1+e}{1-e}$	r_p	e	$\sqrt{\dfrac{\mu}{r_p}\dfrac{(1-e)^2}{1+e}}$	$\sqrt{\dfrac{\mu}{r_p}(1+e)}$	$\sqrt{\mu r_p(1+e)}$	$-\dfrac{\mu}{2r_p}(1-e)$	$2\pi\sqrt{\dfrac{r_p^3}{\mu(1-e)^3}}$
V_a,V_p	$\dfrac{\mu}{V_a V_p}$	$\dfrac{2\mu}{(V_a+V_p)\sqrt{V_a V_p}}$	$\dfrac{2\mu}{V_a(V_a+V_p)}$	$\dfrac{2\mu}{V_p(V_a+V_p)}$	$\dfrac{V_a-V_p}{V_a+V_p}$	V_a	V_p	$\dfrac{2\mu}{V_a+V_p}$	$\dfrac{V_a V_p}{2}$	$\dfrac{2\pi\mu}{\sqrt{V_a^3 V_p^3}}$
V_a,r_a	$\dfrac{\mu r_a}{2\mu-r_a V_a^2}$	$r_a V_a\sqrt{\dfrac{r_a}{2\mu-r_a V_a^2}}$	r_a	$\dfrac{r_a^2 V_a^2}{2\mu-r_a V_a^2}$	$1-\dfrac{r_a V_a^2}{\mu}$	V_a	$\dfrac{2\mu-r_a V_a^2}{r_a V_a}$	$r_a V_a$	$\dfrac{2}{r_a}-\dfrac{V_a^2}{\mu}$	$2\pi\mu\left(\dfrac{r_a}{2\mu-r_a V_a^2}\right)^{3/2}$
V_p,r_p	$\dfrac{\mu r_p}{2\mu-r_p V_p^2}$	$r_p V_p\sqrt{\dfrac{r_p}{2\mu-r_p V_p^2}}$	$\dfrac{r_p^2 V_p^2}{2\mu-r_p V_p^2}$	r_p	$\dfrac{r_p V_p^2}{\mu}-1$	$\dfrac{2\mu-r_p V_p^2}{r_p V_p}$	V_p	$r_p V_p$	$\dfrac{2}{r_p}-\dfrac{V_p^2}{\mu}$	$2\pi\mu\left(\dfrac{r_p}{2\mu-r_p V_p^2}\right)^{3/2}$

*Subscripts a and p refer to apogee and perigee values, respectively.
Source: *Handbook of Mathematical Tables*, CRC Press, Boca Raton, Fla., 1962, p. 423.

Sec. 2.7 Orbital Transfers

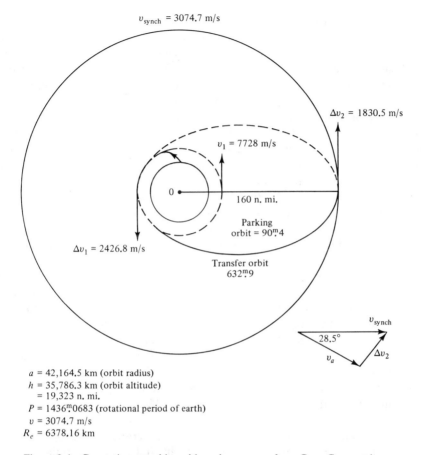

$a = 42{,}164.5$ km (orbit radius)
$h = 35{,}786.3$ km (orbit altitude)
$\quad = 19{,}323$ n. mi.
$P = 1436^m.0683$ (rotational period of earth)
$v = 3074.7$ m/s
$R_e = 6378.16$ km

Figure 2-4 Geostationary orbit and launch sequence from Cape Canaveral.

Canaveral. This transfer maneuver is discussed in the next section. Further refinements on geostationary orbit calculations can be made by considering the nonspherical earth and the effects of lunar and solar gravitation. These effects are discussed to some extent in a later section.

2.7 ORBITAL TRANSFERS

One of the most important applications of orbital mechanics in space communications is planning the transfer of satellites from "parking" orbit to final operational orbit. Indeed, the launch of the satellite into operational orbit is conveniently divided into two phases, the first of which is devised to place the satellite in a low earth orbit. The second is to transfer it from this parking orbit to the desired one. The first phase is essentially a powered flight. Its theoretical analysis has many elements in common with that of aerodynamics of ordinary aircraft. Some insight into this part of the mission is given in Chapter 4. For the discussion in this section, we assume that the spacecraft is already in a low earth orbit (LEO). Typically, the launch of a satellite on the Space Transportation System (STS) or "Shuttle" will place the spacecraft in 160-nautical mile parking orbit inclined at 28.5°. The basic maneuver to a higher-altitude orbit, circular or elliptical, was first studied by the German physicist Hohmann almost 60 years ago, in connection with interplanetary flight. Hohmann showed that the minimum energy transfer was obtained by giving the spacecraft two increments of velocity. The first increment is given to place it in an elliptical transfer orbit whose apogee is at the desired final height. This height can be the altitude of a circular orbit or the apogee of a second elliptical orbit. If nothing further were done, the spacecraft would remain in this orbit, returning to perigee at the same point in inertial space where the first velocity increment was given. On the other hand, if a second incremental velocity is given, equal to the difference between the transfer orbit apogee velocity and the velocity in final orbit at the desired height, the desired orbit will be achieved. If the parking orbit and operational orbit are not coplanar, changes in inclination can also be made, either at perigee, apogee, or both. It is common to make most of the total plane change at apogee because the required changes in velocity, and thus the required expenditure of propellant, are less. When the final orbit is the geostationary one, the perigee "burn" is given when the satellite, in parking orbit, crosses the plane of the equator—a "node" of the parking orbit. The major axis of the elliptical transfer orbit will thus lie in the equatorial plane, even though the plane of the transfer orbit is still inclined. Thus, the satellite will also be in the equatorial plane at apogee, a requirement for its insertion into geostationary orbit. Equations for required velocity increments are easily derived from the basic results that we have already had. In fact, the formulas of Table 2-1 can be used in a straightforward—albeit algebraically complicated—manner, to arrive at the useful relations in Eqs. (2-17) through (2-20).

For circular parking and final orbit:

$$R_E = \text{earth radius}$$

$$h_1 = \text{parking orbit altitude}$$

$$v_1 = \text{parking orbit velocity}$$

$$v_2 = \text{final orbit velocity}$$

$$h_2 = \text{final orbit altitude}$$

$$\Delta v_1 = v_p - v_1$$

$$\Delta v_2 = v_2 - v_a$$

Sec. 2.7 Orbital Transfers

If

$$K = \frac{R_E + h_1}{R_E + h_2} \quad \text{and} \quad v_1 = \sqrt{\frac{\mu}{R_E + h_1}}$$

then

$$\Delta v_1 = v_1 \left(\sqrt{\frac{2}{1+K}} - 1 \right) \tag{2-17}$$

$$\Delta v_2 = v_1 \sqrt{K} \left(1 + \frac{2K}{1+K} - 2 \cos \zeta \sqrt{\frac{2K}{1+K}} \right)^{1/2} \tag{2-18}$$

where ζ is the plane change at apogee as in Figure 2-5.

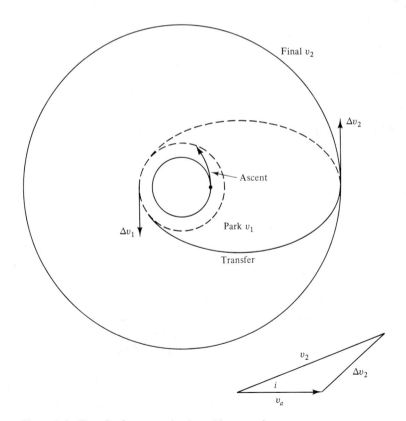

Figure 2-5 Transfer from one circular orbit to another.

If there is a plane change ξ at perigee, Δv_1 is given by

$$\Delta v_1 = v_1 \left(1 + \frac{2}{1+K} - 2\cos\xi \sqrt{\frac{2}{1+K}}\right)^{1/2} \quad (2\text{-}19)$$

If there is *only* a plane change at apogee and no circularization,

$$\Delta v_2 = 2v_0 K \sqrt{\frac{1}{1+K}} \sqrt{1 - \cos\zeta} \quad (2\text{-}20)$$

It is routine to show that for a given total plane change there is an optimum apportionment between apogee and perigee. One takes the expression for the total Δv, given by the sum of the expressions in Eqs. (2-18) and (2-19), and minimizes it by differentiation. The results can easily be found numerically. As an example, the launch into geostationary orbit from Cape Canaveral is optimized if about 2.2° of inclination is removed at perigee and the remainder at apogee.

The diagram of Figure 2-6 is given by Kaplan and Yang (1982) to show a typical vector diagram for injection into geostationary orbit from Cape Canaveral (Eastern Test Range, ETR).

Table 2-2 shows some typical Δv requirements for a set of common satellite orbits. The parking orbit is assumed to be 160 nautical miles. The velocity increments and the characteristics of the transfer orbit assume that the entire plane change is at apogee. As we have already mentioned, this is

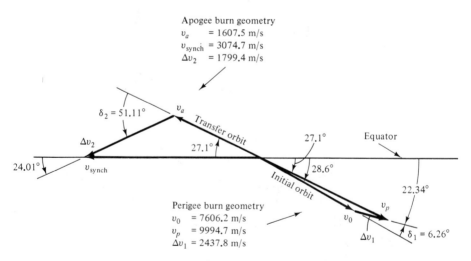

Figure 2-6 ETR launch into 296-km circular orbit at 28.6°. (M. Kaplan and W. Yang, "Finite Burn Effect on Ascent Stage Performance" *Journal of Astronomical Sciences*, Vol. XXX, No. 4 Oct–Dec, 1982, pp. 403–414.)

Sec. 2.8   Transfer Between Coplanar Elliptic Orbits   43

TABLE 2-2 Characteristics of Common Circular and Transfer Orbits

Final Orbit Characteristics				Transfer from 160 Nautical Mile (296.3 km) Parking Orbit			
h (km)	P (min)	v (m/s)	i (deg)	Δv_1 (m/s)	Δv_2 (m/s)	T_{tr} (min)	
296	90.4	7,728	—	—	—	—
500	94.6	7,613	0.0	57.9	57.4	92.5
1,000	105.1	7,350	0.0	191.1	186.4	97.7
5,000	201.3	5,919	0.0	948.6	829.2	142.2
10,000	347.7	4,933	0.0	1,484.0	1,179	205.3
20,000	710.6	3,887	0.0	2,035	1,417	352.4
35,786	1,436.1	3,075	0.0	2,427	1,467	632.9
35,786	1,436.1	3,075	2.0	2,459	1,469	632.9
35,786	1,436.1	3,075	5.0	2,459	1,480	632.9
35,786	1,436.1	3,075	8.0	2,459	1,500	632.9
35,786	1,436.1	3,075	10	2,459	1,518	632.9
35,786	1,436.1	3,075	28.5	2,459	1,837	632.9
35,786	1,436.1	3,075	45.6	2,459	2,263	632.9
35,786	1,436.1	3,075	62.8	2,459	2,742	632.9
60,000	2,836.6	2,451	0.0	2,690	1,403	632.9
150,000	10,257.1	1,597	—	2,975	1,140	1,157.9
Escape	—	—	—	3,201	—	—

not necessarily the minimum fuel mission when a specific launch is considered, and there is often a small change in inclination at perigee. The table shows the variation in velocity increment as a function of apogee plane change. This is important when it is noted that the minimum inclination for the transfer orbit will be the latitude of the launch site. Any lesser inclination requires a powered turn on the part of the launch vehicle. Chapter 4 contains further elaboration on this subject, since the plane change is not the only effect of the launch latitude. The earth's velocity of rotation is also a significant factor. It is interesting to note that the total velocity increments for circular orbits above about 10,000 km are greater than the total velocity increments to "escape." In other words, high-altitude circular orbits are considerably more difficult to achieve than is total escape from the earth's gravitational influence.

2.8 TRANSFER BETWEEN COPLANAR ELLIPTIC ORBITS

It is often practical to accomplish the transfer from parking to operational orbit by a series of velocity increments at perigee or apogee or both. One common reason for doing this is to permit the use of low-thrust engines. In that case, if one attempted to give the entire velocity increment in a single burn, the time taken would be long enough so that the velocity increment could no longer be considered as "impulsive." There would be a loss in efficiency of the maneuver. Relationships among payload thrust, burn time, and propellant characteristics are discussed in Chapter 4. Another practical reason is to permit the careful measurement and control of spacecraft attitude on successive apogees so as to produce an accurate final orbit. We present here a collection of equations that permit the calculation of velocities and incremental velocities in a transfer between any two coplanar elliptical orbits. Once again, these equations can be derived from the formulas in Table 2-1. R_{a1}, R_{a2}, and R_{p1}, R_{p2} are the radii at apogee and perigee, respectively, of the initial and final orbits. e_1 and e_2 are the eccentricities of those orbits. The results are perfectly general, and can be modified to consider the case of noncoplanar transfers. It is necessary only to consider the vector addition of velocities and the vector triangle is solved using the law of cosines, as was done in the case of transfer between circular orbits (see Figure 2-7).

$$\Delta v_1 = \sqrt{\frac{\mu}{R_{p1}}} \left[\sqrt{\frac{2(R_{a2}/R_{p1})}{1 + (R_{a2}/R_{p1})}} - \sqrt{1 + e_1} \right]$$

$$\Delta v_2 = \sqrt{\frac{\mu}{R_{a2}}} \left[\sqrt{1 - e_2} - \sqrt{\frac{2}{1 + (R_{a2}/R_{p1})}} \right]$$

(2-21)

Also:

$$v_{p1} = \sqrt{\frac{\mu}{R_{p1}} (1 + e_1)} \qquad v_{a1} = \sqrt{\frac{\mu}{R_{a2}} \frac{2}{1 + (R_{a2}/R_{p1})}}$$

$$v_{p2} = \sqrt{\frac{K}{R_{p1}} \frac{2(R_{a2}/R_{p1})}{1 + (R_{a2}/R_{p1})}} \qquad v_{a2} = \sqrt{\frac{\mu}{R_{a2}} (1 - e_2)}$$

(2-22)

2.9 CHANGE IN LONGITUDE

The satellite longitude in earth coordinates at the moment of insertion into geostationary orbit is often a matter of launch convenience, determined by many factors, several of which are discussed in Chapter 4. At the same time,

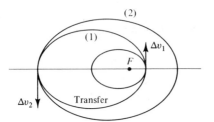

Figure 2-7 Coplanar transfer between elliptical orbits.

the longitude at which the geostationary satellite must be ultimately placed is a matter of operational necessity. It is thus often necessary to move the satellite from its initial position to some other longitudinal location. There are other occasional operational necessities, such as those caused by satellite failures, that require moving a geostationary satellite from one longitude to another.

This maneuver is accomplished by temporarily placing the satellite in a nonsynchronous orbit and permitting the earth to rotate, relatively, under it. At the appropriate moment, it is returned to synchronous orbit at the desired longitude.

The required Δv, for a change of $\Delta\lambda$ degrees/day to the west in longitude, is given by

$$\Delta v = v_s \left[\sqrt{2 - \left(1 + \frac{\Delta\lambda}{360°}\right)^{-2/3}} - 1 \right] \qquad (2\text{-}23)$$

where v_s is the velocity in geostationary orbit. The total requirement is $2\Delta v$ to initiate and then stop the drift. Note that it is dependent only on the rate of longitude change, not on the total change itself.

2.10 ORBITAL PERTURBATIONS

2.10.1 General

So far, we have been concerned with the important but idealized theory of the motion of two bodies when each can be considered as a point mass. In reality, we have dealt with a basic and still simpler case—the *one-body problem*—in which the mass of one body is completely negligible compared to the other. Artificial earth satellites meet that criterion nicely.[1] Unfortunately, the assumption that we have only a two-body problem is not correct because of the presence of the moon and sun, and the assumption that point masses are involved is also not correct since the earth is far from spherical. The earth's distorted shape, the atmosphere, the presence of the sun and moon, and other factors give rise to a series of orbital perturbations that are important. (See Figure 2-8.) They affect both the launch sequence and the final operational orbit and influence the design of the spacecraft in many ways. In a typical geostationary communication satellite, provision for dealing with the gravitational effects of the sun and moon and anomalies of the earth's

---

[1] Note that the motion of a large earth satellite, such as the moon, cannot be handled with that simplfied theory. Lunar motion is one of the most complicated problems in celestial mechanics.

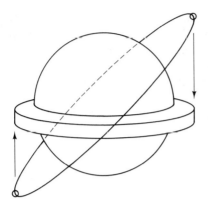

Figure 2-8 Orbital perturbations: sun and moon, oblate earth, triaxial earth.

gravitational field can typically add from 20 to 40% to the total dry mass of the spacecraft in station-keeping propellant. The theoretical treatment of these perturbations is complicated and well beyond the scope of a systems engineering text of this kind. They are properly the problem of a specialist in celestial mechanics, but some appreciation of them is necessary. We state only the most important results of these theories here and give plausible arguments for them, without making any attempt to prove the formulas.

Orbital perturbations of artificial satellites can be placed into three categories:

1. Those due to the presence of other large masses (i.e., the sun and moon)
2. Those resulting from not being able to consider either the earth or the artificial satellite as a point mass
3. Those due to nongravitational sources:
    a. The radiation pressure of the sun
    b. The earth's magnetic field
    c. Micrometeorites
    d. Atmosphere

2.10.2 The Sun and Moon

In the first category, we have the disturbing effects of both the sun and the moon. The precise mathematical analysis of these disturbances is complicated, but the results are simple and can be appreciated physically. One notes that a geostationary satellite is in the plane of the equator, whereas the geocentric orbits of the sun and moon are both inclined positively with respect to the equator. Since the moon is in revolution about the same central body with a period long compared to that of the geostationary satellite, each day the effect of the moon's gravity is to increase the inclination of the satellite slightly (i.e., to pull it out of the plane of the equator). The perturbation is analogous to that of large planets like Jupiter on the plane of the earth's orbit

Sec. 2.10   Orbital Perturbations

in the heliocentric system. The effects of the sun's mass are more subtle inasmuch as the two-body earth–satellite system is already in orbit around the sun. Nevertheless, we can guess that approximately the same thing happens, although to a lesser extent because of the aforesaid orbital motion. The sun produces a kind of gradient effect because of the slightly different distance from the sun to the earth and to the satellite. Although the sun's mass is great, its distance from the earth's orbit is, too. When all the effects are added together, again there is a net tendency to cause the inclination of the satellite to increase slowly. The total effects produce a rotation of the orbital plane of somewhere between $0°.75$ and $0°.95$ per year depending on the launch date. If this effect were not corrected, it would produce a figure-eight apparent oscillation of the satellite at the rate of one per sidereal day. The velocity increment required to correct this rotation is about 50 m/s per year, a large number compared to the effect of other perturbations. It is the major disturbance to be considered in the accurate positional control of a geostationary satellite. There is also a small effect on the initial radius of the satellite, depending on the position of the sun and moon at the time of launch, which can cause a departure of somewhere between -0.5 and $+0.24$ km in the radius from the nominal geostationary value.

A more precise computation of the annual velocity increment for lunar–solar corrections can be made using Kamel's results. Table 2-3 lists the rate

TABLE 2-3 Annual Velocity Increment for Lunar–Solar Corrections

Launch Date (January 1)	$\Delta i/\Delta t$ (0.01°/month)
1981	6.69
1982	6.97
1983	7.25
1984	7.50
1985	7.69
1986	7.82
1987	7.86
1988	7.84
1989	7.74
1990	7.58
1991	7.37
1992	7.12
1993	6.85
1994	6.60
1995	6.40
1996	6.28
1997	6.26
1998	6.36
1999	6.55
2000	6.80

of change of inclination in degrees per month as a function of launch date. Values for dates other than January 1 can be found by interpolation. Over the satellite lifetime T_L, the average value for the years in question is used in the following formula. From the vector diagram for a change in inclination only, Δi, and a constant orbital velocity v_s:

$$\Delta v = 2v_s \sin \frac{\Delta i}{2}$$

or, approximately,

$$\Delta v = v_s \frac{\pi}{180} \left(\frac{\Delta i}{\Delta t}\right)_{av} T_L \qquad (2\text{-}24)$$

Note that the table for $\Delta i/\Delta t$ repeats its values every 18.6 years—the period of a complete revolution of the moon's node around the ecliptic. Thus it can be used for satellite system planning well into the future.

2.10.3 The Oblate Earth

The second category of perturbation includes those due to the non-spherical shape of the earth. The oblate nature of the earth, with its significant equatorial bulge, has been known since Newton's day. (The *geoid*, or gravitational equipotential of the earth, is now known to bulge unsymmetrically and to have approximately a pear shape.) The ellipticity of the equator, giving rise to a completely unsymmetrical (or "triaxial" earth), has been appreciated only since the advent of the artificial satellite era.

The gravitational potential of the earth, instead of being simply μ divided by r (the value for a point source) can be shown to be equal to (Moulton, 1970)

$$U = -\frac{\mu}{r}\left[1 - \sum_{n=2}^{\infty} J_n \left(\frac{R_E}{r}\right)^n P_n(\sin \phi)\right] \qquad (2\text{-}25)$$

where ϕ is the latitude and r is the distance from the earth's center. The earth's equatorial radius R_E is 6378.144 km (Soop, 1983). Taking only P_2, the second-order Legendre polynomial, and neglecting higher-order terms, we have

$$U = -\frac{\mu}{r}\left[1 - J_2\left(\frac{R_E}{r}\right)^2 \frac{3\sin^2\phi - 1}{2} + \cdots\right] \qquad (2\text{-}26)$$

The constant J_2 is sometimes call the oblateness coefficient and is equal to 1.08264×10^{-3}. Subsequent J terms involving both the latitude and longitude (zonal and tesseral harmonics) can be taken if the precision of the analysis justifies it. For instance, a simple modification has been made by adding a term, $J_2^{(2)}$ for the ellipticity (Frick and Garber, 1962):

$$U = -\frac{\mu}{r}\left[1 - J_2\left(\frac{R_E}{r}\right)^2 \frac{3\sin^2\phi - 1}{2} + 3J_2^{(2)}\left(\frac{R_B}{r}\right)^2 \cos^2\phi\cos 2\lambda\right] \qquad (2\text{-}27)$$

Sec. 2.10 Orbital Perturbations

where λ is the longitude.

$J_2 = 1.08264 \times 10^{-3}$ (earth oblateness coefficient)

$J_2^{(2)} = 5.35 \times 10^{-6}$ (Elliptically coefficient from Vanguard measurements)

The analysis of artificial satellite motion under the influence of the oblate earth—that is, using the potential given by Eq. (2-26) (Sterne, 1960)—has been carried out in several places, and the results are important.

There are two principal effects of the J_2 term. They are a rotation of the orbital plane (nodal regression) and a rotation of the major axis within the plane of the orbit.

Nodal regression (Figure 2-9):

$$\dot{\Omega} = -\frac{3}{2}\frac{2\pi J_2 \cos i}{(1-e^2)^2}\left(\frac{R_E}{a}\right)^2 \quad \text{rad/rev}$$

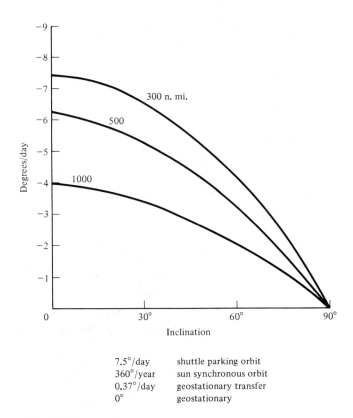

7.5°/day	shuttle parking orbit
360°/year	sun synchronous orbit
0.37°/day	geostationary transfer
0°	geostationary

Figure 2-9 Nodal regression.

or (2-28)

$$\dot{\Omega} = -\frac{9.9641}{(1-e^2)^2}\left(\frac{R_E}{a}\right)^{3.5} \cos i \quad \text{deg/day}$$

The minus sign indicates that the node drifts westward for direct orbits $i < 90°$. The drift is eastward for retrograde orbits ($i > 90°$).

Apsidal rotation (Figure 2-10):

$$\dot{\omega} = \frac{3}{4}\frac{2\pi J_2}{(1-e^2)^2}\left(\frac{R_E}{a}\right)^2 (5\cos^2 i - 1) \quad \text{rad/rev}$$

or (2-29)

$$\dot{\omega} = \frac{4.9821}{(1-e^2)^2}\left(\frac{R_E}{a}\right)^{3.5} (5\cos^2 i - 1) \quad \text{deg/day}$$

An estimate of its pointing error resulting from nodal regression is

$$\varepsilon = 2\sin^{-1}(\sin i \sin \Omega) \quad (2\text{-}30)$$

A third effect of oblateness is that the period of revolution, conveniently

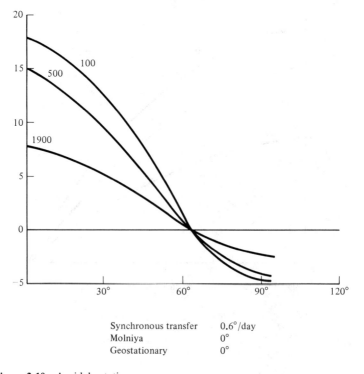

Synchronous transfer	0.6°/day
Molniya	0°
Geostationary	0°

Figure 2-10 Apsidal rotation.

defined as the time between successive passages across the ascending node, is reduced. The nodal period is (Kalil and Martikan, 1963):

$$P_N = \left[1 - \frac{3}{8} J_2 \left(\frac{R_E}{\bar{a}}\right)^2 (1 - e^2)^{-2}(7 \cos^2 i - 1)\right] \bar{P} \quad (2\text{-}31)$$

where $\bar{a}$ is the mean semimajor axis and $\bar{P}$ is the Keplerian period calculated by substituting $\bar{a}$ for a in Eq. (2-11).

Perturbations of the geostationary orbit. The oblate earth produces two effects that cause an error in the satellite period which in principle would require a slight modification to the orbital radius. For a geostationary orbit with $\bar{P} = 86\,164.1$ s, $\bar{a} = 42\,164.1$ km, $e = 0$, and $i = 0$, the ascending node (which may be visualized for very small inclinations) drifts westward by 0.0134 degree in one revolution according to Equation (2-28). On this account the satellite would arrive over the equator late by 3.2 seconds. However, by Equation (2-31) the period is actually shortened by 4.8 seconds and so the satellite actually arrives early by a net 1.6 seconds. To compensate for this, the period should be increased 1.6 seconds by raising the orbit slightly. The new period is thus 86 165.7 s. The corresponding mean orbital radius and height above the equator are 42 164.7 km and 35 786.6 km, respectively. These values may be taken as excellent nominal values for an ideal geostationary orbit and compare well with some representative values for actual satellites listed in Table 2-4. In practice, however, the oblateness correction is negligible in comparison to perturbations caused by the attractions of the sun and the moon, the triaxial component of the earth's shape, and solar radiation pressure and in comparison with orbit injection errors.

Triaxial earth (Figure 2-11). The effect of the noncircular earth's equator can be calculated by considering the longitudinally dependent terms in the expansion for the earth potential. The result is that there are four equilibrium points at the ends of the major and minor axes of the ellipse. The ends of the minor axis are stable equilibrium points where a satellite will stay with no station keeping effort. A satellite at the end of the major axis is in unstable equilibrium and will tend to drift away. It must have its position actively maintained, but the required velocity increment is very slight. Kamel et al. (1973) have analyzed this problem in great detail and give equations for the corrections as a function of longitude and the various coefficients of the earth's potential. The principal result is seen in the curve of Figure 2-12. It is interesting to note that the total annual Δv requirement is independent of the precision with which the satellite is held on longitude. On the other hand, the frequency of required orbital maneuvers will increase notably as the station-keeping accuracy is made tighter.

TABLE 2-4 Examples of Geostationary Satellites

Name	Launch	T (min)	i (day)	h_a (km)	h_p (km)	$\bar{h}$ (km)	a (km)
RCA SATCOM IV	1/16/82	1436.2	0.0	35 798	35 778	35 788.0	42 166.1
WESTAR 4	2/26/82	1436.1	0.0	35 794	35 777	35 785.5	42 163.6
INTELSAT 5 F-4	3/5/82	1436.2	0.0	35 804	35 773	35 788.5	42 166.6
WESTAR 5	6/9/82	1436.2	0.0	35 790	35 787	35 788.5	42 166.6
RCA SATCOM V	10/28/82	1436.1	0.0	36 107	35 466	35 786.5	42 164.6
TDRS 1	4/4/83	1436.1	1.0	35 798	35 776	35 787.0	42 165.1
GALAXY 1	6/28/83	1436.2	0.0	35 795	35 785	35 790.0	42 168.1
TELSTAR 3A	7/28/83	1436.0	0.0	35 792	35 781	35 786.5	42 164.6
INSAT 1B	8/31/83	1436.2	0.2	35 939	35 636	35 787.5	42.165.6
RCA SATCOM VII	9/8/83	1436.2	0.0	35 798	35 778	35 788.0	42 166.1
GALAXY 2	9/22/83	1436.2	0.0	35 796	35 784	35 790.0	42 168.1
INTELSAT 5 F-7	10/19/83	1436.1	0.0	35 802	35 773	35 787.5	42 165.6
BS-2A	1/23/84	1436.1	0.0	35 796	35 777	35 786.5	42 164.6
INTELSAT 5 F-8	3/5/84	1436.1	0.1	35 795	35 778	35 786.5	42 164.6
COSMOS 1546	3/29/84	1436.1	0.8	35 796	35 778	35 787.0	42 165.1
SPACENET 1	5/23/84	1436.2	0.0	35 792	35 783	35 787.5	42 165.6
ECS 2	8/4/84	1436.1	0.0	35 813	35 759	35 786.0	42 164.1
TELECOM 1A	8/4/84	1436.2	0.0	35 802	35 773	35 787.5	42 165.6
TELSTAR 3C	9/1/84	1436.1	0.1	36 157	35 417	35 787.0	42 165.1
GALAXY 3	9/21/84	1436.3	0.2	35 804	35 775	35 789.5	42 167.6

Source: NASA Satellite Situation Report, Vol. 24, No. 2, 1984.

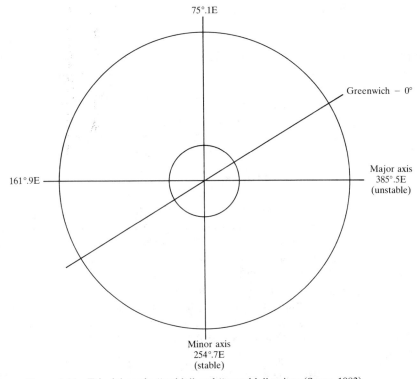

Figure 2-11 Triaxial earth: "stable" and "unstable" points (Soop, 1983).

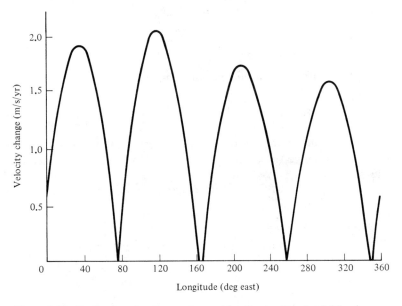

Figure 2-12 Station-keeping Δv requirement for the deadband ± 0.05 rad.

2.10.4 Gravity Gradient Effects

It is sometimes important not only to consider the non-point-mass nature of the earth but also of the satellite. If the satellite is extended geometrically, particularly in a radial direction, the forces of gravity vary along its length, and we have the *gravity gradient effect*. Since the gravitational force diminishes with the square of the distance, the gradient diminishes with the cube. Its effect in geostationary orbit is small and generally negligible. However, as satellites become larger and larger, it can be expected that gravity gradient will be one of the perturbing forces to be dealt with.

2.10.5 Other Perturbations

The principal nongravitational disturbances that have to be considered are the radiation pressure of the sun, the earth's magnetic field, and the influence of micrometeorites. The radiation pressure of the sun is by far the most important in geostationary orbit, since the earth's magnetic field diminishes so rapidly with distance. It can be a serious disturbing force if the satellite area is great and its mass small. Radiation pressure is given by the rate of change of momentum of the impinging photons per unit area. Since the photon energy per unit area, E, is mc^2, where m is the "mass" of the impinging photons, the radiation pressure can be written as

$$P_{rad} = \frac{d}{dt}(mc) = \frac{d}{dt}\left(\frac{mc^2}{c}\right) = \frac{d}{dt}\left(\frac{E}{c}\right) \quad (2\text{-}32)$$

But by definition Φ, the solar constant, is given by

$$\Phi = \frac{dE}{dt}$$

Thus for total absorption and normal incidence,

$$P_{rad} = \frac{\Phi}{c} \quad \text{where } \Phi = 0.137 \text{ W/cm}^2 \quad (2\text{-}33)$$

More generally, remembering that a reflected photon will have double the momentum change of an absorbed photon, for a reflection factor Γ and an angle of incidence i:

$$P_{rad} = \Phi(1 + \Gamma)\cos^2 i$$
$$P_{rad} = 4.57 \times 10^{-6}(1 + \Gamma)\cos^2 i \quad \text{N/m}^2 \quad (2\text{-}34)$$

If the effective surface area is A and the satellite mass is M, the acceleration due to this pressure is

$$a = \frac{P_{rad} A}{M} \quad (2\text{-}35)$$

This acceleration tends to change the orbit shape from circular to slightly elliptical. With area-to-mass ratios typical of most current satellites, the radiation pressure effects are small compared to other effects. As satellites use larger and larger solar panels, it will become a significant perturbation.

2.11 OTHER ORBITS FOR SATELLITE COMMUNICATION

Although the geostationary orbit is of the greatest interest for satellite communication, it is not without its defects (e.g., lack of coverage at the polar caps) and it is a difficult orbit to achieve from northern latitudes.

A variety of other orbits have been both used and proposed for communications-like applications, including synchronous and medium-altitude polar and equatorial orbits.

Of particular interest is the *Molniya orbit*. (See Figure 2-13.) This fascinating special case involves a highly elliptical orbit with apogee remaining over the northern hemisphere and a period synchronized to half a sidereal day. It can also be a full sidereal day (24 hours) if desired. The required inclination in both cases is given by

$$5 \cos^2 i - 1 = 0 \qquad (2\text{-}37)$$

where i has a value of 63.4°, and the resulting apsidal rotation is zero. This permits the apogee to remain over the northern hemisphere.

Figure 2-13 shows the "twelve hour" version of the Molniya orbit. The chosen value of perigee is a compromise between atmospheric drag for low values and loss in payload for higher values. After choice of perigee, the apogee is calculated from the required value of the major axis to yield the half-day period. Station keeping must be provided to allow for gravitational anomalies, the sun and moon, and initial inclination errors. The transfer and injection maneuvers to achieve these orbits are shown in Figure 2-14.

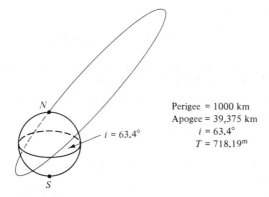

Figure 2-13 Molniya orbit.

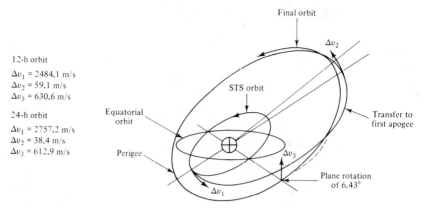

Figure 2-14 Launch into Molniya orbits from Cape Canaveral: 12-h and 24-h orbits at 63.43° inclination; ETR launch into 57° orbit at 296 km.

REFERENCES

FRICK, R. H. and T. B. GARBER, "Perturbations of a Synchronous Satellite," NASA Report R-399, Rand Corp., 1962.

Handbook of Chemistry and Physics, CRC Press, Boca Raton, Fla.

KALIL, F., and F. MARTIKAN, "Derivation of Nodal Period of an Earth Satellite and Comparisons of Several First-Order Secular Oblateness Results," *AIAA J.*, Vol. 1, No. 9, Sept. 1963, pp. 2041–2046.

KAMEL, A., D. EKMAN, and R. TIBBITS, "East–West Stationkeeping Requirements of Nearly Synchronous Satellites Due to Earth's Triaxiality and Luni-Solar Effects," *Celestial Mech.*, Vol. 8, 1973.

KAPLAN, M., and W. YANG, "Finite Burn Effects on Ascent Stage Performance," *J. Astronaut. Sci.*, Vol. 30, No. 4, Oct.–Dec. 1982, pp. 403–414.

MOULTON, F. R., *An Introduction to Celestial Mechanics*, Dover Publications, Inc., New York, 1970.

SMART, W. M., *Textbook on Spherical Astronomy*, Cambridge University Press, Cambridge, 1977.

Additional Readings

BATE, R., D. MUELLER, and J. WHITE, *Fundamentals of Astrodynamics*, Dover Publications, Inc., New York, 1971.

BERMAN, A. I., *Astronautics—Fundamentals of Dynamical Astronomy and Space Flight*, John Wiley & Sons, Inc., New York, 1961.

BROUWER, D. and G. M. CLEMENCE, *Methods of Celestial Mechanics*, Academic Press, Inc., New York, 1961.

KAPLAN, M., *Modern Spacecraft Dynamics and Control*, John Wiley & Sons, Inc., London, 1976.

MUELLER, I. I., *Spherical and Practical Astronomy As Applied to Geodesy*, Frederick Ungar Publishing Co., Inc., New York, 1977.

SOOP, E. M., Introduction to Geostationary Orbits, ESA SP-1053, ESA Scientific and Technical Publications, Noordwijk, The Netherlands, 1983.

3

The Geometry of the Geostationary Orbit

3.0 GENERAL CONSIDERATIONS

There is a wide variety of geometric problems in connection with geostationary satellites. They range from the simple to the extremely complicated. The calculation of radio-frequency link performance, including the effects of atmospheric attenuation, the calculation of antenna coverage and problems in pointing of earth station and satellite antennas, and the prediction of eclipses and sun outages all require the solution of some geometric problem.

3.1 BASIC GEOMETRY

In the first group we consider the most basic problems—ones that can be handled simply and without concern for the nonspherical earth. These involve: given the latitude ϕ of any earth station and the longitude λ taken relative to the subsatellite point, calculating the distance to the satellite and the azimuth and elevation angle of the earth station antenna required for it to point at the satellite. If we assume that the earth is spherical with a radius equal to its mean equatorial radius, we can calculate these quantities using the geometry of Figure 3-1. The basic trigonometric formulas needed are the laws of cosines and sines for plane and spherical triangles, listed in Table 3-1 for convenient reference.

From the spherical triangle *NES* and the plane triangle *OEP*, and using

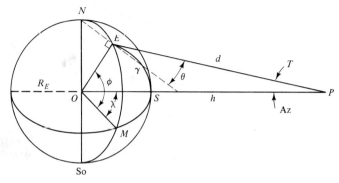

ϕ = goecentric latitude of earth station at E
λ = difference in longitude between E and subsatellite point S (taken positive if earth station is to the west of satellite)
γ = great circle arc ES = ∡ EOS
T = angle of inclination at satellite
R_E = equatorial radius of the earth
h = satellite altitude
θ = angle of elevation at earth station
Az = azimuth angle at earth station = ∡ NES (measured through east from north)
d = slant range from satellite to earth station
P = satellite

Figure 3-1 Basic satellite geometry.

the trigonometric relations of Table 3-1, we can derive the following elementary results in a routine series of trigonometric steps:

Slant range:

$$d = \sqrt{R_E^2 + (R_E + h)^2 - 2R_E(R_E + h)\cos\gamma} \qquad (3\text{-}1)$$

or

$$d = \sqrt{h^2 + 2R_E(h + R_E)(1 - \cos\phi\cos\lambda)} \qquad (3\text{-}1a)$$

Azimuth:

$$\sin Az = \frac{-\sin\lambda}{\sqrt{1 - \cos^2\phi\cos^2\lambda}} = -\frac{\sin\lambda}{\sin\gamma} \qquad (3\text{-}2)$$

Elevation:

$$\cos\theta = \frac{R_E + h}{d}\sqrt{1 - \cos^2\phi\cos^2\lambda}$$

$$= \frac{R_E + h}{d}\sin\gamma \qquad (3\text{-}3)$$

Sec. 3.1   Basic Geometry

TABLE 3-1   Reference Formulas from Plane and Spherical Trigonometry

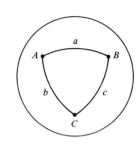

For any spherical triangle ABC whose side lengths a, b, and c are measured by the great circle arcs subtended at the center of the sphere:

$$\frac{\sin A}{\sin a} = \frac{\sin B}{\sin b} = \frac{\sin C}{\sin c} \quad \text{(sine law)}$$

$\cos a = \cos b \cos c + \sin b \sin c \cos A$ (cosine law)

$\cos A = -\cos B \cos C + \sin B \sin C \cos a$

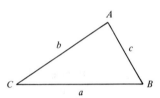

For any plane triangle ABC

$c^2 = a^2 + b^2 - 2ab \cos C$ (law of cosines)

$$\frac{\sin A}{a} = \frac{\sin B}{b} = \frac{\sin C}{c} \quad \text{(law of sines)}$$

and if

$$S = \frac{a + b + c}{2}$$

then

$$\tan \frac{C}{2} = \sqrt{\frac{(S-a)(S-b)}{S(S-c)}} \quad \text{(tangent law)}$$

It is occasionally useful to be able to calculate the slant range given only the angle of elevation:

$$d = (R_E + h)\frac{\cos(\theta + \zeta)}{\cos \theta} \tag{3-4}$$

where

$$\zeta = \sin^{-1}\frac{R_E \cos \theta}{R_E + H}$$

Note that the great-circle arc γ between the earth station and subsatellite point is given simply as

$$\cos \gamma = \cos \phi \cos \lambda \tag{3-5}$$

The limit of optical visibility is given when $\theta = 0$. After some routine algebra, we can derive from Eq. (3-3):

$$\cos \gamma = \frac{R_E}{R_E + h} \tag{3-6}$$

or $\cos \phi \cos \lambda = 0.1513$ for geostationary orbit.

In the calculation of interference between two satellites or two earth stations, it is necessary to calculate the angles subtended at the earth station or satellite in question, respectively. Reference to Fig. 3-2 shows the case of one satellite and two earth stations.

For earth stations at A and B, one calculates the great-circle arc ξ between them, then the chord p and distances d_A and d_B. Finally, using the triangle PAB, one calculates the angle β:

$$\cos \xi = \cos(90 - \phi_A) \cos(90 - \phi_B) \qquad (3\text{-}7a)$$
$$+ \sin(90 - \phi_A) \sin(90 - \phi_B) \cos \lambda$$

$$\cos \xi = \sin \phi_A \sin \phi_B + \cos \phi_A \cos \phi_B \cos \lambda \qquad (3\text{-}7b)$$

$$p = 2R_E \sin \frac{\xi}{2} \qquad (3\text{-}7c)$$

$$\cos \beta = \frac{d_A^2 + d_B^2 - p^2}{2 d_A d_B} \qquad (3\text{-}7d)$$

The angle ψ subtended by satellites at P_1 and P_2 (see Figure 3-3) separated by a longitude λ, at an earth station at E, is calculated in an analogous manner:

$$l = 2(R_E + h) \sin \frac{\lambda}{2}$$
$$\cos \psi = \frac{d_1^2 + d_2^2 - l^2}{2 d_1 d_2} \quad \text{from triangle } EP_1P_2 \qquad (3\text{-}8)$$

The angles can also be calculated from the "tangent" law in those cases where the cosine is so close to unity that numerical accuracy is difficult to achieve with ordinary calculators. For instance, in Eq. (3-8):

$$\tan \frac{\psi}{2} = \sqrt{\frac{(s - d_1)(s - d_2)}{s(s - l)}} \qquad (3\text{-}9)$$

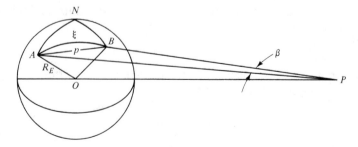

Figure 3-2 Geometry of two earth stations.

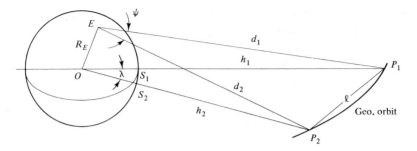

Figure 3-3 Geometry of two satellites.

where

$$s = \frac{d_1 + d_2 + l}{2}$$

3.2 SATELLITE COORDINATES

It is frequently desirable to locate an earth station in satellite-centered spherical coordinates α and β. If ϕ and λ are the earth station latitude and relative longitude, d is the slant range and R_E is the earth's radius, it is easy to show that

$$\sin \beta = \frac{R_E}{d} \sin \phi \qquad (3\text{-}10)$$

$$\sin \alpha = \frac{R_E}{d} \frac{\cos \phi \sin \lambda}{\cos \beta}$$

and as a check equation:

$$\cos \alpha \cos \beta = \cos T \qquad (3\text{-}11)$$

from the spherical right-triangle formula. T is the angle of inclination between the earth station and subsatellite point. It can be calculated from triangle OEP in Figure 3-1 or from Eq. (3-11) with the same result, given in Eq. (3-12):

$$\sin T = \frac{R_E}{d} \sqrt{1 - \cos^2 \phi \cos^2 \lambda} \qquad (3\text{-}12)$$

The relations above are useful in determining antenna pointing angles and in calculating the antenna gain toward a particular earth station. In calculating the coverage of a particular antenna, the reverse transformation is frequently useful. For instance, if the spherical coordinates (α, β) of a constant-gain contour of an antenna are given, that contour can be plotted on the earth as a function of the latitude and relative longitude (ϕ, λ) using the reverse transformation:

$$\gamma = \sin^{-1}\left(\frac{h + R_E}{R_E} \sin T\right) - T$$

$$\sin \phi = \frac{\sin \gamma}{\sin T} \sin \beta \qquad (3\text{-}13)$$

$$\cos \lambda = \frac{\cos \gamma}{\cos \phi}$$

γ is equal to the great-circle arc between the subsatellite point and the point in question on the contour.

3.3 ECLIPSE GEOMETRY

When a satellite is in the shadow of the earth, it is deprived of solar radiation—with two important effects. It is, for almost all communications satellites, without primary power and its temperature balance is changed sharply. The prediction of eclipse duration and onset time are thus of considerable importance. The general geometry of eclipses, with both the satellite and sun of finite size and the satellite in an arbitrary orbit, is extremely complicated. It is dealt with in books on spherical astronomy (Smart, 1977) where eclipses of the moon are calculated. We deal here with the simplified but practical case of a "point" satellite, the extended sun and geostationary orbit. Two separate geometric pictures must be analyzed. Figure 3-4 shows the geometry of the earth's shadow with a finite-size sun. That part of the shadow where the entire sun is blocked is called the *umbra* and the part of the shadow obscured from only part of the sun is rigorously called the *penumbra*. However, in the discussion to follow, *penumbra* is taken to mean the entire shadow (including the umbra). We are interested in calculating the geocentric half-angles ψ_1 and ψ_2, subtended by the umbra and (in the sense defined above) the penumbra, respectively.

$$\sin \alpha = \frac{R_E}{AE} = \frac{R_S}{AE + \rho}$$

where $\rho = SE$ (earth-sun distance)

Thus

$$\sin \alpha = \frac{R_S - R_E}{\rho} \qquad \text{for umbra geometry}$$

Sec. 3.3 Eclipse Geometry

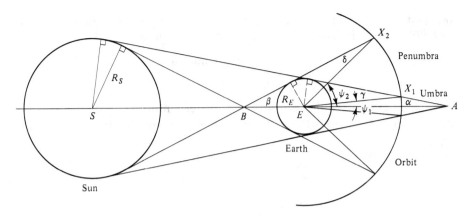

Figure 3-4 Eclipse geometry—umbra and penumbra.

Similarly,

$$\sin \beta = \frac{R_E}{\overline{BE}} = \frac{R_S}{\rho - \overline{BE}}$$

and

$$\sin \beta = \frac{R_S + R_E}{\rho} \quad \text{for penumbra geometry}$$

For a satellite in the plane of the figure at an altitude h, from triangles EAX and BEX_2:

$$\psi_1 = \gamma - \alpha$$
$$\psi_2 = \delta + \beta$$
$$\gamma = \delta = \sin^{-1} \frac{R_E}{R_E + h}$$

Therefore:

$$\psi_1 = \sin^{-1} \frac{R_E}{R_E + h} - \sin^{-1} \frac{R_S - R_E}{\alpha} \quad \text{half-angle of time in umbra} \quad (3\text{-}14)$$

$$\psi_2 = \sin^{-1} \frac{R_E}{R_E + h} + \sin^{-1} \frac{R_S + R_E}{\alpha} \quad \text{half-angle of time in penumbra} \quad (3\text{-}15)$$

We ignore the difference in the value of the distance to the sun, ρ, at the two equinoxes and take it equal to one astronomical unit (AU) or 149.598 × 10⁶ km. The radius of the sun is equal to 698,000 km.

$$\psi_1 = 8.43° \text{ for the umbra}$$

$$\psi_2 = 8.97° \text{ for the penumbra}$$

The durations of the eclipse are calculated simply as a proportion of the mean solar day—automatically accounting for the earth's orbital motion during eclipse:

$$T_1 = \frac{2\psi_1}{360} 1440 = 67^m\!.5 \quad \text{umbra}$$

$$T_2 = \frac{2\psi_2}{360} 1440 = 71^m\!.8 \quad \text{penumbra} \qquad (3\text{-}16)$$

The effect of the varying distance to the sun is less than 2ˢ and negligible for practical spacecraft problems. Note that if the geometry of a point-source sun is used, only the first terms in Eqs. (3-14) and (3-15) are used and an incorrect eclipse duration of 69ᵐ.6 min is found—it is the average of the umbra and penumbra times.

As the sun goes above or below the equator, the eclipse durations become less than the foregoing values and finally goes to zero when the declination of the sun gets high enough. The geometry of this problem is best handled with the geocentric diagram of Figure 3-5.

The circle C represents the earth's shadow (umbra or penumbra) projected on the celestial sphere. The satellite revolves diurnally about the equator while the earth's shadow progresses at the mean rate of the sun (0.985647° per day). The eclipse season starts when the circle C is tangent to the equator. The eclipse reaches its maximum duration when the center of the shadow goes through either equinox, and then declines to zero when the circle is again tangent to the equator. The great-circle arc ΥC is equal to the ecliptic longitude of the sun $\odot$ and the arc $\widehat{CD}$ is equal to the sun's declination δ.

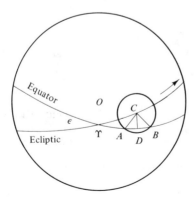

Figure 3-5 Eclipse geometry—seasonal variation.

Sec. 3.3 Eclipse Geometry

Note that the center of the earth's shadow moves relative to the equinox opposite to the location of the sun, but at exactly the same rate. The obliquity of the earth's orbit, ε, and the declination and longitude, are related by the spherical right-triangle relation

$$\sin \delta = \sin \varepsilon \sin \odot \qquad (3\text{-}17)$$

The arc length $\widehat{AC}$ is half the projected angular diameter of the earth's shadow ψ, as calculated for the umbra or penumbra in the preceding section. From the spherical right triangle ACD and the law of cosines:

$$\cos \widehat{AD} = \frac{\cos \psi}{\cos \delta} \qquad (3\text{-}18)$$

$$\widehat{AD} = \cos^{-1} \frac{\cos \psi}{\sqrt{1 - \sin^2 \varepsilon \sin^2 \odot}} \qquad (3\text{-}19)$$

The length of time in eclipse T_e is simply proportional to the arc length $\widehat{AD}$ compared to ψ, its length when the shadow circle C is on the equator. Thus

$$T_e = \frac{T_m}{\psi} \cos^{-1} \frac{\cos \psi}{\sqrt{1 - \sin^2 \varepsilon \sin^2 \odot}} \qquad (3\text{-}20)$$

This equation is plotted in Figure 3-6 for values of ψ corresponding to both the umbra and penumbra. The sun's longitude is taken from the simple *Astronomical Almanac* formula for that quantity as a function of day number. Variations from year to year and from one equinox to the other are negligible for any practical spacecraft purpose.

The moment at which the eclipse "season" starts can be calculated from $T_e = 0$ when

$$\sin \odot = \pm \frac{\sin \psi}{\sin \varepsilon} \qquad (3\text{-}21)$$

The moment of entering eclipse is when the satellite passes the point A. The arc $\widehat{\Upsilon A}$ is given by

$$\widehat{\Upsilon A} = \alpha - \psi \frac{T_e}{T_m} \qquad (3\text{-}22)$$

where α is the right ascension of the sun. It is related to the sun's longitude by the spherical right-triangle expression:

$$\tan \alpha = \cos \varepsilon \tan \odot \qquad (3\text{-}23)$$

Astronomical Almanac tabulations of α, or formulas for its computation as a function of d, the day number during a year, are given with α in time

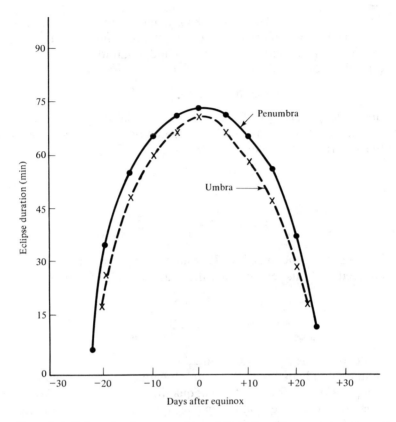

Figure 3-6 Eclipse duration versus time relative to equinox.

units (hours). The conversion from time units to arc is done simply at the earth's rotation rate of 15° per hour.

Eclipse onset time on any particular day is of interest in many satellite applications because that time on the clocks, in the area being served by the satellite, often determines the required service and thus the battery requirements on board the spacecraft.

The easiest way to calculate the clock time at the onset of satellite eclipse is to start from the geometrically evident fact that the eclipse peak is at local

TABLE 3-2

	Spring Equinox	Autumnal Equinox
Enter umbra	23^h56	23^h31
Enter penumbra	23^h52	23^h28

Sec. 3.3 Eclipse Geometry

solar midnight at the subsatellite point. The clock or "mean" time is given by

mean time = apparent solar time − equation of time[1]

$$MT = 24^h - E \qquad \text{at eclipse peak} \qquad (3\text{-}24)$$

$$MT = 24^h - \frac{T_e}{2} - E \qquad \text{at eclipse onset} \qquad (3\text{-}25)$$

where $E = (L - \alpha)$ in time units (15° per hour) and α is the right ascension of the sun as defined and calculated above. L is the mean longitude of the sun and is calculated from the expression

$$L = L_0 + 0.98564736 d \qquad (3\text{-}26)$$

where d is the number of the day starting from January 1 = 1. The coefficient of d, approximately the rate of the mean sun, is simply the quotient of 360° and the number of days in a tropical year and L_0 is the reference value at the start of the year, very close to 280°. Its exact value is found in the *Astronomical Almanac*—for 1984 it was 278.864. The same book each year will also supply simple formulas for calculating the apparent longitude ⊙, the right ascension, and the declination δ. Such formulas are useful for the sun interference problem discussed in Section 3.4.

The most important special case is eclipse onset at the equinoxes when the eclipse duration itself is a maximum. The right ascension is zero at the vernal equinox and 180° at the autumnal. L is calculable from Eq. (3-26) using the day numbers of the equinox from the *Almanac* (typically March 21 and September 23). To a quite satisfactory approximation for most purposes, $d = 80$ for the vernal equinox and 266 for the autumnal equinox. We already have the maximum eclipse durations, so we can calculate Table 3-2. Note that these are local mean times at the *subsatellite* point. Local mean time is the same as standard or zone time only on the meridian that defines the time zone (i.e., Eastern Standard Time is equal to local mean time only at 75°W). Clock rates are equal to that of mean time and are set in any region to the assigned zone. If the satellite is serving an area to the west of the subsatellite point by an amount λ, the earliest clock time T_e, at the point in question, is given by

$$T_c = 23^h\!28 - \frac{\lambda}{15} \qquad (3\text{-}27)$$

---

[1]The equation of time E is the difference between the solar time on any day, as determined from meridian crossings of the sun and the solar time averaged throughout the year. In effect, this mean solar time is calculated assuming that the sun's orbital rate is constant and that it moves in the equatorial plane. It is the line kept by a fictitious "mean sun" which moves at a constant annual rate in the plane of the equator. The longitude of the mean sun is L in Eq. (3-26).

If it is desirable to have eclipse onset as late in the night as practical—say 1:00 A.M.—then λ must be equal to −25°.8. That is, the satellite must be to the *west* of the point in question by this amount. This can be a particularly important consideration in the choice of satellite location for some services where it is possible to consider not operating all or part of the communications system during eclipse because of reduced demand. Note well that the expression in Eq. (3-27) ignores the presence of standard time lines. The second term would not apply, for example, if the service area and subsatellite point were in the same time zone. On the other hand, it would have to be 1 h if these two points were on opposite sides of a zone boundary, even if the value of λ was very small. The geometry of any particular case should be looked at carefully, making allowance for the "standard" time boundaries as they actually are. These boundaries are as much in accord with local convenience as they are with the idealized 15°-wide zones.

On rare occasions the *moon* can eclipse the satellite. In a detailed eclipse analysis this effect must also be taken into account.

3.4 SUN INTERFERENCE

The sun-interference problem is important in satellite communication work, particularly for narrow-beam antennas. The earth station antenna beam must be pointed at the satellite, and when the sun crosses this line of sight the apparent noise temperature of the earth station receiver increases dramatically. In fact, if the earth station beam is narrow compared to the 0.5° angle subtended by the sun, the apparent antenna temperature of the earth station will increase to that of the sun. The interference creates an *outage*. The communication aspects are discussed in detail in Chapter 6; here we will concern ourselves only with the orbital geometry.

A reference to Figure 3-7 shows both the simplicity and awkwardness of the calculation. The small circle *CPD* is the path traced out by the earth station antenna beam on the celestial sphere as the earth rotates. It rotates once a sidereal day around the small circle *CD*. At the same time, the sun moves along the ecliptic. In general, the center of the sun will not be at the intersection of the ecliptic and the small circle at the same moment as the center of the antenna beam is. For a wide-beam antenna, this small difference can be ignored, and one can proceed simply to find the time that the antenna beam center crosses the ecliptic and then assume that this is the moment of peak sun interference. That is done in accordance with the routine shown in the next paragraph. Usually, it is easiest to do for a particular earth station in connection with an *Almanac* for the appropriate year. If the antenna beam is narrow compared to the sun, a more refined procedure is desirable. What is done is to calculate the position of the sun as a function of time in both right ascension and declination. The *Astronomical Almanac* will routinely

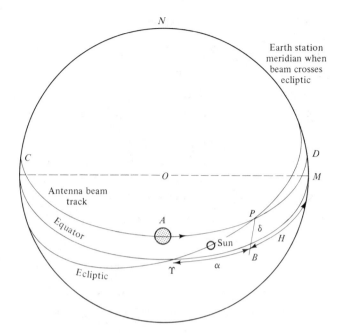

Figure 3-7 Sun interference geometry.

give a simple set of formulas for calculating this to an accuracy of a minute of arc. The right ascension and declination of the beam is calculated from its azimuth and elevation and the sidereal time in accordance with the equations of the next paragraph. The angle between the beam and the sun is then calculated using the cosine formula. Some assumption or separate estimation must be made of the minimum acceptable value of this angle without incurring a *sun outage*. Typically, for large earth station antennas, it is assumed that there will be a sun outage for angles between the two radius vectors of less than 1°.

3.4.1 Sun-Interference Geometry

The procedure for a simplified calculation follows (see Fig. 3-7).

- A The shaded circle represents a region in which the presence of the sun will cause unacceptable deterioration in performance. It rotates along the small circle CD once a sidereal day.
- ☉   The sun; it moves eastward along the ecliptic one revolution per year.
- P The point at which the antenna beam center crosses the ecliptic. Its coordinates are declination δ and right ascension α.
- H The hour angle of the point P at the moment the beam center passes through P.

Calculations are based on an earth station at

$\phi = 39°\!.2906$ N
$\lambda_E = 280°\!.2629$ E   (earth station)
$\lambda_S = 325°\!.50$ E   (satellite)
$\lambda = -45.2371$ (east longitude of earth station relative to satellite)

Since the antenna beamwidth may not be much larger than the errors in angular calculations that ignore the nonspherical earth, we must calculate the azimuth and elevation angles using the methods of Section 3.6. We first calculate the earth's radius, ρ, and geocentric longitude, ϕ':

$\rho = 6396.657$ km   from Eq. (3-56)
$\phi' = 39°\!.0921$ from Eq. (3-57)

These values are then used to calculate the azimuth and elevation:

Az $= 122°\!.0207$ from Eq. (3-59)
$\theta = 25°\!.2424$ from Eq. (3-60)

The hour angle H and declination can be calculated from the foregoing values of Az and θ using the following transformation equations:

$$\sin \delta = \sin \phi' \sin \theta + \cos \phi' \cos \theta \cos \text{Az} \qquad (3\text{-}28)$$

$$\cos H = \frac{\sin \theta - \sin \phi' \sin \delta}{\cos \phi' \cos \delta}$$

$\delta = -5°\!.9306$
$H = -50°\!.4459$
$\quad = -3^h 21^m 47^s$ in time units at 15°/h

The negative value of the angle is taken for H since it is conventionally measured as positive to the west of the observer's meridian. This is the case when the satellite is to the east of the earth station and the azimuth is less than 180°. The geometry of any case should be examined to avoid errors in sign, which are so easy to make with inverse trigonometric functions.

At this point, the *Almanac* can be consulted to determine the moment at which the sun has the given declination. For 1978, the sun had this declination some time on March 5. The right ascension α can be determined at the same moment by interpolation in the table or by the relation

$$\tan = \frac{\cos \varepsilon}{\sqrt{(\sin^2 \varepsilon / \sin^2 \delta) - 1}} \qquad (3\text{-}29)$$

Sec. 3.4 Sun Interference

where ε is the obliquity of the earth's orbit. For 1978 it was $23°.4420$. It changes slowly and the correct value for any year will be given in the *Almanac*:

$$\alpha = 13°.8612$$

In time units of 15° per hour:

$$\alpha = 0^h 55^m 27^s + 24^h = 23^h 04^m 33^s$$

The 360° (or 24 h) is added because right ascension is reckoned to the east of the vernal equinox, whereas the formula gives the usual first quadrant value of the angle. The local sidereal time (LST) is calculated from the universal relation.[2]

$$LST = H + \alpha \qquad (3\text{-}30)$$

The geometric part of the sun-interference problem is completed with the calculation of local sidereal time. This time is the angular position of the rotating earth relative to the universal astronomical reference point, the vernal equinox, Υ. Note that this local sidereal time does not change from year to year. It depends only on the geometry of the problem.

To determine local clock time, we must convert to mean solar time. This conversion requires a knowledge of the sidereal time on the date in question—it can be found from tables or formulas in the *Almanac*. We set up the schedule shown in Table 3-3.

TABLE 3-3 Sun Outage: Local Mean Time Calculation

Hour angle H	$-\ 3^h\ 21^m\ 47^s$	Calculated from earth station
$+$ Right ascension α	$+\ 23^h\ 04^m\ 33^s$	azimuth and elevation and obliquity of ecliptic
Local sidereal time LST	$19^h\ 42^m\ 46^s$	$H + \alpha$
$+$ Earth station *west* longitude (time units)	$+\ 05^h\ 18^m\ 57^s$	Given earth station location
Greenwich sidereal time (GST)	$25^h\ 01^m\ 43^s$	
$-$ GST at 0^h UT on 3/5/78	$-\ 10^h\ 49^m\ 33^s$	From *Almanac* or formula
Sidereal time interval after midnight	$14^h\ 12^m\ 10^s$	
$-$ Correction to mean solar interval (ratio of 0.0027379) after midnight	$-\ \ \ \ \ \ 02^m\ 20^s$	$\dfrac{\text{Mean sidereal time}}{\text{Mean solar time}} = 1.00273379$
Greenwich Mean Time GMT	$14^h\ 09^m\ 50^s$	
$-$ Time zone correction	$-\ 05^h\ \ 0^m\ \ 0^s$	Location of earth station
Eastern Standard Time	$09^h\ 09^m\ 50^s$	Time on local earth station clock

---

[2]This simple formula is rather useful. It applies to any celestial object: the sun or moon, stars, and artificial satellites alike.

3.4.2 Duration of Interference

The duration of the interference can be estimated from the scanning velocity of the beam (i.e., the earth's rotation of 15° per hour or $4^m\!.0$ per degree) and the simplifying assumption that the outage lasts through a "tolerance" angle $\pm \beta$:

$$T = 8\beta \text{ (minutes)}$$

A fair value for β is

$$\beta = 0°\!.54 + 2\beta_0 \tag{3-31}$$

β_0 is the half-power beamwidth of the antenna and $0°\!.54$ is the apparent diameter of the sun. The formula is better for narrow beams than for wide beams, but the effects of the sun on a wide beam are much less severe (see Chapter 6). Since the solar declination changes about $0°\!.4$ per day, the number of interferences N is an equinoctial season is also estimable from β.

$$N = \frac{2\beta}{0.4} = 5 \text{ days/season} \tag{3-32}$$

It must be emphasized that we have calculated the time at which the earth station beam crosses the ecliptic on the day that the sun is closest to that declination. The sun changes its declination about 23" per day at that time of year, and itself subtends an angle of about 30". Thus two consecutive daily passes of the earth station beam could intersect near the upper and lower limits of the sun without intersecting the center. Nonetheless, it is a good method for finding the day of the year and the clock time within a few minutes of the peak interference.

It is also a starting point for the more elaborate routine in which the angle between the radius vector from the earth station to the sun and from the earth station to the satellite is calculated as a function of time. If it is found to be less than some preset value, which depends on the beamwidth of the antenna and the acceptable deterioration in performance, there is an outage. The procedure for doing this is straightforward, albeit complicated.

One calculates the right ascension and declination of the sun as a function of Greenwich Mean Time, or more properly, *universal time* (UT), for several days before and after the day calculated as above and for some hours before and after the outage. *Almanac* simplified formulas are used. The hour angle of the sun is calculated from the sidereal time and the relation LST = H + α. Both hour angle and declination are now given for the sun and the satellite as seen from the earth station. The "cosine" formula (Table 3-1) is used to calculate the angle between the radius vectors.

3.5 CALCULATION OF GROUND TRACES AND SATELLITE TOPOCENTRIC COORDINATES AS A FUNCTION OF TIME

The tracking of satellites from an earth station normally requires the computation of azimuth (Az) and elevation (θ) as a function of calendar date and time. The orbital equations of Section 2.3 are sufficient to do this but require coordinate transformations and a procedure for relating clock time to the rotational position of the earth.

A special case of the theory of Section 2.5 is the tracking of an "almost synchronous" satellite. The procedure is first to calculate the satellite position in its own orbit, which is fixed in inertial space, and then to transform these polar coordinates to nonrotating geocentric coordinates (right ascension and declination). We then correct for parallax between the earth's center and the surface and convert the corrected right ascension and declination into azimuth and elevation. Right ascension and declination are *nonrotating* coordinates, whereas azimuth and elevation are *rotating*. Declination is not affected by the rotating coordinates, but the right ascension α is related to the rotating hour angle (the angle at any instant between the local meridian and the object) by Eq. (3-30). Thus, in terms of the Greenwich sidereal time (GST) and the east longitude λ of the earth station, we can write

$$H = \text{GST} + \frac{\lambda}{15°} - \alpha \quad \text{in time units} \quad (3\text{-}33)$$

since the earth rotates 15° per hour. Note the following general relations:

local mean solar time = universal time + east longitude

local mean sidereal time = Greenwich mean sidereal time + east longitude

local apparent sidereal time = local mean sidereal time + equation of equinoxes

local hour angle = local apparent sidereal time − apparent right ascension

α is found from the coordinate transformations and GST must be found from the sidereal time, which is the rotational position of the earth relative to the vernal equinox. The expression (*Astronomical Almanac*, 1984) is

$$\text{GMST}_0 = 24110^s\!.54841 + 8640184^s\!.812866 T_u$$
$$+ 0^s\!.093104 T_u^2 - 6^s\!.2 \times 10^{-6} T_u^3 \quad (3\text{-}34)$$

GMST_0 is the Greenwich mean sidereal time at midnight universal time

(UT) of the start of the day in question. T_u is the time from 2000 January 1.5, the astronomer's way of writing noon, January 1, 2000 UT.

T_u is the elapsed time since 2000 January 1.5, and is measured in Julian centuries of 36,525 days of universal time. It is taken in increments of 1 day and is given by

$$T_u = \frac{JD - 2{,}451{,}545}{36{,}525} = \frac{d}{36{,}525} \tag{3-35}$$

JD is the Julian day number[3] and d is the number of Julian days after the reference date, which itself is Julian day number 2,451,545 at noon.

Equation (3-34) can be rewritten for any particular year by changing the reference date to January 0.0 of that year (astronomical shorthand for midnight start of December 31 of the previous year). d becomes the number of whole days after January 0.0. The new reference value of GMST is found from Eq. (3-34). Some tedious arithmetic produces the following expression for 1984:

$$GMST_0 = 6\!\overset{h}{.}5905966 + 0\!\overset{h}{.}0657098242 d \tag{3-36}$$

The T^2 and T^3 terms have been ignored as negligible for this purpose. For a universal time UT, after midnight on the day d, the Greenwich sidereal time is

$$GMST = GMST_0 + 1.002737909 UT \tag{3-37}$$

The coefficient of the second term is the ratio of sidereal to universal time. The hour angle (in degrees) from Eq. (3-33), is finally written as

$$H = 98\!\overset{\circ}{.}858949 + 0\!\overset{\circ}{.}98564736 d + 15.041068\ UT + \lambda - \alpha° \tag{3-38}$$

The conversion from time units (usual for sidereal time and right ascension) to degrees (usual for hour angles) is done at 15° per hour.

We are now in a position to calculate the right ascension α and declination δ of the satellite. We first calculate either the period P or a, the semimajor axis of the orbit, dependent on which is given, from the usual

$$P = 2\pi \sqrt{\frac{a^3}{\mu}} \tag{3-39}$$

where $\mu = 3.986018 \times 10^{14}$ m^3/s^2, t is the time in question and τ is the time at perigee. Both are given as Julian dates. $(t - \tau)$ permits the calculation of the mean anomaly M from the definition of M in Eq. (2-10):

$$M = \frac{2\pi}{P}(t - \tau) \tag{3-40}$$

---

[3] Julian days begin at noon and are counted starting from 4713 B.C. Conversion tables to Julian days from calendar date are found in the *Almanac*.

Sec. 3.5 Calculation of Ground Traces and Satellite Coordinates

The eccentric anomaly E is found from Kepler's equation,

$$M = E - e \sin E \tag{3-41}$$

Since this equation is not explicitly solvable for E (except by a not very useful infinite series), it must be solved using any convenient method of successive approximations. Newton's method (see any text on advanced algebra) or other numerical methods work quite well.

The true anomaly v is finally calculated from E using the expression

$$\tan \frac{v}{2} = \frac{1+e}{1-e} \tan \frac{E}{2} \tag{3-42}$$

The frequently useful magnitude of the radius vector r is given as

$$r = \frac{a(1-e^2)}{1+e \cos v} \tag{3-43}$$

Right ascension and declination are found from

$$\alpha = \Omega + \tan^{-1}[\cos i \tan(\omega + v)] \tag{3-44}$$

$$\delta = \tan^{-1}[\sin(\alpha - \Omega) \tan i] \tag{3-45}$$

or

$$\delta = \sin^{-1}[\sin(\omega + v) \sin i]$$

The declination and hour angle (found from α) are also the terrestrial coordinates of the subsatellite point (the locus of which is called the *ground trace*). Declination is equal to terrestrial latitude and hour angle to longitude relative to the reference meridian—usually Greenwich. Ground traces are of great interest in planning nongeostationary orbits for such purposes as remote sensing and navigation.

To calculate azimuth Az and elevation θ at an earth station, given the latitude ϕ and hour angle H, we must convert from geocentric to topocentric coordinates.

Our calculations of H and δ are referred to the center of the earth. For many astronomical calculations, the error due to the finite distance of the body involved compared to the radius of the earth is negligible. In the case of an artificial satellite, this is not the case and we must make a parallax correction. We can follow the method used by astronomers to correct for the moon's parallax (*Explanatory Supplement*, 1961). If H and δ are the corrected values (see Section 3.6 for formulas) of the hour angle and declination at the earth's surface, the standard spherical trigonometric formulas can be used to find azimuth and elevation:

$$\tan A = \frac{\sin H}{\cos \phi \tan \delta - \sin \phi \cos H} \tag{3-46}$$

$$\sin \theta = \sin \phi \sin \delta + \cos \phi \cos \delta \cos H \tag{3-47}$$

An alternative and simpler method can be developed for a spherical earth by recognizing that the geometry of Fig. 3-1 is not fundamentally altered if the point P representing the satellite is not in the equatorial plane (Fig. 3-8). In particular, we can take the subsatellite point P to be at a latitude equal to the satellite's geocentric declination δ_0 and a longitude relative to the earth station equal to the negative of the geocentric hour angle H_0 (note that hour angles are conventionally measured to the west and longitudes to the east).

The geometry is as before with the reference plane now OMS rather than OMA. The arc $\widehat{ES}$, which we call γ, as before, can be calculated from the spherical triangle ESM and from the spherical right triangle SAM. Both the cosine and sine formulas from Table 3-1 are used. After a little trigonometric manipulation, we have

$$\cos \gamma = \cos \phi \cos \delta_0 \cos H_0 + \sin \phi \sin \delta_0 \tag{3-48}$$

and

$$\frac{\sin Az}{\sin 90 - \delta_0} = \frac{\sin H_0}{\sin \gamma} \tag{3-49}$$

We now can use the expressions already derived [Eqs. (3-1) through (3-3) and (3-12)] with the new expression for γ to derive the following:

Slant range:

$$d = \sqrt{R_E^2 + (R_E + h)^2 - 2R_E(R_E + h)(\cos \phi \cos \delta \cos H_0 + \sin \phi \sin \delta_0)} \tag{3-50}$$

Elevation angle:

$$\cos \theta = \frac{(R_E + h) \sin \gamma}{d} \tag{3-51}$$

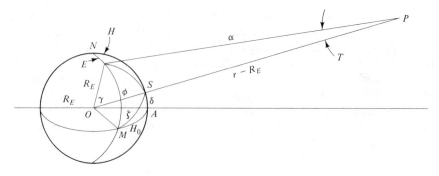

Figure 3-8 Geometry for a satellite out of the equatorial plane.

Sec. 3.6   The Nonspherical Earth

Azimuth:

$$\sin Az = \frac{-\cos \delta_0 \sin H_0}{\sin \gamma} \quad (3\text{-}52)$$

Inclination:

$$\sin T = \frac{R_E \sin \gamma}{d} \quad (3\text{-}53)$$

These expressions can be used to derive the topocentric coordinates and slant range as long as the needed accuracy does not require consideration of the nonspherical nature of the earth. For more precise corrections, the methods of the following section should be used.

3.6 THE NONSPHERICAL EARTH

3.6.1 General

Most of the geometric calculations in satellite communication can assume a perfectly spherical earth with negligible loss in accuracy. The computation of slant range for space loss, elevation angles for rain losses, discrimination angles for interference, and coverage patterns, among others, are in that category where angular errors of tenths of a degree are of no importance. There are cases, however, where a more careful computation must be made.

At a location E on the earth's surface (Figure 3-9), the latitude is ϕ, as determined by measurements to a local vertical. It is called the *geodetic* or *geographic latitude* and is the value normally given for an earth station. Because of the nonspherical earth, this angle is different from ϕ', the geocentric latitude. For our calculations, the earth's section is taken to be an ellipse, with the following parameters (the "Hayford" ellipse):

$$a = R_E = 6378.16 \text{ km} \quad \text{(equatorial radius)}$$

$$f = \frac{a-b}{a} = \frac{1}{298.25} \quad \text{(flattening factor)}$$

The eccentricity e of this ellipse is defined as

$$e^2 = \frac{a^2 - b^2}{a^2} \quad (3\text{-}54)$$

and is therefore given by

$$e^2 = 2f - f^2 = 0.0066946 \quad (3\text{-}55)$$

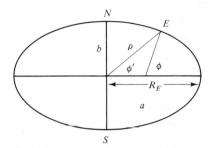

Figure 3-9 The earth as an ellipse.

The polar coordinate equation for the ellipse can be approximated adequately by

$$\rho = R_E(1 - f \sin^2 \phi') \tag{3-56}$$

Also, exactly, from the equations for an ellipse and the fact that the negative reciprocal of d_y/d_x at any point on a curve is the tangent of the slope angle of the normal to the curve at the same point:

$$\tan \phi' = (1 - e^2) \tan \phi$$

or (3-57)

$$\tan \phi' = (1 - f)^2 \tan \phi$$

Thus, given the latitude, one can calculate the geocentric angle ϕ and the earth's radius ρ at that point. The corrected distance to the satellite d can be shown to be given by

$$d = \sqrt{\rho^2 + (R_E + h)^2 - 2\rho(R_E + h) \cos \phi' \cos \lambda} \tag{3-58}$$

If the value for d along with the value of ϕ' instead of ϕ is used, the corrected values of azimuth and elevation can be found from Eqs. (3-2) and (3-3), modified as follows:

$$\sin Az = \frac{\sin \lambda}{\sqrt{1 - \cos^2 \phi' \cos^2 \lambda}} \tag{3-59}$$

$$\cos \theta = \frac{R_E + h}{d} \sqrt{1 - \cos^2 \phi' \cos^2 \lambda} \tag{3-60}$$

3.6.2 Parallax Conversion on a Nonspherical Earth

To convert geocentric coordinates (H and δ) to topocentric coordinates (θ and Az), one must correct for both the parallax due to the earth's radius and the nonspherical earth. Both Smart (1977) and the *Explanatory Supplement* (1961) give detailed and exact procedures for this correction as applied to the moon. An artificial earth satellite presents identically the same problem and we use those results. If H_0 and δ_0 are the geocentric values of hour angle and declination—as normally calculated from the equations for orbital motion—then H and δ, the topocentric values, are

$$\tan H = \frac{\sin H_0 \cos \delta_0}{\cos \delta_0 \cos H_0 - (\rho/r) \cos \phi'} \tag{3-61}$$

$$\tan \delta = \frac{\cos H [\sin \delta_0 - (\rho/r) \sin \phi']}{\cos \delta_0 \cos H_0 - (\rho/r) \cos \phi'} \tag{3-62}$$

ϕ' is the geocentric latitude and ρ the earth's radius at the observer's location (calculated as in 3.6.1). r is the radius vector to the satellite. Equations (3-61) and (3-62) are derivable from either Smart (1977) or the *Explanatory Supplement* (1961) through routine but lengthy manipulations.

3.7 THE APPARENT POSITION OF AN ALMOST GEOSTATIONARY SATELLITE

One of the useful applications of the methods of Section 3.5 is to plot the apparent motion of an almost synchronous satellite. There is always some residual eccentricity and inclination in a nominally geostationary orbit and these lead to a cyclic change in the apparent (topocentric) position of the satellite. Calculating this motion is necessary in designing the earth station tracking system.

Using the equations of Section 3.5, for zero eccentricity ($M = E = v$), the hour angle h is

$$h = v - \alpha \tag{3-63}$$

and

$$\sin h = \sin v \cos \alpha - \cos v \sin \alpha$$

Using

$$\cos \alpha = \frac{\cos v}{\cos \delta}$$

and

$$\cos \delta = \sqrt{1 - \sin^2 v \sin^2 i}$$

as identities for the spherical right triangle, substituting Eq. (3-63), and after some reduction, we obtain

$$\sin h = \frac{1}{2} \frac{\sin 2v(1 - \cos i)}{\sqrt{1 - i^2 \sin^2 v}} \tag{3-64}$$

$$\sin h \simeq \frac{i^2}{4} \sin 2v \tag{3-65}$$

and

$$\sin \delta = \sin i \sin v \qquad (3\text{-}66)$$

$$\simeq i \sin v \qquad (3\text{-}67)$$

3.8 THE TRACK OF AN ALMOST SYNCHRONOUS SATELLITE

Equations (3-65) and (3-66) are the parametric equations for a figure eight, and are shown in Figure 3-10. Note that the maximum change in declination is equal to the orbit inclination and the maximum change in hour angle is very much less. The figure eight, shown here about the equator, is simply transformed to the nominal elevation angle for earth stations not located on the equator, using the equations relating azimuth and elevation to hour angle and declination. Often, the maximum dimensions of the figure eight are all that are needed for planning.

Another important case is for a perfectly equatorial orbit with some residual eccentricity. A good approximation to the equation of the center is

$$v \simeq M + 2e \sin M \qquad (3\text{-}68)$$

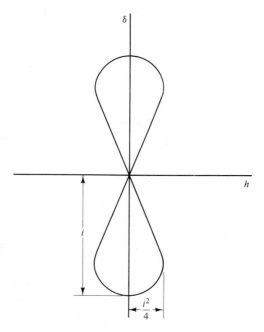

Figure 3-10 Apparent position of a synchronous satellite inclined i degrees to equator.

If the satellite is synchronized so that M is equal to nt, where n is the mean residual motion, then

$$h = nt - v \qquad (3\text{-}69)$$

$$h = -2e \sin nt \qquad (3\text{-}70)$$

That is, the satellite will vary cyclically about its nominal longitude at the sidereal rate and with an excursion of $\pm 2e$.

REFERENCES

The Astronomical Almanac, U.S. Government Printing Office, Washington, D.C. (any year).

Explanatory Supplement to the Astronomical Ephemeris and the American Ephemeris and Nautical Almanac, Nautical Almanac Offices of the U.K. and U.S.A., Her Majesty's Stationery Office, London, 1961.

SMART, W. M., *Textbook on Spherical Astronomy*, Cambridge University Press, Cambridge, 1977.

Additional Readings

See the References in Chapter 2 for additional readings on orbits.

4

Launch Vehicles and Propulsion

4.0 INTRODUCTION

The launching of a satellite into orbit is an extraordinarily complex and costly operation. A launch vehicle as a system includes structure, engines, propellant storage and pumps, guidance, and control. Several stages are often used. Every known engineering discipline is involved and a complete grasp of all the details is beyond anyone's reach. Even an understanding of launch vehicles at the systems level, to the depth necessary, for instance, to supervise the design and development of such vehicles, requires considerable study. Certainly, any such undertaking is beyond the scope of satellite communication systems engineers. Yet a reasonable familiarity with the principal ideas in launch vehicles is also a necessity for them.

Typically, the launch vehicle costs as much as the satellite. Indeed, in the early days of satellites, launch vehicles usually cost more than the spacecraft. Furthermore, the launch vehicle and mission dictate some of the important compromises that must be made in spacecraft design and in the communications mission itself. In addition to the obvious questions of payload mass and dimensions, there are other less obvious relations between the capabilities of the launch vehicles and the design of the communications payload. It is the purpose of this chapter to acquaint the satellite communication systems engineer with the notions involved in launching a satellite into orbit, the subsequent orbital maneuvers necessary to change locations and correct for orbital anomalies, and something of the launch vehicles that will be available for communications satellites during the rest of this century. The physical

Sec. 4.2 The Rocket Equation   83

ideas involved in propulsion are useful, not only to understand the launch but also the spacecraft design itself because of its need to carry its own propulsion system.

4.1 LAUNCH MISSIONS

Launch missions may be divided into two phases. The first takes the payload from the ground, overcomes the earth's gravity and resistance of the atmosphere, and places the payload in a low-altitude "parking" orbit. Normally, this orbit is as low as possible, consistent with the residual atmospheric drag and the duration in that orbit. The second phase comprises a series of transfer maneuvers necessary to take the satellite from parking into final operational orbit and, in the case of a geostationary satellite, to its final longitudinal location. The first phase is best characterized as conceptually similar to ordinary powered flight. A launch vehicle "flies" from the surface of the earth into parking orbit. Both inertial and aerodynamic forces are involved. After the aerodynamic, powered-flight phase of the mission is completed and the spacecraft is in parking orbit, the next phase maneuvers the satellite into final orbit. The standard routine for so doing was first proposed by a German physicist, Hohmann, in connection with his research on interplanetary travel in the early 1920s. This mission plan is simple and basic. It is shown schematically in Figure 4-1. Two separate increments in velocity must be given to the satellite: one to place it in a transfer orbit whose apogee is at the desired final altitude, the second to circularize the orbit at the final altitude. Changes in inclination can be made at perigee, at apogee, or at both. If the final orbit is geostationary, the apogee impulse may not place the satellite at the desired terrestrial longitude and subsequent maneuvers are necessary. In Sections 4.4 and 4.5 we discuss maneuvering from parking orbit into final position, and in Section 4.3 we discuss, at least briefly, the main problems of powered flight. Because both powered flight and subsequent orbital maneuvers rely on rocket engines, the elementary physics of such engines are described in Section 4.2 and a simple derivation given of the widely useful basic rocket equation. Finally, in Section 4.7, a brief summary of the capabilities of existing launch vehicles, both for powered flight injection into parking orbit and later orbital maneuvers, is presented.

4.2 THE ROCKET EQUATION

4.2.1 The Basic Equation

A rocket engine develops its thrust F by expelling propellant (such as gas molecules) at a constant velocity v_e relative to the vehicle. This is shown diagrammatically in Figure 4-2 for a liquid rocket engine in which a fuel and

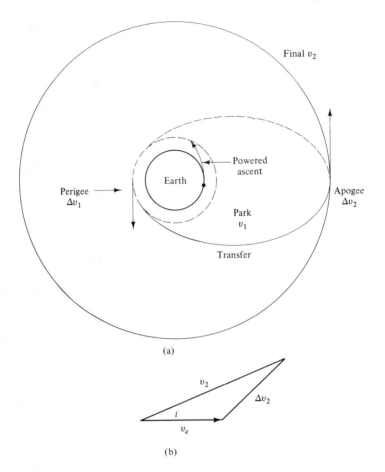

Figure 4-1 Transfer from one circular orbit to another: (a) coplanar maneuvers; (b) plane change at apogee.

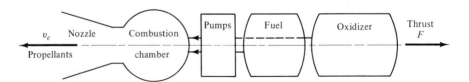

Figure 4-2 Basic liquid rocket engine.

Sec. 4.2 The Rocket Equation

an oxidizer are burned in a combustion chamber to produce high-temperature, high-pressure gas.[1]

Rigorous derivations of the basic law of motion for a rocket are not difficult and are based on the principle of conservation of momentum (Berman, 1961). A simplified derivation follows: v is the vector velocity of the vehicle in an inertial coordinate system, M-dm_e is the instantaneous mass of the vehicle and remaining propellant, and dm_e is the mass of propellant expelled during the period in which the velocity changes by dv. We assume that the momentum in inertial coordinates Mv is constant for the entire system and is equal to the sum of the momenta of the expelled propellant and of the vehicle including the remaining propellant. Thus

$$(M - dm_e)(v + dv) + (v - v_e)\, dm_e = Mv \tag{4-1}$$

where $dM = -dm_e$. This becomes, after simplification, the ordinary differential equation

$$M\, dv + v_e\, dM = 0 \tag{4-2}$$

It is directly integrable, subject to the initial condition that when $v = 0$, $M = M_0$,

$$v = v_e \log \frac{M_0}{M} \tag{4-3}$$

Equation (4-3) is widely applicable. It gives the final value v of a rocket's velocity after an amount of propellant $(M_0 - M)$ has been expelled. M_0/M is called the *mass ratio*. It is equal to the ratio of the initial total mass of rocket plus propellant to the final mass after the propellant is used. v_e is the exhaust velocity and a measure of the efficacy of the propellant. The required value of the final velocity v depends on the orbit to be attained and the final mass M is the sum of the payload and the residual structure—hence the magnitude of v_e is the characteristic of the propellant that largely determines the payload capacity of the rocket.

It is important to note that the equation is applicable to any rocket propulsion system, regardless of the means used to impart velocity to the ejected propellant. It applies equally well to chemical rockets, ion propulsion, and photon engines.

4.2.2 Chemical Propulsion

If a chemical propellant is used, the velocity of the molecules expelled is related to the absolute temperature T of the burning reaction by the classical

---

[1] The term *propellant* is used both for the fuel/oxidizer combustion and for the discharged gas. The context will usually make the intended meaning clear.

relation between the kinetic energy of a monatomic gas molecule and its absolute temperature, assuming that the energy is all stored in translation:

$$\tfrac{1}{2}mv_e^2 = \tfrac{3}{2}kT \tag{4-4}$$

where k is Boltzmann's constant and m is the absolute molecular mass. Thus

$$v_e = \sqrt{3\frac{kT}{m}} \tag{4-5}$$

The relation for more complicated molecules, which may store energy in translation, rotation, and vibration, is more involved. Nevertheless, the same general dependence of exhaust velocity on temperature and molecular weight exists. This simple relation reveals at a glance the need for high-temperature reactions (limited ultimately by material technology) and light molecules. Hydrogen (H_2) and oxygen (O_2) constitute virtually the ideal chemical rocket propellants. They burn energetically and nontoxically to produce water, itself a desirably light molecule, and at a high temperature. However, they are not burned stoichometrically, but an excess of hydrogen is used to produce an average molecular mass of propellant as low as possible. Only enough oxygen is used to sustain the reaction at the desired temperature and pressure. Their drawback as propellants is that to store them in sufficient total mass, they must be liquefied at low temperature and stored cryogenically. This is significant when the upper stages of a launch, for instance from the NASA Space Transportation System (STS) or Shuttle, must remain in parking orbit for long times. Various other combinations of propellant and oxidizer [e.g., hydrazine (N_2H_4) and nitrogen tetroxide (N_2O_4)] can be stored at room temperature, but give a lesser impulse and have the further disadvantage of being toxic. A wide variety of propellants, liquid and solid, are in use depending on the specifics of the application.

4.2.3 Ion Engines

Another way of achieving high exhaust velocities is through the use of ion engines. Metallic molecules such as mercury and cesium can be ionized and then accelerated to high velocities by an electrical field. If E is the accelerating voltage and e the electron charge,

$$\tfrac{1}{2}mV_e^2 = eE \tag{4-6}$$

The velocities achievable with ordinary accelerating voltages are many times higher than those realizable with chemical propulsion. The thrusts produced, however, are low, since the rate of total mass expulsion is low. Chemical rockets, because of their high thrusts, can be used in both powered flight and orbital maneuvers, but the ion engine is restricted to low-thrust maneuvers in free space.

Equation (4-3) also leads to the inference that the ultimate rocket engine would use photons—the exhaust velocity being the velocity of light—and the payloads, say for an interstellar mission, would be a maximum. The thrusts and accelerations would be extremely low and usable only after escape from the solar system had been accomplished by conventional rockets. No such technology is available today.

4.2.4 Specific Impulse and Thrust

Although exhaust velocity is the propellant characteristic that determines the payload, it is not alone in determining the thrust of a rocket engine. The thrust is important in determining the time required to achieve a given velocity increment and it depends on the rate of mass expulsion. In powered flight thrust is necessary to overcome gravitational resistance and atmospheric drag.

The propellant exhaust velocity is normally not specified explicitly, but instead a related parameter called *specific impulse*, I_{sp}, is used. In engineering usage it is defined as

$$I_{sp} = \frac{\text{thrust (units of force)}}{\text{rate of propellant burning (in weight units)}} \tag{4-7}$$

which is equivalent to

$$I_{sp} = \frac{F}{\dot{m}g}$$

where $\dot{m}$ is the burning rate of propellant in mass units. It can be related to the exhaust velocity by applying Newton's second law to a restrained rocket ($v = 0$). If propellant is expelled at a rate $\dot{m}$ and with a velocity v_e, then $\dot{m}v_e$ is the rate of change of momentum and thus equal to the force exerted.[2] Measurements of the thrust exerted by rocket engines are made by restraining them in exactly that manner. From the definition of I_{sp}, we have

$$\dot{m}gI_{sp} = \dot{m}v_e$$

and

$$v_e = gI_{sp}$$

We can then rewrite Eq. (4-3) for the final velocity v in its most common scalar form:

$$v = gI_{sp} \log \frac{M_0}{M} \tag{4-8}$$

---

[2]Actually, an additional factor accounts for the behavior of the rocket engine nozzle. This is discussed in Section 4.3.3.

TABLE 4-1 Impulse and Density for Propellants

(a) Combinations[a]

		Specific Gravity	Specific in Vacuum (seconds) Impulse (s)
(C) Oxygen (O_2)	(C) Hydrogen (H_2)	0.326	430
(C) Oxygen	(S) RP-1	1.011	328
(C) Oxygen	(S) Hydrazine (N_2H_4)	1.067	338
(C) Oxygen	(S) UDMH	0.976	336
(C) Oxygen	(S) Ammonia (NH_3)	0.835	319
(C) Fluorine (F_2)	(C) Hydrogen	0.539	440
(C) Fluorine	(S) Hydrazine	1.332	388
(C) Fluorine	(S) Ammonia	1.181	385
(C) Fluorine	(S) RP-1	1.210	350
(C) Fluorine	(S) UDMH	1.194	368
(C) Fluorine	(S) 0.5UDMH–0.5N_2H_4	1.267	376
(C) Fluorine	(C) Diborane (B_2H_6)	1.122	397
(C) Fluorine	(S) Pentaborane (B_5H_9)	1.202	384
(C) Oxygen difluoride (OF_2)	(S) RP-1	1.319	379
(C) Oxygen difluoride	(S) Pentaborane	1.220	395
(C) Oxygen difluoride	(C) Hydrogen	0.385	436
(C) Oxygen difluoride	(S) Hydrazine	1.275	371
(C) Oxygen difluoride	(C) Diborane	0.995	399
(S) Nitrogen tetroxide	(S) Hydrazine	1.223	314
(S) Nitrogen tetroxide	(S) UDMH	1.172	309
(S) Nitrogen tetroxide	(S) 0.5UDMH–0.5N_2H_4	1.202	312
(S) Nitrogen tetroxide	(S) MMHN$_2$H$_3$ (OH)	1.000	325
(C) Chlorine trifluoride (ClF_3)	(S) Hydrazine	1.501	312
(S) Nitric acid (IRFNA)	(S) 0.5UDMH–0.5N_2H_4	1.275	297
(S) Nitric acid	(S) Pentaborane	1.149	321
(S) Hydrazine	(S) Pentaborane	0.796	358
(C) Fluorine–oxygen (FLOX—20%)	(S) RP-1	1.154	338

[a] (C), cryogenic; (S), storable.
[b] Specific impulse given at standard conditions: Sea level, and chamber pressure of 1000 psi.
[c] Nozzle-area expansion ratio taken at 40.

Besides I_{sp}, other factors, such as the density of the propellants, whether or not they are toxic, and whether or not they are cryogenic, are of considerable practical importance. Table 4-1 lists specific impulses and densities for a wide variety of chemical propellant combinations. In connection with solid

TABLE 4-1 (*continued*)

(b) Solids

	Percent Oxidizer	Specific Impulses		Specific Gravity
		Sea Level[b] seconds	Vacuum[c] seconds	
Plastisols:				
polyvinyl chloride				
A	74	225	266	1.63
B	74	220	260	0.059
C	81	241	284	1.72
D	40	225	266	2.4
E	60	259	306	1.80
F	66	283	334	1.72
Polymers				
Polyurethane				
A	62	235	277	1.66
B	65	227	268	1.66
C	69	232	274	1.66
Polybutadiene				
A	70	245	289	1.77
B	70	247	291	1.77
C	69	244	288	1.77
D	70	244	288	1.77
Polysulfide				
A	76	221	261	1.72
B	68	241	284	1.72
C	74	226	267	1.72
Double base	% Nitroglycerin			
A	1.4	263	310	1.66
B	41	231	273	1.60
C	39	240	283	1.69

rocket motors, which have a fixed charge of propellant, one often speaks of the total impulse. It is the product of the thrust F (approximately constant) and the burning time t_b and is equal to the specific impulse times the weight of propellant consumed gM_p. This is written as

$$Ft_b = gI_{sp}M_p \qquad (4\text{-}9)$$

and M_0 is given by the final mass M' plus the mass of propellant consumed:

$$M_0 = M + M_p \qquad (4\text{-}10)$$

g, the standard acceleration of gravity, appears in Equation 4-9 merely because of the definition of I_{sp} in units involving the *weight*, rather than *mass*, of propellant. The standard value of g is 9.80665 m/s^2.

4.3 POWERED FLIGHT

4.3.1 General Picture: Aerodynamic Forces

The launch phase from earth to parking orbit is best understood by appreciating its similarity to ordinary powered flight. The equations for powered flight are Newton's laws. The thrust of the engine and the aerodynamic forces combine to produce an equilibrium with the gravitational force of the earth and the inertial forces resulting from the vehicle acceleration. Figure 4-3 shows these forces. A glance shows that the situation is complex. None of the forces is constant. The mass m is diminishing rapidly as propellant is consumed. The gravitational pull mg diminishes also with altitude. The aerodynamic forces tend to diminish with altitude because of the decreasing density of the air, but increase generally with the square of the rocket speed, which is increasing rapidly with altitude. Both lift and drag are proportional to the dynamic pressure q, equal to $\rho v^2/2$, where ρ is the density of the air and v is the rocket speed. q is multiplied by an effective cross-sectional area A and a

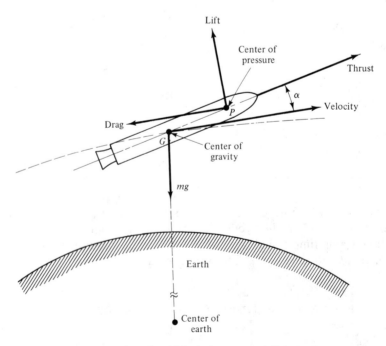

Figure 4-3 Forces on a launch vehicle during powered flight in the atmosphere.

4.3.2 Simple Equations

If we take the considerably simplified picture shown in Figure 4-4, where we assume that the thrust vector **T** is tangent to the flight path and the center of pressure and gravity are coincident, we can write the set of equations (4-11) through (4-14). h is the altitude of the vehicle center of gravity. These equations represent, respectively, Newton's second law, the aerodynamic drag, the inverse-square law of gravitation, and the reduction in mass because of the expenditure of propellant.

$$m \frac{dv}{dt} = T - D - mg \sin \phi \qquad (4\text{-}11)$$

$$D = \tfrac{1}{2}\rho v^2 C_D A \qquad (4\text{-}12)$$

$$g = \frac{g_o R_E^2}{(R_0 + h)^2} \qquad (4\text{-}13)$$

$$m = M_0 - \dot{m}t \qquad (4\text{-}14)$$

Note that all the variables m, v, t, and h are interrelated. The simultaneous solution of these equations, particularly when they are generalized for the forces in Figure 4-3 and put in a nonrotating set of coordinates, is a complicated exercise in numerical analysis. Equation (4-11) can be integrated in form as

$$v = \int_0^t \frac{T}{M} \, dt - \int_0^t \frac{D}{m} \, dt - \int_0^t g \sin \phi \, dt \qquad (4\text{-}15)$$

The first term on the right is equal to the free-space-velocity increment as given by the rocket equation (4-8). The net increment in velocity given to the rocket is thus seen to be representable as the free-space value diminished

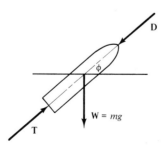

Figure 4-4 Simplified force diagram.

by two terms—which we designate as v_d and v_g—to allow for drag and gravity losses, respectively. We can turn this notion around and state that the velocity increment for which propellant must be supplied in accordance with Eq. (4-8) can be written as the sum of the orbital velocity and the losses attributable to gravity and atmospheric drag. This approach can be further generalized to allow increments for maneuvering, thrust not being in the same direction as velocity, variations in atmospheric pressure, and other effects (Section 4.3.5).

4.3.3 Expansion Ratio

To understand the impact of changes in atmospheric pressure in the course of a mission, we must consider the factors affecting rocket motor thrust. Figure 4-5 illustrates the physical parameters involved. In addition to the component of thrust attributable to the rate of change of propellant momentum, there is a component representing the force on the exit section area A_e resulting from any difference in the pressures P_e and P_0 on the two sides of the exit section. It can be shown (Seifert and Brown, 1961) that the total thrust F is

$$F = \dot{m} v_e + A_e(P_e - P_0) \tag{4-16}$$

This equation may expressed alternatively as

$$F = \dot{m} c \tag{4-16a}$$

where c is the *effective* exhaust velocity. (In the discussion prior to Equation (4-8) we assumed the condition of optimum expansion, i.e. $P_e = P_0$. In this special case $c = v_e$.) By introducing the concept of *thrust coefficient* C_F, we can restate the thrust as

$$F = C_F A_t p_c \tag{4-16b}$$

where A_t is the throat area and p_c is the chamber pressure. The thrust coefficient is the figure of merit for the nozzle design. The figure of merit for the propellant is the characteristic exhaust velocity

$$c^* = A_t p_c / \dot{m} = c / C_F \tag{4-16c}$$

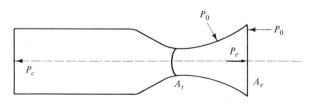

Figure 4-5 Nozzle expansion.

Therefore, in terms of C_F and c^* the thrust is

$$F = \dot{m}\, C_F\, c^* \qquad (4\text{-}16d)$$

and the specific impulse is

$$I_{sp} = c/g = C_F c^*/g \qquad (4\text{-}16e)$$

P_e is affected by the thermodynamic behavior of the propellant in the combustion chamber and by the expansion ratio ε

$$\varepsilon = \frac{A_e}{A_t} \qquad (4\text{-}16c)$$

Further analysis reveals that the maximum value of C_F, and thus the maximum thrust, occurs for the value of ε which brings P_e equal to P_0. Since P_0, the atmospheric pressure, varies over a wide range from sea level to orbital vacuum, it is impossible to pick a single optimum value of ε for an entire flight. Different values are used for the successive stages more nearly to approach an optimum. Typically, values of ε around 6 to 8 are used for the first stage, 20 to 30 for the second stage, and 30 to 45 for upper stages. There is nevertheless some loss in thrust due to expansion ratio mismatch, especially during the low altitude parts of the mission.

4.3.4 The Earth's Rotation

Another effect that influences rocket performance during the launch into parking orbit is that of the earth's rotational velocity. This velocity, if in the correct direction, can be considered as "free" to the launch vehicle. It is sometimes called the *sling effect*. With its benefit the vehicle's propellant systems have less work to do. Since the earth's rotational velocity is a maximum at the equator, launch sites close to the equator have this advantage over those more remote. Table 4-2 shows this effect for some representative launch sites. The advantages of the sling effect are small but definite.

Note that launching into any specific orbital plane orientation also requires precise timing because of the rotation of the earth. The injection into orbit must take place when the vehicle passes through the target plane. To

TABLE 4-2 Earth's Rotational Velocities at Various Launch Sites

	Latitude	Earth's Velocity (m/s)
Equatorial site	0.00	465
Kourou, French Guiana	5.23	463
Cape Canaveral, Florida	28.5	409
Baikonur, USSR	45.6	325
Plesetsk, USSR	62.8	213

permit launching before or after this precise moment, it is necessary to allow for a small plane change. Three hundred meters per second will allow about a 2-h "window" for launch into a specific 28.5° orbit from Cape Canaveral. It is important to emphasize that the savings in incremental velocity as a result of the launch site location and the earth's rotational speed is of advantage only to the rocket designer. It is academic to the spacecraft designer once the vehicle has achieved parking or transfer orbit since those orbits are fixed in inertial space. Orbital transfer maneuvers are quite independent of the powered-flight mission into parking orbit. However, any subsequent change in inclination is very much of concern to the spacecraft designer and is discussed in Section 4.4.

4.3.5 Composite Velocity Requirement

We can summarize the results of these effects in Eq. (4-17), which is Eq. (4-15) generalized somewhat:

$$v_{\text{total}} = v_0 + (v_m + v_D + v_w + v_g + v_n) \pm v_R \qquad (4\text{-}17)$$

where v_0 = orbital velocity to be achieved
v_m = loss due to maneuvering
v_w = launch window allowance
v_D = loss due to atmospheric drag
v_g = loss due to gravity
v_n = loss due to non-optimum nozzle expansion ratio
v_R = correction for velocity of earth's rotation

Equation (4-17) can be used to estimate powered-flight payloads. A typical velocity budget might be:

v_0 (150 km parking orbit)	7800 m/s
v_g and v_D (gravity, drag loss)	1500 m/s
v_m (maneuvering)	200 m/s
v_w (launch window allowance)	300 m/s
v_R (earth's rotation)	$-$ 400 m/s
	9400 m/s

Changes in payload, as a result of mission changes, can be estimated with surprising accuracy using Eqs. (4-8) and (4-17).

The combined velocity losses for gravity and drag are dependent on the mission flight time. A high thrust-to-weight ratio reduces this time and thus reduces the integrated gravitational and drag terms in Eq. (4-15). A plot of

Sec. 4.3   Powered Flight

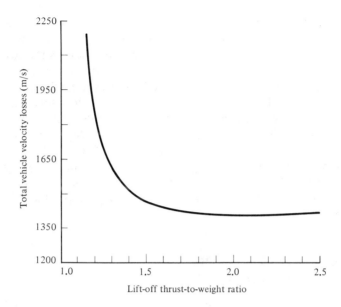

Figure 4-6 Gross launch velocity losses versus thrust-to-initial weight ratio.

typical combined losses as a fraction of thrust-to-initial weight ratio is shown in Figure 4-6.

4.3.6 Parking Orbit

The vehicle lifts off vertically and after a transitional maneuver is usually programmed to take a *gravity turn*. In such a trajectory the thrust vector remains in the direction of the velocity, minimizing the aerodynamic effects. In some missions, the engines are used right up to orbital insertion and in others part of the flight is coasting.

When the vehicle reaches a velocity sufficient for the inertial and gravitational forces to be in equilibrium, it is said to be in *parking orbit*. Such orbits are usually near circular and at altitudes between 150 and 300 km. The velocities and other orbital parameters are related by the equations of Chapter 2. Although the aerodynamic forces in such orbits are small, they are not completely negligible and will cause the spacecraft slowly to reenter the atmosphere if means are not taken to prevent it—such as occasional additional propulsion. For purposes of launching into final orbits, the decay times from parking orbit are generally long compared to the times needed for orbital determination and maneuvering.

During the ascent phase, the earth's rotation is an analytical complication, but when the vehicle is in parking orbit this orbit can be considered as constant in inertial space—that is, independent of the earth's rotation. It will be affected slowly by the earth's gravitational anomalies and by the sun and moon, as discussed in Chapter 2. The most important effect is that attributable to the earth's equatorial bulge, which produces the two classical effects of nodal regression and apsidal rotation. In particular, for the Shuttle parking orbit of 160 nautical miles, the nodal regression will be 7.5° per day, as shown in Figure 4-7. This can be an important effect for spacecraft designers. Its immediate significance is that the orientation of a parking orbit relative to the sun will change this amount every day in addition to the approximately 1° per day motion of the sun itself in the same coordinate system. Some Shuttle multiple-payload conflicting launch constraints based on different desired sun angles can be resolved by waiting between successive payload ejections for the composite 8.5° per day to produce the desired difference.

4.4 INJECTION INTO FINAL ORBIT AND ORBITAL MANEUVERS

After the aerodynamic or powered-flight part of the mission is completed and the spacecraft is in parking orbit, the next operation is to maneuver into the final operational orbit, as shown in Figure 4-1. At a node of in the parking orbit (that is, as the satellite crosses the equatorial plane), an increment in velocity Δv_1 is added, usually in a period of time short compared to the orbital rotational period, sufficient so that the resulting elliptical orbit has its apogee at the desired final height. The apogee will lie in the equatorial plane when the spacecraft arrives at apogee, either for the first time or at any subsequent apogee; it is given a second increment in velocity Δv_2, the purpose of which is to circularize the orbit and to produce any necessary change in inclination. Note that the inclination can also be changed at perigee. In fact, for most missions, the optimum overall (maximum payload) mission calls for the removal of a small amount of inclination at perigee—about 2°. This optimum is ascertained by differentiation of the total Δv equations in the Appendix with respect to this chapter versus ζ, the plane change at perigee, and solving the resultant equation numerically. There are occasionally other operational reasons for some plane change at perigee.

The choice of which apogee to use depends on the available thrust, how long it takes to make accurate orbital and attitude measurements, its final longitude, and other practical problems. It is common for launching into geostationary orbit to fire the apogee engines at about the third apogee. The velocity increments required for any perigee or apogee firing are calculated using the equations of Chapter 2. The most important of these equations are collected for convenience in the Appendix to this chapter. Some interesting results are shown in Table 4-3.

The reduction to Δv value at apogee required in an Ariane launch is

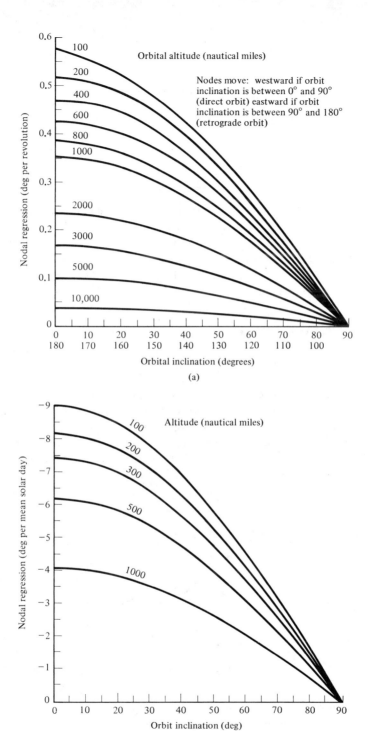

Figure 4-7 Nodal regression: (a) degrees per revolution; (b) degrees per mean solar day.

TABLE 4-3 Apogee Kick Velocity Increments

Launch Site	v_2 (m/s)
Equator (100 n. mi.)	1479
Shuttle launch (28.5°)	1831
Ariane (5.29°)	1493

not negligible. Figure 4-8 shows, for solid and liquid rocket motor propellants, the percentage of mass in transfer orbit that can be injected into final orbit as a function of the required change in inclination.

Note that the disadvantages for high latitudes are definite but by no means insuperable. The Soviet Union indeed launches into geostationary orbit from rather high latitudes.

4.5 MISSION POSSIBILITIES

The overall mission into geostationary orbit can be accomplished by several configurations of rocket stages. The simplest for the spacecraft designer is to have available a launch vehicle in which the entire third stage with spacecraft is injected into geostationary orbit. It is convenient but inefficient. The only system that we know of capable of doing this is the Titan 3–Transtage series. It can inject several thousand kilograms of useful payload into geostationary orbit, depending on the upper stage. No apogee kick engine is necessary—spacecraft designers need design the satellite only for operation in final orbit

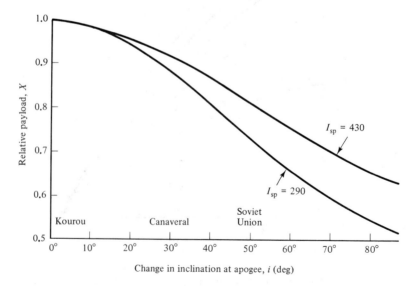

Figure 4-8 Loss in payload versus change in inclination at apogee.

and not concern themselves with getting there. This deluxe way of going to geostationary orbit is, unfortunately, expensive. To date it has been used only for military vehicles.

The most common method is typified by that used by the Atlas Centaur and Thor Delta launch vehicles during the last 15 years in the United States and envisioned by Ariane in the future. That is, the launch vehicle goes to parking orbit and inserts a composite spacecraft into geostationary transfer orbit (GTO). The composite spacecraft carries an apogee kick motor for insertion into operating orbit (GEO). This is less convenient than inserting the upper stage into GEO but more efficient. The greatest payload is achieved by "staging" as much as possible so as not to carry any more dead weight than necessary.

A third method is that envisioned by the Shuttle, in which the reusable vehicle or Orbiter goes to parking orbit only; the spacecraft designer must be concerned with both the perigee and apogee injections. In return for these complications and spacecraft costs, it is hoped that there will be substantial compensation to the spacecraft designer. First, it is envisioned that the overall cost, even allowing for the extra upper stage, will be less because of the lower costs of the reusable Shuttle. Second, the Shuttle will permit a much larger spacecraft physically, that is, a greater diameter for spinning satellites, a substantial advantage for primary power and rotational stability, and longer lengths for deployable structures. Third, the manned Shuttle offers the spacecraft designer and operator the possibility for a recheck of the satisfactory performance of the spacecraft after the hazardous primary ascent phase of the mission and before continuing with the transfer orbit operations. It will be possible to return to earth with a payload no longer suitable for operation. The Shuttle will eventually offer the possibility of assembling large platforms in parking orbit, refueling and replenishing them, and launching very large spacecraft into operational orbits. Perhaps most important, a Shuttle-maintained space station, supplemented by an orbital transfer vehicle (OTV), could refuel satellites in geostationary orbit and replace failed transponders. We are early in the Shuttle era and designers have not begun to exploit these possibilities, but we expect that they will as time goes on.

4.6 LOW-THRUST VARIATIONS

Several variations on the basic Hohmann maneuver are possible and probably will be employed during the Shuttle era. Perigee and apogee stages used up to now have been of the kind in which the "burn" time is short or "impulsive." One assumes that the time needed to give an increment in velocity is negligibly short compared to the orbital period. It is possible to configure transfer maneuvers using restartable liquid engines and lower thrusts in which the burning time is an appreciable fraction of the transfer orbit period. If this is the case, there will be a loss in efficiency since the thrust is not always tangential to the orbital direction. Usually, in missions of this type, the total required velocity is supplied in increments in successive apogee or perigee burns rather than all at once. To compensate for the loss in efficiency compared to impulsive maneuvers, there will be improvements in orbital injection accuracy because of the longer time in which to refine the measurements and the lesser errors because of thrust misalignment. This is because the thrusts are lower and the spacecraft attitude can be recorrected after each burn.

Gimballed engines can also be used with continuous correction of the thrust vector so as to have it in the same direction as the velocity—a "gravity turn" again. If there is no steering of the spacecraft during a burn maneuver and it is held in an inertially fixed attitude, the loss in efficiency of a low-thrust nonimpulsive burn can be at least approximated (Figure 4-9). We assume that the orbit in the vicinity of perigee (or apogee) is a circle with a radius equal to its value at that moment. We define an efficiency of propellant utilization η_p as the ratio of the energy imparted to the spacecraft by a force F at an angle φ to the direction of motion to the energy imparted if φ were always zero:

$$\eta_p = \frac{\int_{-\varphi_m}^{\varphi_m} F \cos \varphi \, d\varphi}{\int_{-\varphi_m}^{\varphi_m} F \, \varphi \, d\varphi} = \frac{\sin \varphi_m}{\varphi_m} \qquad (4\text{-}18)$$

φ_m in turn is approximated by

$$\varphi_m = 2 \frac{vt_b}{r} \qquad (4\text{-}19)$$

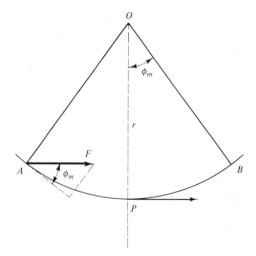

Figure 4-9 Approximate effect of burn duration.

Sec. 4.6 Low-Thrust Variations   101

where v and r are taken at perigee or apogee, whichever is appropriate, and t_b is the burn time of the rocket engine. Note that the angle φ_m will be greater and the consequent loss more serious at perigee because the velocity is higher and the radius lower than at apogee.

Burning time can be calculated from Eq. (4-9). A variation in the approach is to set a minimum efficiency, then calculate a permissible value of φ_m and thus burning time. Equation (4-9) is then used to determine how much propellant is to be used.

Note that for burning times of the order of a few minutes or less (typical of most "solid" missions), the efficiencies are close to unity. As the thrust and acceleration go below about 1 g, the burning time can be long enough to cause an annoying loss in efficiency and prompt the consideration of multiple-burn maneuvers. The group of equations in the Appendix, together with those of this chapter, are sufficient for preliminary mission planning. An illustration of a typical launch sequence is seen in Figure 4-10.

Exact calculations require the integration of the differential equations or orbital motion with external forces applied. The procedure, because of the nonlinear nature of the equations and the several coordinate systems involved, is very complicated (Porcelli, 1982). Its use is needed only in a precise mission plan and comparison.

There are other implications in low-thrust missions for the system designer, such as longer times to operation and more passages through the Van Allen radiation belts. The low-thrust bipropellant systems can even be used after orbital injection for stationkeeping in operation. We expect many such missions as a result of extensive Shuttle use, and designers will learn to adapt to these idiosyncracies. Many all-liquid and combined solid–liquid engines will be designed for transfer from STS to geostationary orbit.

To summarize some of the possibilities available to spacecraft designers during the 1980s, we have prepared Table 4-4 to show the main characteristics of primary launch vehicle combinations and Table 4-5 to describe the payload capabilities of certain upper stages. It is important to emphasize that the numbers are all approximate. There are literally dozens of factors that cause the payload capabilities of a launch vehicle to vary from model to model and from one production unit to the next. Among the many considerations are launch windows, telemetry and tracking requirements, mass of attachment and separation fittings, and fairing size and mass. Final mission planning must be done in close cooperation with the manufacturer of the launch vehicle. Every detail must be considered, including such factors as the weather at the moment of lift-off. Launch-vehicle capacities are being continuously upgraded by manufacturers. Fuel tanks are enlarged, structural masses are reduced, engines are improved, and booster rockets are added. The numbers are, nevertheless, sufficiently accurate to be illustrative and suggestive of which vehicles are applicable to a particular mission.

TABLE 4-4 Launch Vehicle Combinations

Vehicle	Upper Stage	Stage No.	Description	Propellants	Stage Initial Mass	Burn-out Mass (kg)	Thrust (kN)	Burn Time (s)
Atlas SLV-3D		1/2	Liquid booster	LO_2/RP-1			1,646[a]	
		1	Liquid sustainer	LO_2/RP-1	121,582	—	267	
	D1-A Centaur	2	Liquid sustainer	LO_2/LH_2	13,952	1,905	131	
Titan III E		0	Two solid boosters	NH_4ClO_4/Al	693,000[b]		10,053[b]	
		1	Core 1 liquid	N_2O_4/UDMH-N_2H_4			2,300	
		2	Core 2 liquid	N_2O_4/UDMH-N_2H_4			450	
	D1-T Centaur	3	Liquid	LO_2/LH_2		2,023	131[c]	
	AKM	4	TEM 364-4	Solids[d]				
Thor Delta 3900 Series		0	Nine Thiokol TX526-2; nine boosters	Solids	192,000[b]		3,412	
3910 Series		1	RS27 liquid	LO_2/RJ-1			912	
3910		2	TR201 liquid	N_2O_4/VOMH-N_2H_4			43.8	
	PAM	3	Star 48	Solids			65.4	
3913		3	TE-364-3	Solids	718	59	42.169	44.1
3914		3	TE-364-4	Solids	1,122	76.6	66.586	43.6
3920 Series		2	AJ10-118K	N_2O_4/UDMH-N_2H_4			44.5	
3920	PAM	3	Star 48	Solids			65.4	
3923		3	TE-364-3	Solids	718	59	42.169	44.1
3924		3	TE-364-4	Solids	1,122	76.6	66.586	43.6

Vehicle	Stage	Engine	Propellant	Thrust		Mass		
Ariane 1	1	4 × Viking V	NO₄/UDMH	207,000[b]		2,437		
	2	1 × Viking IV	N₂O₄/UDMH	36,600		700		
	3	HM7	LO₂/LH₂	9,400		60		
Ariane 2	1	4 × Viking V	N₂O₄/UDMH			2,700		
	2	1 × Viking IV	N₂O₄/UDMH	188,000		800		
	3	HM7	LO₂/LH₂			62		
Ariane 3	Same as Ariane 2	Two strap-ons	Solids	234,000[b]		1,320		
Ariane 4	Same as Ariane 2	Four strap-ons	Solids			2,640		
Shuttle (STS)	0	SRB	Solids	2,000,000[b]		23,600[c]		
	1	SSME	LO₂/LH₂			6,270		
	2	OMS	N₂O₄/MMH			53		
SSUS-A		Thiokol–Minute Man III PKM	Solids	3,792 (7,891)[f]	318	199.27		59.5
SSUS-D		Thiokol–Star 48 PKM	Solids	1,917 (4,567)[f]	191	57.82		84
IUS	1	Large solid rocket motor (SRM-1) PKM	Solids	16,894	1093	185.1		151
	2	Small solid rocket motor (SRM-2) AKM	Solids	3,815 (16,996)[f]	984	76.3		100

[a] Total Atlas lift-off thrust, including vernier engine 1917 kN.
[b] Total lift-off value.
[c] Three-burner Centaur.
[d] Two-burn plus AKM.
[e] Recoverable.
[f] Total installed mass, including spacecraft and excluding NASA-provided tiedown fittings, SC-ASE, and mission-specific hardware.

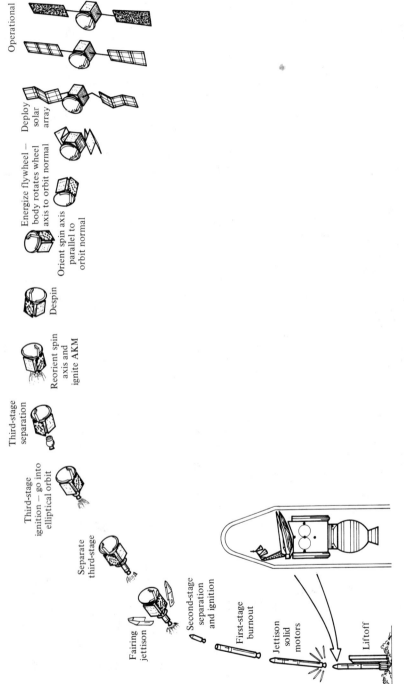

Figure 4-10 Advanced RCA Satcom launch sequence. (Reprinted by permission of RCA Astro-Electronics.)

Chap. 4   Appendix

The currently available Shuttle and its upper stages are shown in Tables 4-4 and 4-5. Once again, it is important that the numbers be taken only as approximations, especially when they pertain to payload. There are several versions of the Boeing Interim Upper Stage (IUS)—some for geostationary orbit and others for interplanetary missions. We have shown the two-stage vehicle intended to launch the NASA Tracking and Data Relay Satellite (TDRS) into geostationary orbit. We have also shown the mass budgets and specific impulses of the constituent solid rocket motors (SRM-1 and SRM-2) for perigee and apogee injection. These data can be convenient in more detailed mission planning or in considering the separate use of these motors. The most energetic stage shown is the Centaur, a modified version of that now used with the expendable launch vehicle, Atlas Centaur. This is the result of the high-impulse hydrogen–oxygen restartable engines. Note that it uses just about the entire payload capability of the Shuttle, so it is suitable only for the launch of large platforms. The Centaur stage itself could inject still more mass into geostationary orbit—it has the propellant capacity to inject as much as 8500 kg, but the Shuttle cannot yet lift the necessary total mass into parking orbit. The Centaur also has a low-thrust mode (5%) suitable for injection of very large and delicate structures into geostationary orbit by six or eight successive perigee and apogee burns.

APPENDIX

The calculation of velocity increments Δv_1 from parking orbit into transfer orbit and Δv_2 from transfer into final orbit is done using the following formulas:

$$\Delta v_1 = v_1 \left[1 - 2 \cos \zeta \sqrt{\frac{2}{H + K} + \frac{2}{1 + K}} \right]^{1/2}$$

$$\Delta v_2 = v_1 \sqrt{K} \left[1 - 2 \cos i \sqrt{\frac{2K}{1 + K} + \frac{2K}{1 + K}} \right]^{1/2}$$

where

$$K = \frac{R_E + h_1}{R_E + h_2}$$

$$v_1 = \sqrt{\frac{\mu}{R_E + h_1}} \quad \text{for a circular orbit}$$

TABLE 4-5 Upper Stage Payload Capacity

Vehicle	Upper Stage	Fairing Diameter (m)	Parking Orbit Altitude
Atlas SLV-3D	D1-A Centaur	3.05	185.2
Titan III-E	DI-T Centaur	3.73	185.2
	TE-M-364-4	3.73	185.2
Thor Delta	—	2.18	185.2
3910	PAM	2.18	
3913	TE 364-3	2.18	
3194	TE 364-4	2.18	
3920	—	2.18	185.2
3920	PAM	2.18	
3923	TE 364-3	2.18	
3924	TE 364-4	2.18	
Ariane 1	HM 7	2.90	200
Ariane 2	HM 7	2.90	200 (5.2°)
Ariane 3	HM 7	2.90	200 (5.2°)
Ariane 4	HM 7	2.90	200
Shuttle (STS)	SSUS-A	15.85 × 4.57 (with SSUS)	296
Shuttle (STS)	SSUS-D	4.5 × 3.0 (with SSUS)	296
Shuttle (STS)	IUS	12.8 × 4.11 (with IUS)	296

[a] With solid AKM.
[b] Lunar escape.
[c] Three-burn.
[d] Two-burn and AKM.
[e] Venus escape.

Also,

$$T = 2\sqrt{\frac{a^3}{\mu}} \qquad \text{for any orbit}$$

where R_E = earth's equatorial radius

h_1 = initial orbit altitude

TABLE 4-5 (continued)

Low Earth Orbit	Payload (kg) Synchronous Transfer Inclination		Geostationary	Escape
5,200	28.5°	1906	1000[a]	1200[b]
15,400	28.3°	—	3290	4500[c]
2,930	28.7°	7120	4000[d] 3750[a,d]	
3,400	28.7°	1102 845 936 1272 1032 1107	577 442 490 666 540 579	
4,850	8.0°	1800	1050	817[e]
5,100	8.0°	2175	1280	1450
5,900	8.0°	2580	1510	1700[f]
	8.0°	3300	1940	
29,500	28.5°	1994	1050	
29,500	28.5°	1247	656.3	
29,500	28.5°	4316 (SMR-1)	2272 (SRM-2)	

[f] Payloads include adapter fittings.
[g] Restricted by Shuttle cargo bay's allowable envelope (18.13 m long × 4.77 m in diameter unloaded). SSUS-A and IUS are horizontally configured and SSUS-D is vertically configured. IUS is nonspinning solid upper stage.

h_2 = final orbit altitude

i = plane change at apogee

ζ = plane change at perigee

μ = 3.98603 × 10^{14} m³/s²

a = semimajor axis of elliptical orbit

a = $R_E + h$ for a circular orbit

REFERENCES

BERMAN, A. I., *The Physical Principles of Astronautics*, John Wiley & Sons, Inc., New York, 1961.

HAVILAND, R., and C. M. HOUSE, *Handbook of Satellites and Space Vehicles*, D. Van Nostrand Company, Princeton, N.J., 1965.

PORCELLI, G., "Shuttle to Geostationary Orbital Transfer by Mid Level Thrust," AIAA 9th Commun. Satellite Syst. Conf., San Diego, Mar. 7–11, 1982.

SEIFERT, H. S., and K. BROWN, *Ballistic Missiles and Space Vehicle Systems*, John Wiley & Sons, Inc., New York, 1961.

5

Spacecraft

5.0 INTRODUCTION

The spacecraft provides a platform on which the communications equipment can function and maintains this platform in the chosen orbit. The design of such spacecraft is a complicated exercise involving just about every branch of engineering and physics. The interrelations among the requirements for communication performance, the need to provide a congenial environment for the communications equipment, and the problems of launching into the desired orbit constitute space systems engineering. For satellite communications systems engineers, who are concerned with the performance of the communications system, it is important to have an adequate understanding of this process. The costs of the communications package itself are normally only about 20% of the total costs. The remaining costs are dedicated to the platform and launch. Because satellite communications systems costs run into the hundreds of millions of dollars, the decisions made by systems planners are critical. It is not the intent of this chapter to go in depth into the problems of spacecraft design, since these are the provinces of the various engineering disciplines, but rather to give the systems planner an appreciation of these problems and how they interact with each other. Each of the main subsystems is discussed and the most important decisions concerning them and their relationships to the others are highlighted. The chapter contains practical procedures for estimating the primary power consumption of a spacecraft and its mass in orbit. Such estimates are required by most cost models for economic planning. The mass and power models in this chapter are quite satisfactory for such cost predictions and for estimating launch vehicle needs.

5.1 SUBSYSTEMS

In addition to the communications gear, the typical satellite contains the following support subsystems: (1) structure, (2) attitude control, (3) primary power, (4) thermal control, (5) propulsion, and (6) telemetry, tracking, and command. Table 5-1 is a simple chart listing these systems, their purposes, and the principal parameters that characterize them quantitatively. We define a utilization factor, u, as the ratio of the communications package weight to the dry mass in orbit of the spacecraft. For most spacecraft, u varies from 0.25 to 0.30. Larger, body-stabilized spacecraft tend to have the higher values of u.

5.2 STRUCTURAL SUBSYSTEM

The structure to hold the spacecraft together must be designed to withstand a variety of loads. During launch and transfer, there are accelerations, vibration and aerodynamic loads, centrifugal stresses, operating thrusts, and separation shocks. In operating orbit, we again find operating thrusts, centrifugal stresses, radiation pressure, and micrometeorite impacts.

A wide variety of materials and techniques have been used for spacecraft structures. Figure 5-1 shows some common structural types, mostly derivative from aeronautical practice, and Table 5-2 lists some common structured materials. Typically, the percentage of the total spacecraft mass represented by the structure is something less than 20%. This percentage tends to decrease as the spacecraft gets larger if it is body-stabilized, and to increase slowly if it is spin-stabilized and the diameter increases. Some empirical data based

Figure 5-1 Typical structures.

TABLE 5-1 Satellite Subsystems

System	Function	Principal Quantitative Characteristics
Communications Transponders Antennas	Receive, amplify, process, and retransmit signals; capture and radiate signals	Transmitter power, bandwidth, G/T, beamwidth, orientation, gain, single-carrier saturated flux density
Structure	Support spacecraft under launch and orbital environment	Resonant frequencies, structural strengths
Attitude control	Keeps antennas pointed at correct earth locations and solar cells pointed at the sun	Role, pitch, and yaw tolerances
Primary power	Supply electrical power to spacecraft	Beginning of Life (BOL) power, End of Life (EOL) power; solstice and equinox powers, eclipse operation
Thermal control	Maintain suitable temperature ranges for all subsystems during life, operating and nonoperating, in and out of eclipse	Spacecraft mean temperature range and temperature ranges for all critical components
Propulsion	Maintain orbital position, major attitude control corrections, orbital changes, and initial orbit deployment	Specific impulse, thrust, propellant mass
Telemetry, tracking, and command (TT&C)	Monitor spacecraft status, orbital parameters, and control spacecraft operation	Position and velocity measuring accuracy, number of telemetered points, number of commands
Complete spacecraft	Provide satisfactory communications operations in desired orbit	Mass, primary power, design lifetime, reliability, communications performance—number of channels and types of signals

on many spacecraft are shown in the curves of Figure 5-2. They are shown to illustrate trends and orders of magnitude only, since individual spacecraft may depart noticeably. As a practical matter for mass estimating, it can be assumed that the structure will be somewhere between 15 and 20% of the dry mass of the spacecraft, regardless of the method of attitude control.

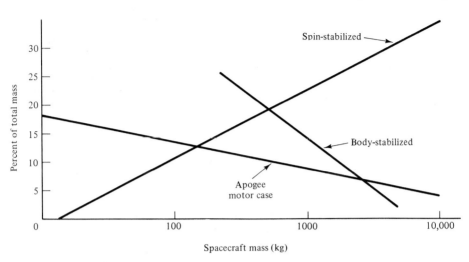

Figure 5-2 Structure and apogee motor factors.

TABLE 5-2 Common Structural Materials

Aluminum
Magnesium
Stainless steel
Invar
Titanium
Graphite-reinforced phenolic (GFRP)
Fiberglass epoxy
Beryllium

5.3 ATTITUDE CONTROL

The attitude control subsystem determines the whole character of the spacecraft (Kaplan, 1976). It can be considered as the "lead" subsystem, and virtually every spacecraft design decision is related to the choice of attitude-control technique. This system must accomplish two things. First, it must keep the antennas pointed in the proper direction (i.e., toward the region to be communicated with) on the surface of the earth or perhaps another satellite, and second, it must keep the solar cells pointed toward the sun. Note that both functions require a double action on the part of the attitude-control system. It must pitch the satellite 15° per hour to maintain earth pointing. At the same time, it must correct for attitude changes resulting from orbital disturbances and from upsetting torques generated when making station-keeping maneuvers.

Attitude control systems can be divided into two categories: active and passive. The only completely passive attitude control that has been used with any success is the gravity gradient method. In this, the difference in the attractive forces of gravity on the elements of the satellite closest and farthest from the center of the earth creates a correcting couple that tends to maintain the satellite aligned with the local vertical. Although this system has worked satisfactorily for low-orbit satellites, the gravity gradient itself diminishes with the cube of the distance from the center of the earth and is consequently too low to stabilize satellites in geostationary orbit. To have adequate torque, it is necessary to make the satellite extremely long, using extendable booms, and to find some method to damp the oscillation (libration) about the vertical axis. Several experiments have been tried, with unsatisfactory results. In the absence of some radically new idea, this method will not be used for geostationary satellites.

The active methods are divided into two categories: spinning and three-axis-stabilized satellites. These are colloquial descriptions and both are partially misnomers. The earliest *spin-stabilized satellites* indeed spun in their entirety with the axis oriented normally to the orbital plane. The attitude was fixed in inertial space and the method could be called quasi-passive. The antenna patterns were omnidirectional in the plane normal to the spin axis and matched the visible earth in the other. This was an impossibly extravagant use of transmitter power. The clear requirement to keep one or several narrow antenna beams pointed at the earth must be met in this kind of satellite by "despinning" or counterrotating the antenna. As the number of beams and the transponder complexity increase, it has been recent practice to despin an entire platform. This complicates the situation considerably, but does achieve the purpose of effectively stabilizing the antenna beams in three axes.

The *three-axis* or *body-stabilized* satellite comes in several varieties, including some in which there is only one large momentum wheel spinning on the inside of the satellite. In a sense, it is simply a spinner turned inside out. The distinction between the two classes is probably best emphasized by noting the use of a spinning drum to carry solar cells in one case and the carrying of solar cells on flat panels in the other.

It is important to emphasize that all attitude control systems function in accordance with the block diagram of Figure 5-3. Any perturbation in the

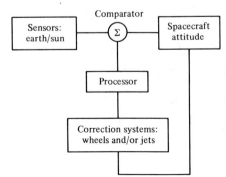

Figure 5-3 Structure and apogee motor factors.

attitude or position of the satellite is detected by sensors, compared to a reference, an error signal derived that is used to command corrections. The corrections are achieved either by varying the speeds of spinning wheels or by thrusters, or by some combination.

All attitude control systems require sensors. It is the resolution of these sensors that limits the ultimate pointing accuracy of the spacecraft. Sensors can be optical either in the visible or infrared regions of the spectrum or they can be radio-frequency sensors to work in conjunction with ground-based transmitters. The accuracies of some important kinds of available sensors are shown in Table 5-3.

Very important is the fact that all active attitude control systems, whether they use single inertial wheels, multiple wheels, or despun platforms, require thrusters to correct large errors. This is sometimes referred to as *dumping momentum* since the use of a jet to expel propellant does change the angular momentum of the spacecraft itself, whereas changing wheel speeds does not. Figure 5-4 shows schematically the four principal kinds of active attitude control systems, and Figures 5-5 through 5-8 show the layouts and orientations of these kinds of satellites in orbit.

The choice between the two classes, best characterized as *spinning-drum-stabilized* and *body-stabilized* satellites, is the principal choice of the spacecraft designer. It involves many complicated and interrelated considerations. Here we can only highlight the advantages and disadvantages of both types.

The advantage of the drum spinner started off as that of simplicity; that is, if no attempt is made to despin any part of the satellite, it is simplicity itself and remains quasi-passive. As more and more of the interior of the spacecraft is counterrotated, this advantage evanesces. For a spinning satellite to be absolutely stable dynamically, the moment of inertia about the desired

TABLE 5-3 Sensor Summary

	Accuracy	Mass (kg)	Power (W)
Earth sensors			
Pulse generator	0.1–0.5°	0.05–1.0	1.0
Passive scanners	0.5–3.0°	1.0–10.0	0.5–14.0
Active scanners	0.05–0.25°	3.0–8.0	7.0–11.0
Sun sensors			
Pulse generator	~0.2°	0.1	None
Solar aspect	0.01–1.0°	0.05–0.2	None
Null seekers	0.01–0.2°	0.05–1.5	None
Magnetometers	1.0–5.0°	1.0–2.0	2
Star sensors	0.02–0.1°	1.5–10.0	1.5–20
Inertial	0.01°/h–0.05°/h	0.1–1.0	0.3–8 (requires a.c. source)

Sec. 5.3 Attitude Control   115

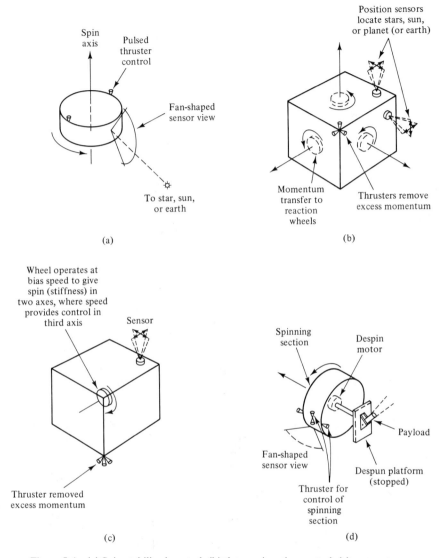

Figure 5-4 (a) Spin-stabilized control; (b) three-axis active control; (c) momentum bias control; (d) dual spin control.

spin axis must be greater than that about any orthogonal axis; that is, for absolute stability, this ratio should be greater than 1. Normally, the diameter of the solar cell-carrying drum is limited by the launch vehicle aerodynamic fairing size, or, in the case of the STS (Shuttle), by the cargo bay dimensions. As the requirement for primary power increases, it is necessary to increase the length of the drum and, in so doing, the ratio of the desired to the undesired moments of inertia diminishes. The problem is aggravated by the

116   Spacecraft   Chap. 5

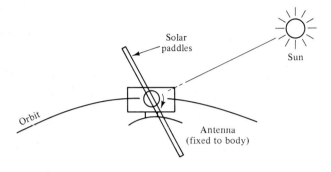
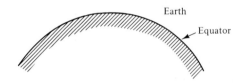

Figure 5-5 Body-stabilized satellite.

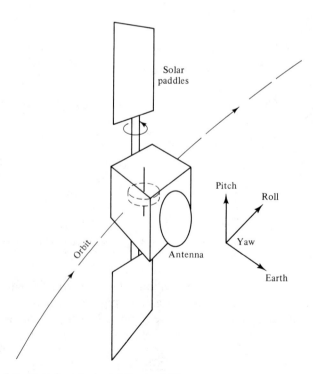

Figure 5-6 Axes for a body-stabilized satellite.

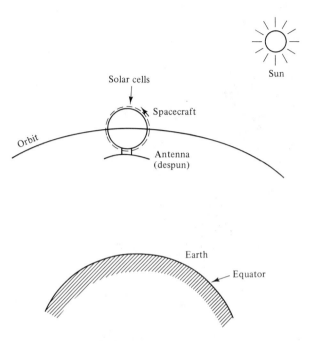

Figure 5-7 Drum-stabilized or "spinner" satellite.

counterrotating platform, which contributes nothing to the angular momentum about the desired axis, but considerably to the angular momentum about the undesired axes. The situation is complicated by changes during the spacecraft lifetime because of fuel depletion. Spinning fuel tanks, which at the beginning of the mission are almost full, slowly empty and create a problem through fuel "sloshing." Since sloshing represents a loss of energy in the spinning part, it is a serious destabilization influence. The net result is that two-body spinners (as this type of satellite is often called) require active nutation control, both in transfer and in final orbit; that is, the sensors must be coupled with the thrusters so as actively to maintain the satellite spinning about the desired axis.

In addition to possible simplicity, the spinning satellite has several other advantages. The cells absorb solar energy only throughout about one-third of a revolution, whereas they radiate to space through the entire cycle. This is a disadvantage in gathering solar energy, but the average temperature is noticeably lower. If only the geometry were involved, the loss in effective energy gathering would be equal to π, but as a result of the lower temperature the factor seems only to be about 2.3 to 2.5. The future use of gallium arsenide instead of silicon for solar cells will cause this factor again to approach π since it is less sensitive to temperatures.

In a spinner, the propellant tanks have no difficulty in feeding propellant to the reaction control system. They merely exploit the centrifugal force of

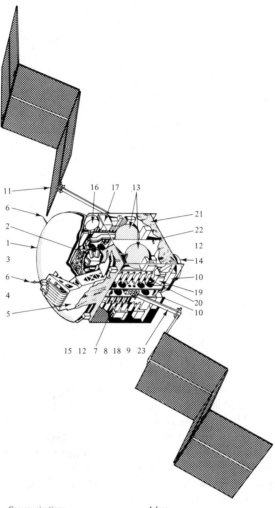

Communications
1. C-band antenna reflector
2. Communications receivers
3. Antenna feed horns
4. Antenna waveguide
5. Antenna tower
6. Omni antenna
7. Transponders and solid-state
8. Electrical power conditioner

Power
9. Power supply electronics
10. Battery packs
11. Solar-array panel

Propulsion
12. Reaction engine
13. Propulsion tanks
14. Apogee kick motor

Adacs
15. Earth scanner
16. Momentum wheel
17. Attitude control electronics

CT&T
18. Central logic processor
19. Command logic decoder
20. Transponder control electronics

Thermal
21. Thermal blanket structure
22. Central core
23. Solar-array boom

Figure 5-8 Advanced RCA Satcom. (Reprinted by permission of RCA Astro-Electronics.)

the spinning fuel tanks. In contrast, a body-stabilized satellite requires separate pressurization systems to manage the propellant.

The number of thrusters in a spinner is greatly reduced. Basically, only three are required—one in each axial direction and one radially. The radial thruster is controlled to fire at the correct moment during the rotation, thus achieving any desired torque orientation. Again, the body-stabilized satellite typically requires at least six thrusters and sometimes more to produce impulses and torques in all the desired directions.

On the other hand, the body-stabilized satellite has noticeable advantages as the primary power requirements increase. It is necessary only to provide solar cells for the desired power, not for three times as much. As the power increases, the panels are simply made larger. Their size is not inhibited by the Shuttle cargo bay or fairing limitations, since the panels are typically made unfoldable or deployable. The corresponding technique with spinning satellites is that of using telescoping solar arrays. Since such mechanisms are difficult and complicated to achieve reliably, they again detract from the initial simplicity advantage of the spinner. Other proposals using extended flat panels that also spin have been made but not implemented. Such a *tri-spun* satellite would solve the power and moment of inertia problems, but at a high price in increased complexity.

The final choice between the two types is complex and judgmental. The simple assignment of qualitative advantages and disadvantages is not sufficient, but rather each characteristic must be quantified and a final assessment of spacecraft mass, performance, and reliability made before the choice is possible. Table 5-4 summarizes the highlights of both systems of attitude control.

As a final interesting comparison, Figure 5-9 shows schematically the total spacecraft mass as a function of primary power for both kinds of attitude control systems. It can be said, in general, that beyond a certain primary power requirement, the body stabilization system will be superior. This crossover point is the intersection of two lines, not varying much from each other in slope. Therefore, the location of the crossover will continue to be a sensitive function of solar cell and other technologies, launch vehicle dimensions, and engineering ingenuity.

5.4 PRIMARY POWER

5.4.1 General Considerations

There are two possible sources of primary power for a satellite today—nuclear and solar. Nuclear supplies can be further divided into two categories. The first includes systems in which a small nuclear reactor heats a boiler with a working fluid such as mercury and the vapor is used to drive a turbine–alternator combination, typically in a Brayton cycle. It is a steam generating

TABLE 5-4 Spinning Satellites

Drum-Stabilized

Rotating drum provides inertial stiffness

Antennas, and usually transponder electronics, are counterrotated to point continuously in the desired direction

Spin-axis alignment is maintained with thrusters

Propellant feed is accomplished by centrifugal force

Only one-third of solar array is exposed to the sun at any time

Solar array diameter is limited by launch vehicle dimensions

Dynamic stability with any appreciable counterrotating mass is conditional only and must be supplemented with active attitude control

Number of thrusters (ignoring redundancy) is only three

Body-Stabilized

Motor-driven flywheels exchange angular momentum with body of spacecraft (it is possible to use only a single flywheel)

Redundant flywheels are possible to avoid single-point failures

Great flexibility in communications configuration is achievable

Large primary power can be made available with deployable solar arrays, not limited by launch vehicle constraints

Typically, a large number of thrusters required (at least six) to permit losing total momentum and reorienting spacecraft in any desired way

Separate pressurized propellant management systems are necessary to feed propellant to thrusters since there is no centrifugal force

---

station in miniature. The second type includes a single radioisotope thermoelectric generator (RTG) that heats lead telluride thermocouples to generate electricity. This latter type tends to be used more frequently for smaller power supplies. Both kinds have the advantage of needing no batteries during eclipse and the disadvantage of requiring substantial shielding to protect the spacecraft electronics from radiation damage. A deep-space mission, where the solar energy will be feeble, often has no option but the use of some kind of nuclear supply, whereas communications satellites do have the choice. At present, the nuclear fuels that are both easy to handle and require little shielding (e.g., curium-244 and plutonium) are expensive. On the other hand, cheap and easily available fuels such as strontium-90 are dangerous, difficult to handle, and a grave environmental hazard. In summary, although nuclear power supplies are practical and in use when necessary, commercial satellite communication from earth orbit seems to have no need that yet justifies the present high cost. To date, all communications satellites have used solar energy with ever-improving solar cell efficiency. The remainder of this section will be devoted to the conventional solar primary power systems.

5.4.2 Estimating Primary Power Needs

We consider a reasonably general case in which the solar array can be either on flat panels or a spinning drum. There may be a number of different types of transmitters, multiple receivers, housekeeping power, and battery

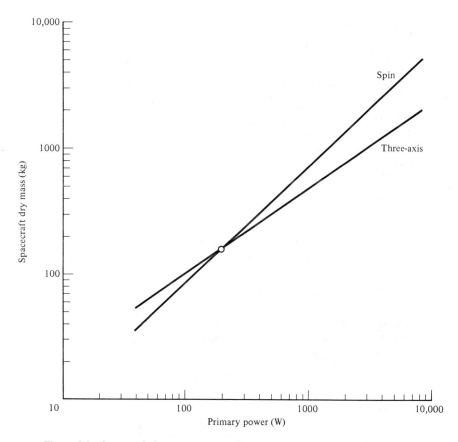

Figure 5-9 Spacecraft dry mass versus primary power.

service during eclipse. Figure 5-10 is the block diagram of a generalized solar primary power system. If n_i, RF_i, and η_i are the number, RF transmitted power, and efficiency, respectively, of the ith transmitter type, the total transmitter power is

$$P_t = \sum_{i=1}^{n} n_i \frac{RF_i}{\eta_i} \quad (5\text{-}1)$$

The efficiencies should be "overall" and taken to include driver amplifier primary powers.

Receiver power P_r may be known separately and added to P_t or estimated as a factor a to allow for other transponder needs not specifically added. The factor a seems to be about 1.05 for large satellites and higher, perhaps 1.10, for smaller or more complicated transponders. The factor a or P_r must include the remaining primary power for elements such as frequency converters and local oscillators. The housekeeping power P_h includes the power for the telemetry, tracking, and command subsystem (TT&C), attitude control, pro-

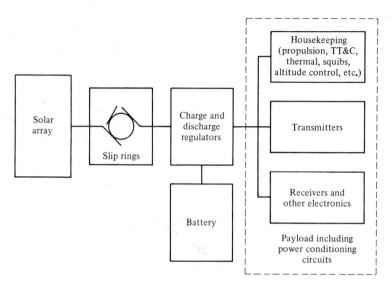

Figure 5-10 Basic satellite primary power system.

pulsion, and heaters for use during eclipse. It has a constant component P_{h0}, an eclipse heater power P_{he}, and a component hP_T proportional to the transponder power P_T, where

$$P_T = P_t + P_r = aP_T$$

During eclipse, the required power P_e is

$$P_e = \frac{(e + h)P_T + P_{h0} + P_{he}}{\eta_d} \tag{5-2}$$

where η_d is the battery-discharging efficiency, including regulator and conditioner circuits. The energy U that must be stored in the battery is

$$U = \frac{P_e t_e}{d} \tag{5-3}$$

where t_e is the eclipse time and d is the allowable depth of discharge. The charging power P_c, if the battery is to be charged in a time t_c, is then

$$P_c = \frac{U}{\eta_c t_c} = \frac{P_e t_e}{d \eta_c t_c} \tag{5-4}$$

The total primary power to be supplied by the array during equinox is given by the sum of the transponder charging and housekeeping powers. Thus

At equinoxes:

$$P_{EQ} = a(1 + h)P_t + P_{h0} + \frac{t_e}{d \eta_c \eta_d t_c}[a(e + h)P_t + P_{h0} + P_{he}] \tag{5-5}$$

At solstices:

$$P_{\text{SOL}} = a(1 + h)P_T + P_{h0} \tag{5-6}$$

5.4.3 Solar Array Requirements and Size

The solar array itself must be sized to provide the power called for by Eqs. (5-4) or (5-5), allowing for both the deterioration due to radiation damage and for the seasonal variation in available solar power. The degradation is handled by an exponential factor e^{-kN}, where N is the design lifetime. k is typically about 0.025 to allow for about a 20% deterioration in seven years. Greater or smaller values are used depending on the nature of the solar cells.

The solar flux density G varies during the year because of the varying distance to the sun and the varying declination. The composite effect yields a maximum near the vernal equinox and a minimum near the summer solstice (northern hemisphere) when the earth–sun distance is a maximum and the sun's declination is 23.5° above the equator. In theory at least, the solar arrays could be designed to track the sun's declination, but the small improvement would hardly justify the addition of another degree of freedom to the drive mechanism with the attendant loss in reliability and increase in cost.

The seasonal variations in the solar flux density can be calculated from the factor $\cos \delta / d^2$, where δ is the declination of the sun and d is the relative distance in astronomical units (1 AU is equal to the semimajor axis of the earth's orbit). If we take the value of a seasonal factor F as equal to 1 at the mean distance to the sun of 1 AU, we have the results shown in Table 5-5.

If the power demand of the spacecraft during eclipse is typical (i.e., housekeeping and transponder operation but not much heater power), the array size is determined by the summer solstice since the extra battery-charging power required at equinox is less than the extra solar power available. On the other hand, a satellite with substantial power requirements during eclipse may have its array size determined by the autumnal equinox. Eclipses during that season are more severe since the available solar power is less then than during the spring eclipse season. It is interesting to note that the power balance in the spacecraft can change during spacecraft lifetime for several

TABLE 5-5 Seasonal Variation in Solar Flux Density

Date	Relative Distance (AU)	Declination (deg)	F
Vernal equinox	0.996	0	1.008
Summer solstice	1.016	23.4	0.889
Autumnal equinox	1.0034	0	0.993
Winter solstice	0.984	−23.4	0.948

reasons. Components deteriorate and thermal coating characteristics change; for example, the absorptivity of surfaces often increases with time.

The nominal required array power at the beginning of life P_A, allowing a margin m_A—typically about 5%—and the previously mentioned radiation degradation factor, can be written as

$$P_A = m_A e^{kN} \frac{P_{EQ}}{F_{EQ}}$$

or (5-7)

$$P_A = m_A e^{kN} \frac{P_{SOL}}{F_{SOL}}$$

whichever is greater. Figure 5-11 shows a typical case of available array power versus years. In this case, the margin is always greater at equinox, but both equinox and solstice available array powers degrade exponentially.

The array size in square meters, A, is calculated from

$$A = \frac{P_A}{G \eta \eta_A S} \qquad (5\text{-}8)$$

where G = solar constant 1370 W/m² at a mean distance from the sun of 1.0 AU
η = solar cell efficiency
η_A = factor to allow for various losses, such as those due to cover glasses, panel wiring, packing, etc. (0.85 is typical)
S = shadowing factor from antenna masts and other structures (0.9 is typical)

This size A is the useful array size. If the satellite is stabilized as a drum spinner, the array size must be increased in geometric theory by a factor S to allow for not all the cells facing the sun. This is a drawback to such stabilization methods as the power requirement increases. The loss is partly offset by less cell efficiency deterioration due to solar heating since the revolving drum tends to stabilize cell temperatures at a lower value. The practical reduction factor seems to be about 2.5 rather than π because of this.

Solar cell efficiencies are improving all the time. Early cells were typically about 8%, but more recently cells have been developed to exploit the substantial amount of energy in the ultraviolet part of the sun's spectrum, together with better coatings for the cover glasses and other techniques. Cells are now made with values of η from 13 to 15%. Gallium arsenide cells promise still higher efficiencies. Substantial improvements in resistance to radiation damage have also been made. Thus a careful estimate of array size and mass must be made using up-to-date and relevant cell efficiencies and degradation factors.

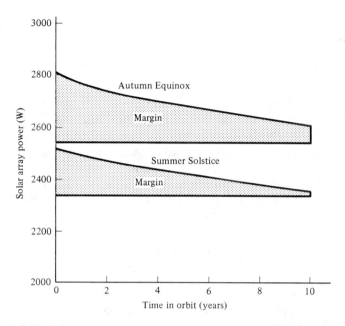

Figure 5-11 Solar array margin.

5.5 THERMAL SUBSYSTEMS

It is necessary that the mean spacecraft temperature and the temperature of all the subsystems be maintained within limits suitable for satisfactory operation. The performance and reliability of everything in the spacecraft is more or less sensitive to temperature. Some devices, such as valves, thrusters, bearings, and deployment mechanisms, can fail to operate completely if the temperature gets too low or too high.

We can construct a simple and informative model for the overall spacecraft thermal equilibrium. As a closed system, we must have a balance between the heat added (by absorption from the sun and earth and internal generation) and the heat lost through radiation. There is no mechanism for losing heat other than radiation. We consider only energy absorbed from the sun, a spacecraft absorptivity α and area A_a, an emissivity ε and radiating area A_r, and an internal heat generation Q in watts. At equilibrium

$$\begin{array}{c}\text{solar energy} \\ \text{absorbed}\end{array} + \begin{array}{c}\text{heat generated} \\ \text{internally}\end{array} = \text{heat radiated}$$

$$\text{or} \quad \alpha A_a G \quad + \quad Q \quad = \quad \varepsilon \sigma T^4 A_r \qquad (5\text{-}9)$$

The right-hand term is the radiation in accordance with the Stefan–

Boltzmann law and σ is a constant equal to 5.67×10^{-16} W/(m²·K⁴). The equilibrium temperature is simply

$$T^4 = \frac{1}{\sigma}\left(\frac{\alpha A_a}{\varepsilon A_r} G + \frac{Q}{\varepsilon A_r}\right) \qquad (5\text{-}10)$$

At any particular temperature the absorptivity α and emissivity ε of a body are the same. In the interesting case of a spacecraft, however, this is not so since the absorption is from the sun at a temperature of many thousand of kelvins whereas the radiation to space is at temperatures near 300 K. Thus the ratio of α/ε in the appropriate temperature range becomes a thermal subsystems designer's choice to control the equilibrium. The ratio of effective absorbing to radiating areas is a geometric property. A plane sheet has a ratio of 1:2 since one surface absorbs but both radiate. For a sphere in sunlight the factor is 1:4, the ratio of the projected or cross-sectional absorbing area to the total surface radiating area. Equilibrium temperature as a function of α/ε is shown in Figure 5-12. Some representative values of α and ε are listed in Table 5-6. Note the wide range of possibilities available to the spacecraft designer, depending on the choice of materials and finish.

Since the solar constant drops to zero during eclipses, the heat balance can change dramatically depending on how much electronic equipment is kept operating. Active control in the form of louvers to vary A_r and α/ε is sometimes used and heaters, especially for components sensitive to cold such as batteries and thrusters, are often used. The temperature of individual elements is

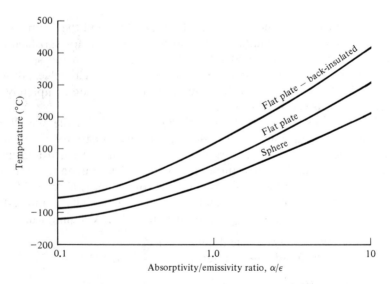

Figure 5-12 Equilibrium temperatures (solar absorption only). (Adapted from R. Haviland and C. House, *Handbook of Satellites and Space Vehicles*, D. Van Nostrand Company, Princeton, N.J., 1965.)

TABLE 5-6 Selected Values of Absorptivity and Emissivity for Preliminary Design

Material	Condition	Absorptivity for 6000-K Solar Radiation	Emissivity, Body at 0–40° C	α/ε
Aluminum 24-ST	Polished	0.29	0.09	3.2
	Oxidized	0.55	0.20	2.7
Copper	Polished	0.35	0.03	11.7
	Oxidized	0.76	0.78	1
Graphite	Milled	0.88	—	~1
	Pressed	—	0.98	~1
Stainless steel 301	Polished	0.38	0.16	2
Clear silicon on stainless	—	0.91	0.91	~1
Magnesium, Dow metal	—	0.31	0.15	2
Molybdenum	Polished	0.36	0.02	18
Nickel	—	0.36	0.12	3
Plastic laminate × 12,100	As received	0.85	0.79	1
Rhodium	As received	0.26	0.01	26
Tantalum	—	0.45	0.03	15
Titanium	—	0.50	0.08	6.0
K-Monel	—	0.42	0.23	1.8
Silver	Polished	0.07	0.03	2.3
Zinc	Polished	0.55	0.25	2.2
Oil paint, lampblack	—	0.96	0.97	1
Oil paint, white lead	—	0.25	0.93	~0.26
Oil paint, carbonate	—	0.15	0.91	~0.16
Quartz	—	Transparent	0.89	—
Glass	—	Transparent	0.95	—

controlled by heat sinks and conducting paths to radiators, internal radiation, and sometimes "heat pipes." The latter device exploits the high heat of evaporation of fluids at a constant temperature to absorb large amounts of heat from a high-temperature source such as a transmitter tube, and take it away by fluid convection to a radiator. Heat pipes are highly effective for moving heat on a satellite.

5.6 PROPULSION SUBSYSTEM

The spacecraft requires propulsion in many missions to inject itself into final orbit, maintain itself in the correct orbit, and to assist in attitude control. The subsystem for accomplishing these maneuvers is called the *propulsion subsystem* or, often, the *reaction control system* (RCS). The latter designation is intended to refer to the liquid engine part of the propulsion system—usually used for attitude control and stationkeeping. Final orbit injection is commonly done with a solid rocket engine, but we are seeing more and more liquid engines used for that purpose.

A common method of injecting into geostationary orbit is by the use of a solid apogee kick motor (AKM). If M_0 is the initial on-orbit mass of the spacecraft, including propellant and residual AKM case (as estimated in Section 5.9), the propellant required (M_p) is

$$M_p = M_0 \left(1 - e^{-\Delta v/gI_{sp}}\right)$$

using the results of Chapter 4. Δv depends on the launch latitude and other mission details. For an Ariane launch using an 8.5° inclination, the value of Δv is 1515 m/s, whereas from Cape Canaveral with 28.5 it will be around 1837 m/s. These values are also affected by transfer orbit perigee and the equations of Chapter 3 are used to determine exact values. With solid AKMs and specific impulses in the vicinity of 290 s, the value of M_p for an Ariane launch is about 70% of M_0, whereas from Cape Canaveral it would be about 91%. Typically, the spacecraft attitude is maintained in transfer orbit and during injection by spinning about the AKM axis. This is often done regardless of the operational spacecraft attitude control system. Three-axis control in transfer is also used but is not as common.

Table 5-7 lists some typical solid rocket motors suitable for use as apogee or perigee kick engines. Note that the specific impulses are given in N·s/kg rather than in seconds. The ratio of the two values is simply the acceleration of gravity. Total impulse is often quoted for solid motors. For liquid engines, the total impulse is usually not, since it depends on the tankage (amount of propellant carried).

On-orbit propulsion is required for north-south and east-west stationkeeping, the correction of initial launch errors, longitudinal relocation, and attitude control. The velocity requirements for all the orbital maneuvers are determined by the methods of Chapter 4. The attitude control requirements

TABLE 5-7 Solid-Propellant Space Motors[a]

Manufacturer	Designation	Total Impulse[b] (10^6 N·s)	Maximum Thrust (N)	Mass (kg)	I_{sp} (n·s/kg)	I_{sp} (s)
Thiokol	Star 62	7.8231 (a)	78,320	2,890	2,857	291
Thiokol	Star 75	13.2655 (p)	143,691	4,798	2,907	296
Thiokol	Star 31	3.7380 (a)	95,675	1,398	2,876	293
Thiokol	Star 37XE	3.7607 (a)	62,678	1,400	2,822	288
Thiokol	Star 48	5.6960 (a)	67,818	2,114	2,835	289
CSD	IUS large	27.9393 (p)	192,685	~10,660	2,871	293
CSD	IUS small	7.7163 (a)	75,650	~2,900	2,862	292
Aerojet	62-KS-33,700	9.3103 (a)	149,965	3,606	2,803	286
Aerojet	66-KS-60,000	17.5962 (p)	268,335	7,033	2,813	287

[a] PKM impulse range is 1.29×10^7 to 1.73×10^7 N·s; AKM impulse range is 3.75×10^6 to 5×10^6 N·s.

[b] (a), AKM candidate; (p), PKM candidate.

Sec. 5.6 Propulsion Subsystem

are generally small and must be estimated from the disturbing torques, mostly from station-keeping thrust and radiation-pressure unsymmetries. They are typically about 10% of the maneuvering requirement. Some common values are given in Table 5-8.

Early attitude control and propulsion (RCS) systems used cold gases such as nitrogen and hydrogen peroxide. They had very low specific impulses and were quickly supplanted by monopropellant hydrazine (N_2H_4). The hydrazine, under the influence of a catalyst, decomposes exothermally into ammonia and nitrogen and provides a specific impulse of about 225 s. Bipropellant systems, typically using monomethyl hydrazine (MMH) and nitrogen tetroxide as an oxidizer, are in increasing use. Although they are more complicated, they provide impulses around 290 s and can provide enough thrust to be used as apogee engines, as discussed in Chapter 4. The characteristics of some typical systems are shown in Table 5-9.

Monopropellant systems can also have their impulses increased by raising the temperature with resistive heating (electrothermal thrusters). Since I_{sp} is proportional to the square root of the absolute temperature (Chapter

TABLE 5-8 Attitude Control Requirements

N-S station keeping	46–48 m/s per year, depending on launch date
E-W station keeping	0–2 m/s per year, depending on location
Initial errors	85 m/s to correct 600-km low apogee and 1° plane error
Relocation	28 m/s 5°/day and stop 17 m/s 3°/day and stop
Attitude control	20–50 m/s 7-year life

TABLE 5-9 Liquid Bipropellant Space Engines

Manufacturer	Nominal Thrust (N)	Fuel	Oxidizer	O/F Ratio	I_{sp} (s)
Marquardt					
R-40A	4,005	MMH	N_2O_4	1.6	299
R-IE-3	111	MMH	N_2O_4	1.6	272
Bell	4,210	MMH	N_2O_4	1.6	289
Rocketdyne	445	MMH	N_2O_4	1.65	299
Rocketdyne	1,335	MMH	N_2O_4	1.57	287
Aerojet	445	MMH	Mon. 8	1.65	304
M.B.B.	400	MMH	N_2O_4	—	307
S.E.P.	61,668	LH_2	LOX	—	432
Aerojet OMS	26,750	MMH	N_2O_4	1.65	313

4), the increase must be substantial and this is not an easy technology. Operating temperatures of about 2000° C are needed.

Figures 5-13 and 5-14 show typical schematics for mono- and bipropellant reaction control systems. It is common in modern monopropellant systems to use the thermally heated high-performance thrusters only for north-south stationkeeping. These thrusters are less reliable but at the same time the north-south stationkeeping requires the greater I_{sp} and is often less critical to mission success.

Gaseous nitrogen (GN_2) is shown as the pressure-producing method to provide propellant feed in a weightless environment. Spinners can use cen-

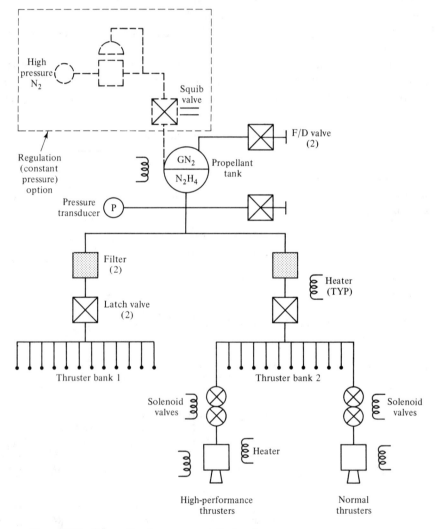

Figure 5-13 Schematic of a monopropellant system.

Sec. 5.7   Telemetry, Tracking, and Command   **131**

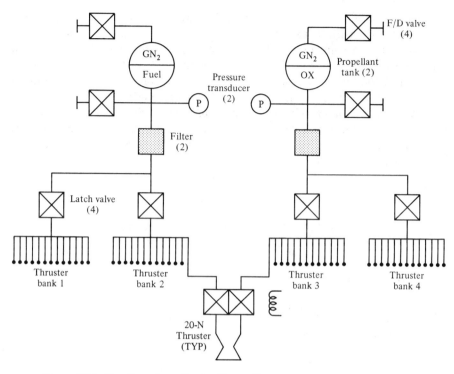

Figure 5-14 Baseline schematic of a bipropellant system.

trifugal force to accomplish the same function and also use noticeably fewer thrusters because of the possibility of timing the rotating radial thruster.

Note that the specific impulse of a thruster may vary with lifetime. Typically it drops slowly with reduced tank pressure. Such reductions in tank pressure are typical of "blow-down" pressurizing systems. In a careful calculation of spacecraft fuel consumption, one can assume that the volume of gas increases to replace the consumed liquid propellant, and the pressure drops in accordance with Boyle's Law. Empirical data on the variation of I_{sp} are required to complete the computation and are usually available from the thruster manufacturers.

5.7 TELEMETRY, TRACKING, AND COMMAND

The three related functions—telemetry, tracking (including range measurements), and command—are usually grouped into one subsystem called *telemetry, tracking and command* (TT&C) or alternatively *telemetry, tracking, command and ranging* (TTC&R). All three are essentially communications functions. Thus the computations of link performance, signal-to-noise ratios, error rates, and other communications parameters are identical in principle to those for telephone, TV, and data. The transmission rates are generally much lower than those for the payload, whereas the antenna patterns and link power budgets must be configured both for transfer orbit and operational orbit. Early satellites used VHF and S-band for TT&C regardless of the communication band, but more recently there has been a tendency to use a segment of the communication band itself for those services.

VHF and S-band links, in common use for all three services, typically have characteristics along the lines of Table 5-10. A simplified block diagram of a spacecraft TT&C system is shown in Figure 5-15. No redundancy is shown.

TABLE 5-10 VHF and S-Band Characteristics

	VHF	S-band
Uplink (MHz)	148–149.9	2025–2120
Downlink (MHz)	136–138	2200–2300
Number of sinusoidal tones	6	7
Ranging accuracy (m)	20–140	5–175
Propagation errors (m)	100–2000	0–300

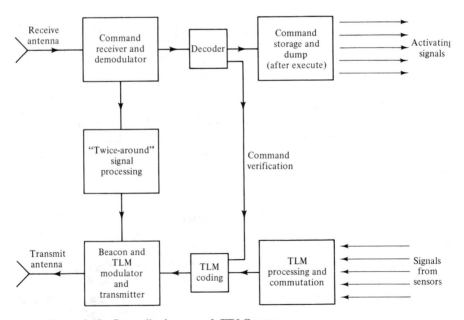

Figure 5-15 Generalized spacecraft TT&C system.

5.7.1 Telemetry

The satellite condition must be known on the ground at all times. It is usual to choose some hundreds of points around the spacecraft and measure such quantities as voltages, currents, temperatures, pressures, and the status of switches and solenoids. Sensors for these quantities are provided together with analog-to-digital (A-D) converters and their outputs are sampled in a commutation system. PCM, time-division multiplexing, and phase-shift keying are the usual telemetry transmission modes. FM analog telemetry is still used occasionally (Stiltz, 1966).

5.7.2 Tracking

Beacon transmitters are usually provided on the spacecraft for tracking during launch and operations. This transmitter can also carry telemetry signals and range signal "turnaround" and command verifications. Angular measurements are done by conventional terrestrial methods using large antennas and *monopulse* or conical scanning systems developed years ago for radar.

Ranging is done by one of two methods. A standard is to phase modulate the uplink or "command" carrier with pairs of low-frequency tones, detect these signals on board, and remodulate the telemetry carrier on the downlink. The earth station compares the transmitted and received phases to calculate the range. By using the tones in pairs, it is possible to resolve the range ambiguities otherwise present.

Another method is to transmit pulsed signals on the uplink and retransmit them on the downlink, measuring the range by time difference in the usual radar manner.

5.7.3 Command

A wide variety of such systems has been developed and the choice depends on the number of commands, the rest of the TT&C, and the "security" required. Digital systems and low-frequency tones are both used.

Most command systems take a similar sequence of operations to protect against unauthorized or fake commands and errors. The sequence is:

1. An *enabling* signal is transmitted to permit command system operation.
2. The specific command is sent and stored.
3. The command is *verified* by transmitting to the earth the telemetry link.
4. An *execute* signal is transmitted and the command is carried out.

Commands are necessary for many functions during manual operation—specifically, transponder switching, stationkeeping, attitude changes, gain control, and redundancy control. During launch there may be others, for example, separation commands, antenna and solar panel deployment, and apogee motor firing. There is a growing tendency to encrypt command signals so as to prevent unauthorized tampering with the satellite—or even malicious mischief. This encryption is done by one of the pseudorandom coding systems and requires both terrestrial and on-board equipment. It is an undesirable but increasingly necessary complication.

5.7.4 Antennas

TT&C antennas are usually quite different from those used for the payload. They have the necessity of remaining in communication during transfer orbit. If the spacecraft is spinning during that phase, these antennas should be omnidirectional since the spacecraft attitude relative to the earth will change continuously. Such an antenna is not theoretically possible and can be approximated only by antennas such as biconical horns and turnstiles. In addition, depending on the location of the terrestrial facilities, the spacecraft may not always be in view of a tracking station. This is, in fact, usually the case since most systems operate only over part of the earth and have no need for worldwide facilities except during launch. It is thus common for a regional satellite operator to make arrangements with the operators of international systems (e.g., European Space Agency, NASA, and Intelsat), who perforce maintain worldwide tracking and command facilities. Once the satellite is in operation, stabilized, and located in orbit, it is possible to use narrow-beam horn antennas for TT&C.

5.8 ESTIMATING THE MASS OF COMMUNICATIONS SATELLITES

Estimating the mass of a communications satellite is necessary early in program planning, because both the cost of the satellite and that of launching it are dependent on that mass. Ultimately, it is possible to do this precisely only by the systematic addition of the masses of individual components. This tedious procedure is practical only after the system design has been set and all the characteristics of the satellite established. At this point, spacecraft manufacturers can calculate the in-orbit mass, typically with a tolerance of a few percent. Unfortunately, the setting of the principal system and spacecraft characteristics, which in turn permits manufacturers to do careful weight and power estimating, is itself dependent on the economics of the system and therefore on the aforementioned masses and powers. The circle is evaded by the development of simplified models for estimating weight and power. These models are useful to assess the effects of changes in systems parameters such as transmitter power and efficiency, eclipse operation, redundancy, and life.

Sec. 5.8 Estimating the Mass of Communications Satellites

The creation of such a model in itself is not simple since it depends on empirical relations, the specific nature of such system designs, and above all, certain technologies that determine such critical characteristics as the mass and efficiency of high-power amplifiers, the specific impulse of propulsion systems, and the weight of batteries. The utility of any such method—and many have been developed during the past 20 years—is limited to estimating the effect of proposed system changes, helping to ensure that the spacecraft is within the range of a particular launch vehicle, and highlighting the features that cost money. In our opinion, no such general model is good enough for making firm fixed-price quotations or absolute commitments to launch vehicle limits.

The method we have outlined starts from the parameters usually known from system requirements (e.g., RF transmitter power, number of transponders, life and eclipse operation, etc.). They are functions of needed channel capacity, operational coverage, and orbital geometry.

5.8.1 Mass of Primary Power Subsystem

The problem is approached by starting with a calculation of the primary power to be produced by the solar array.

The methods of Section 5.4 are used to determine P_A, typically at the summer solstice and the beginning of life in accordance with Eq. (5-7). The mass of the primary power subsystem itself M_{pp} is then given by

$$M_{pp} = M_a + M_b + M_c \tag{5-11}$$

where M_a = array mass
M_b = battery mass
M_c = power control mass

M_a and M_c can be found to a good approximation by quasi-empirical formulas, and M_b using Eq. (5-3):

$$M_a = \frac{SP_a}{\gamma} + 10 \tag{5-12}$$

$$M_c = 0.01 P_a + 10 \tag{5-13}$$

$$M_b = \frac{u}{\beta} \frac{P_e t_e}{\beta_d} \tag{5-14}$$

where γ is the array factor in W/kg; and

$$\begin{aligned} M_{pp} &= \frac{SP_a}{\gamma} + 10 + 0.01 P_a + 10 + \frac{P_e t_e}{\beta d} \\ &= \left(\frac{S}{\gamma} + 0.01\right) P_a + t_e \frac{(e+h)P_T + P_{h0} + P_{he}}{\beta \eta_d d} + 20 \end{aligned} \tag{5-15}$$

where β = battery factor, Wh/kg
s = "spin" factor = $2.0 < s < \pi$
   = 1.0 for three-axis
e = eclipse transponder operation
t_e = eclipse time = 1.2 h
d = depth of discharge
η_d = charge efficiency = 0.9

The factor γ was about 30.0 W/kg at Beginning of Life (BOL) (in 1985) and is improving steadily. The battery factor β is about 30 Wh/kg for both NiH and NiCd batteries and has improved slowly over the years. A great advantage today of NiH batteries is the possibility of 0.7 to 0.8 depth of discharge.

5.8.2 Payload Mass M_{PL}

The mass of the payload, considered as the communications antenna and transponder, is given by the total transponder and antenna masses. If there are n_i transmitters of each of i types with an individual mass M_i and if the mass of each antenna is M_{aj}, then

$$M_{PL} = bR \sum_i n_i M_i + \sum_j M_{aj} \qquad (5\text{-}16)$$

where R is the transmitter redundancy and b is a factor to allow for the mass of the receivers, switches, up and down converters, filters, and all the remaining transponder electronics. Values of b vary from 1.1 to 2.0, depending on complexity. If specific data on the masses of the receivers, filters, switches, and so on, are available, they should be used in preference to the factor b and simply added to the two terms of Eq. (5-16).

Antenna mass estimating, because of the possibly complex shapes and multiple feed structures, is very difficult. For simple parabolic reflectors with average f/D ratios, the mass will be proportional to the surface area of the reflector, which in turn depends on the square of the diameter. Because of questions of stiffness, surface tolerances, and feed accommodation, the composite antenna masses seem to vary more rapidly. For relatively simple reflector and feed combinations, we can use the following relations, in the absence of any other information:

$$M_{ant(1)} = 12 + 4D^{2.3} \quad \text{at C-band} \qquad (5\text{-}17)$$

$$M_{ant(2)} = 9 + 3D^{2.3} \quad \text{at K-band} \qquad (5\text{-}18)$$

Because of the extremely wide variation in antenna designs, materials, and techniques, it is best to use a specific estimate.

5.8.3 Support Subsystems

The mass M_s of the support systems can be related to the dry mass M_D in several ways. Data on 12 spinning and 18 body-stabilized communication satellites, almost all designed during the decade 1972–1982, yielded the results of Table 5-11. We must be aware that there is some lack of consistency in available data on mass breakdowns.

These ratios can be used for quick estimating, but this method is not recommended. It takes no cognizance of the changing ratio of subsystem-to-dry spacecraft mass as the total mass increases. Nor does it allow for variation in transponder mass as a function of primary power mass. These ratios should be used as guides only. A better way to proceed is to use the rather well correlated results of a regression analysis on the aforementioned typical systems but without the primary power subsystem and without the payload. It produces the straight lines of Figure 5-16 and the related equations (5-21) and (5-22). Both equations fit with correlation coefficients in excess of 0.98.

We note that M_D comprises payload M_{PL}, primary power M_{pp}, and support subsystems M_{ss}, the mass of which is linearly related to the dry spacecraft. Thus

$$M_D = M_{PL} + M_{pp} + M_{ss}$$

$$= M_{PL} + M_{pp} + kM_D + M_1 \quad (5\text{-}19)$$

$$M_D = \frac{M_{PL} + M_{pp} + M_1}{1 - k} \quad (5\text{-}20)$$

TABLE 5-11 Relative Masses of Satellite Subsystems

	Body Stabilized		Spin Stabilized	
	% Total Dry Mass	% Sub-systems	% Total Dry Mass	% Sub-systems
Cable-harness	4	8.2	4	8.2
TT&C	4	8.2	4	8.2
Structure (includes mechanical integration and balance weights)	18	36.7	21	41.2
Attitude control (ACS)	7	14.3	5	9.8
Propulsion (RCS)	5	10.2	3	5.9
AKM case	7	14.3	8	15.7
Thermal	4	8.2	5	9.8
Support subsystems	49%	100%	51%	100%
Communications	28%		25%	
Primary power	23%		24%	
Total	100%	100%	100%	100%

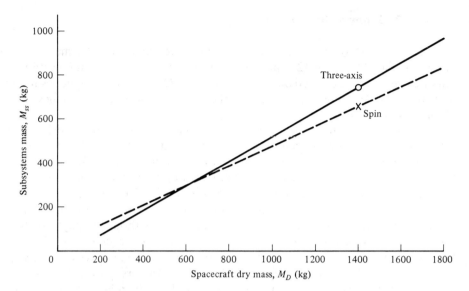

Figure 5-16 Subsystems mass compared to spacecraft dry mass.

M_{PL} and M_{pp} are calculated in accordance with the methods of the previous paragraphs. The subsystems are found from:

Three-axis:
$$M_{ss} = -10 + 0.50 M_D$$

Spin: (5-21)

$$M_{ss} = 30 + 0.45 M_D$$

(5-22)

We can then write for the dry mass:

Three-axis:
$$M_D = 2.00(M_{PL} + M_{pp} - 10) \text{ kg}$$

Spin:
$$M_D = 1.82(M_{PL} + M_{pp} + 30) \text{ kg}$$

The wet mass at the beginning of life in orbit, M_0, is found from the dry mass by calculating the amount of propellant needed in accordance with the methods of Section 5.6. A good general expression, absent a more detailed calculation, is to use 50 m/s for N years of design life:

$$M_0 = M_D e^{(100 + 50N)/gI_{sp}} \quad (5\text{-}23)$$

This assumes average north-south station-keeping requirements, about 100 m/s for correcting initial errors and repositioning and about 10% extra for attitude control. The 50 m/s term is only about 10 m/s without north-south station keeping. The methods of Section 5.6 should be used if more careful estimates are needed.

A solid apogee motor case has been assumed. In those cases where a liquid AKM is used, the residual mass of the engines in orbit is generally a little higher than that of a solid case—say 50%—and the structure and attitude control systems should be increased about 10% to accommodate liquid engines. This can easily be made up for by the more energetic liquid propellants.

5.9 SYSTEM RELIABILITY AND DESIGN LIFETIME

System reliability is an enormously complicated subject (Bazovsky, 1961; Feller, 1950; Sandler, 1963). It is based on the theory of probability, and some of it is of dubious applicability since the populations are small. Nonetheless, it is all that we have and much preferable to nothing. Again, here we can only highlight the main concepts as applied to system planning. The extensive applications of the theory to component and subsystem quality control will not be considered here.

Spacecraft system life depends on random failures, wear-out, and the exhaustion of expendables, notably propellant. Each of these three factors is predicted differently. The different failure rates during different parts of a total mission are seen in Figure 5-17. Fig. 5-18 distinguishes between random failures and wear-out.

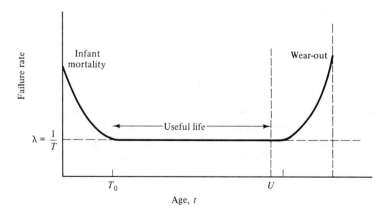

Figure 5-17 Failure rate versus satellite age.

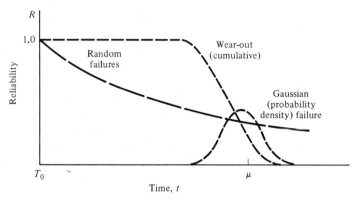

Figure 5-18 Wearout and random failures.

5.9.1 Random Failures and Design Life

Rare events, such as peoples' deaths and component failures, are generally *Poisson distributed*. That is, the probability of exactly n events, if the average number of such events in a given time is a, is given by

$$w(n) = \frac{a^n}{n!} e^{-a} \quad \text{where } a = n \tag{5-24}$$

If λ is the failure rate equal to a/t, the probability of no failures $w(0)$ or the system reliability R is given by

$$R = w(0) = e^{-a} = e^{-\lambda t} \tag{5-25}$$

The probability of at least one failure in any period t is $(1 - e^{-\lambda t})$ and the probability density as a function of time $w(t)$ is the first derivative $\lambda e^{-\lambda t}$. The mean value of the time to first failure T is simply the integral of $tw(t)$ over all possible values:

$$T = \int_0^\infty \lambda t + e^{-\lambda t}\, dt = \frac{1}{\lambda} \tag{5-26}$$

Following Maral et al. (1982), we can define a useful or mission life U at the end of which the service is truncated. In this case the average life τ is given by the sum of two integrals, the second of which is a delta function normalized by the factor $e^{-u/T}$, so that the cumulative probability of failure to infinity is equal to 1. For this truncated case, the average life is

$$\tau = \int_o^u \lambda t\, e^{-\lambda t}\, dt + e^{-u/T} \int_o^\infty ts(t - u)\, dt \tag{5-27}$$

$$\tau = T(1 - e^{-u/T}) \tag{5-28}$$

where T is the mean time to failure, as already defined, for a constant failure rate. τ/T is the probability of a failure during the useful life. This equation is useful for planning the number of launches required over a long period. For a basic system of one satellite with no spare, over a system lifetime L, the number of launches needed n (if each launch has a probability of success p) is

$$n = \frac{L}{pT(1 - e^{-u/T})} \qquad (5\text{-}29)$$

If the time to replace a satellite is T_R, the average time of unavailability during L years is LT_R/pT and the system availability A is given by

$$A = 1 - \frac{T_R}{pT} \qquad (5\text{-}30)$$

If there is an "in-orbit" spare, twice as many launches are necessary, but the availability of the system becomes

$$A = 1 - \frac{2T_R^2}{P^2 T^2} \qquad (5\text{-}31)$$

5.9.2 Wear-Out

Components such as bearings, solenoid-operated valves, and TWT cathodes wear out in accordance with a normal or Gaussian distribution rather than failing at random. In such cases, the reliability is given by

$$R = 1 - \int_{-\infty}^{t} \frac{1}{\sigma\sqrt{2\pi}} e^{-(t-n)^2/2\sigma^2} \, dt \qquad (5\text{-}32)$$

where m is the mean life and σ is the standard deviation. If it is necessary to consider wear-out, the reliability due to wear-out must be multiplied by the random failure reliability to determine the composite. Systems are normally designed so that mean wear-out times are long compared to the design lifetimes. The exact calculation, considering random failures, wear-out, and truncation due to exhaustion of expendables involves all three probability densities and is not necessary for most system planning.

Some numbers are interesting. See Table 5-12, which describes a system of one operating satellite, taking three months to launch, and a design lifetime U of seven years. Note that the mean time to failure for random failures must be substantially longer than the design life if reasonable system availabilities are to be achieved.

Systems using more than one satellite to cover the operating territory (e.g., Intelsat and wide-area direct broadcasting systems) can use the same basic ideas but must consider each satellite as a *Bernoulli trial*. If the satellites are interchangeable and we have a system of N operational satellites and S

TABLE 5-12 Satellite System Design

Design life, U	7 years	10 years
Meantime to failure, T	20 years	20 years
Average lifetime	5.9 years	7.9 years
Probability of failure during life, τ/T	0.29	0.39
Time to replace	0.25 year	0.25 year
Probability of launch success	0.9	0.9
Annual launch rate (no spare)	0.19	0.14
Availability, A	0.986%	0.986%
Annual launch rate (with spare)	0.38	0.28
Availability	0.9996%	0.9996%

spares, the probability of at least n operating satellites is found from the Bernoulli formula for exactly k successes in n tries:

$$P(k) = \binom{n}{k} p^k (1-p)^{n-k} \qquad (5\text{-}33)$$

where the binomial coefficient is equal to $n!/k!\,(n-k)!$.

In this case, $k = N$ and $n = N + S$. The probability of at least N successes is the system reliability R and is found by calculating $P(k)$ for all values of k from N to $N + S$ and summing them.

An important special case is when $S = 1$ (one spare satellite). Then there are only two terms in the sum: that is, either all but one satellite works or all the satellites work and the system still functions. In that case

$$R_N = R^N [N(1-R) + 1] \qquad (5\text{-}34)$$

5.9.3 System Modeling

The overall system reliability is a series model in which the composite probability of success (no failure) is the product of the constituent subsystem reliabilities. The subsystem reliabilities themselves are usually a network of series–parallel paths depending on the redundancy. In principle, these networks are all extensions of that shown in Figure 5-19. R is always the reliability

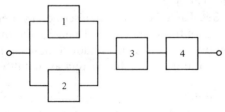

Redundancy 2:1

Figure 5-19 Basic subsystems reliability model.

of a constituent block or probability of success and Q is $(1 - R)$ or the probability of a failure.

We then have overall reliability R_o of the series parallel combination given by

$$R_o = (1 - Q_1 Q_2) R_3 R_4$$
$$= (R_1 + R_2 - R_1 R_2) R_3 R_4 \qquad (5\text{-}35)$$

Arbitrarily complicated systems, with "live" or "standby" redundancies at any level, can be built up in a straightforward manner using the notion above. Real systems are complicated and must be programmed on a computer. If the redundancy switching itself is a problem, it must be considered as another element in the model, usually in series with the parallel elements. Standby failure rates are usually different from operating failure rates, and this must also be considered.

At high levels of complexity, subsystems comprising electronic and mechanical components can be considered to have a constant failure rate. Some elements are "time" dependent and some, switches for instance, tend to be "cycle" dependent. For series elements

$$R = e^{-\Sigma_i \lambda_i t_i} \qquad (5\text{-}36)$$

where t_i is that part of the total time period or number of cycles in question over which the ith component is accumulating time. λ_i is the relevant failure rate, per hour or per cycle depending on which is involved. "One-shot" probabilities must be multiplied into the above if applicable.

In the series–parallel cases, the equations of the form Eq. (5-35) are simply multiplied out, yielding the composite reliability function—a combination of exponentials. The mean time to failure can be found by numerical integration. This composite mean time to failure is set equal to the time at which R is equal to e^{-1}. This composite value can be used in system planning as discussed in Section 5.9.2.

REFERENCES

BAZOVSKY, I., *Reliability Theory and Practice*, Prentice-Hall, Inc., Englewood Cliffs, N.J., 1961.

FELLER, W., *An Introduction to Probability Theory and Its Applications*, John Wiley & Sons, Inc., London, 1950.

HAVILAND, R., and C. HOUSE, *Handbook of Satellites and Space Vehicles*, D. Van Nostrand Company, Princeton, N.J., 1965.

KAPLAN, M., *Modern Spacecraft Dynamics and Control*, John Wiley & Sons, Inc., London, 1976.

MARAL, G., M. BOUSQUET, and J. PARES, *Les Systèmes de Télécommunications par Satellites*, Masson et Cie, Éditeurs, Paris, 1982.

MIYA, K., *Satellite Communications Engineering*, Lattice Company, Tokyo, 1975.

SANDLER, G., *Systems Reliability Engineering*, Prentice-Hall, Inc., Englewood Cliffs, N.J., 1963.

STILTZ, H., *Aerospace Telemetry*, Vols. I and II, Prentice-Hall, Inc., Englewood Cliffs, N.J., 1961, 1966.

6

The RF Link

6.0 GENERAL CONSIDERATIONS

The performance of a satellite communication link is conveniently considered in two parts. The first part, commonly called the RF link and considered in this chapter, deals with the calculation of an available carrier-to-noise-density ratio (C/N_0). The second part, considered in Chapters 7 and 8, deals with the calculation of channel performance and the number of channels available as a function of the available carrier-to-noise density. The results of this second part calculation depend on the modulation and multiple access systems to be used.

    The performance of the radio-frequency link is determined from the characteristics of the radio terminals (i.e., transmitters, receivers, and antennas), the characteristics of the propagation medium, and the possible interference. The first two aspects of the RF link are considered here, while the important question of interference is treated separately in Chapter 11.

    Basic to the performance of the link is noise—which can be generated by active electron devices, received from outer space, or be simply the thermal noise inherent in the random motion of the electrons. Electron devices also, by virtue of their nonlinearities, generate another class of noiselike transmission impairment, loosely called intermodulation—considered in more detail in Chapter 7. In this chapter we consider only how its effects can be combined with the thermal and active device noise.

6.1 NOISE

The most convenient starting point for RF link discussions is with the definition of *available thermal noise*—the noise due to the random fluctuation of electric currents. Nyquist proved, from thermodynamic considerations, that the mean-squared voltage across a resistance R and measured in a bandwidth B was given by

$$\overline{e_n^2} = 4kTBR \quad \text{volts} \tag{6-1}$$

where e_n = noise voltage
k = Boltzmann's constant, 1.38×10^{-23} J/K
R = resistance, Ω
B = noise bandwidth, Hz
T = absolute temperature, K

The available noise power P_n into a matched load is then given as

$$P_n = \frac{\overline{e_n^2}}{4R} = kTB \tag{6-2}$$

P_n is independent of frequency (hence the term "white"). The Nyquist formula is valid over most of the communications spectrum, but in the higher "millimeter wave" and infrared regions the more exact quantum formula valid at all frequencies must be used.

$$P_A = \frac{hf}{e^{\frac{hf}{kT}} - 1} + hf \quad \text{W/Hz} \tag{6-3}$$

where e is the base of natural logarithms.

Note that at very high frequencies this exact expression shows that the *thermal* noise vanishes, leaving only the quantum noise (second term). We define a noise density N_0 so that $N = N_0 B$ for nonquantum thermal noise where $N_0 = kT$. If we take the ratio of P_A to N_0, we have

$$\frac{P_A}{N_0} = \frac{hf/kT}{e^{hf/kT} - 1} + \frac{hf}{kT} \quad \frac{hf}{kT} \ll 1 \tag{6-4}$$

$$\lim_{f/T \to 0} \frac{P_A}{N_0} = 1$$

$\dfrac{hf}{kT} = 1$ can be considered "transitional" between the two expressions.

This occurs for

$$\frac{f}{T} = 21 \quad \text{if } f \text{ is in GHz and } T \text{ is in K}$$

This illustrates that the lower the system temperature, the lower the frequency at which one should use the full quantum expression. It is not necessary to use it for any satellite communication problem today—even in the high gigahertz range—but for future, still higher frequencies, and in the infrared and optical regions, the quantum expression must be utilized.

6.2 THE BASIC RF LINK

6.2.1 The Inverse-Square Law Applies to Any Link

Starting with the isotropic radiation of a power P_T in concentric spherical waves and defining the gain G_T as the power ratio of the signal in the desired direction to that which would have radiated isotropically, we can write the signal (carrier) power intercepted by a receiver at distance R, having an antenna with effective aperture A_{eff} as

$$C = \frac{P_T}{4\pi R^2} G_T A_{eff} \tag{6-5}$$

since

$$\text{flux density} \times \text{effective aperture} = \text{carrier level}$$

Thus, the parameters which affect the actual received power in any link are *transmitter power, transmitter antenna gain, distance between transmitter and receiver*, and *receiver antenna (effective) size*. Eq. (6-5) can be restated in terms of receiver antenna gain, a useful engineering term, by applying the *universal antenna formula* (Jasik, 1961)

$$A_{eff} = \frac{G\lambda^2}{4\pi}$$

where G is the gain of the antenna. Then

$$C = \frac{P_T G_T}{4\pi R^2} \left(\frac{G_R \lambda^2}{4\pi}\right) \tag{6-6}$$

The required power for information transmission can be determined to be some carrier-to-noise ratio (C/N) above the available thermal noise. That value of (C/N) depends on the required information rate, the required signal-to-noise ratio (for analog signals) or bit error rate (for digital signals), and the modulation system and associated bandwidth. Its calculation is examined in detail in Chapter 7.

The available thermal noise is given by

$$N = k\,T_s\,B$$

where T_s, the *system temperature*, is a measure of receiver system noise performance, and B is the bandwidth required by the desired information rate and modulation scheme. Thus the required carrier power is

$$C = \frac{C}{N} N = \frac{C}{N} k T_s B$$

and thus we can write

$$\frac{C}{N} = \frac{(P_T G_T)}{\left(\frac{4\pi R}{\lambda}\right)^2} \left(\frac{G_R}{T_s}\right) \frac{1}{kB} \quad (6\text{-}7)$$

The following factors appearing in that equation are defined as follows for engineering convenience

$$P_T G_T = \text{e.i.r.p.} \quad (\textit{equivalent isotropic radiated power})$$

$$\left(\frac{4\pi R}{\lambda}\right)^2 = L_s \quad (\textit{free space loss})$$

where λ is the wavelength of the signal.

Free space loss is a traditional term arbitrarily defined to allow radio link power system power calculations in terms of the gains of both antennas. The actual relationship by which electromagnetic radiation density diminishes with distance, called "spreading loss," is just the inverse square relationship $\frac{1}{4\pi R^2}$ and not at all dependent on wavelength.

Rewriting Eq. (6-7) using these engineering terms, we get:

$$\left(\frac{C}{N}\right) = \frac{\text{e.i.r.p.}}{L_s} \left(\frac{G_R}{T_s}\right) \frac{1}{kB}$$

We thus see that the performance of the receiver, as it determines (C/N), is characterized by (G_R/T_s), a factor which (called G/T) is widely cited for satellite receiving systems. Like many ratios, G/T is often stated in dB form, using the "unit" dB/K. This is spoken "dB per K (kelvin)," but in fact must be thought of as "dB with respect to a reference of 1 K^{-1}."

Eq. (6-7) may be recast in a more general form, independent of bandwidth, as

$$\left(\frac{C}{N_0}\right) = \frac{\text{e.i.r.p.}}{L_s} \left(\frac{G}{T}\right) \frac{1}{k} \quad (6\text{-}8)$$

This is the fundamental relationship indicating RF link performance, and will be extensively cited in the work to follow.

6.2.2 The Significance of C/N_0

C/N_0 is an interesting parameter, the significance of which is best appreciated from a look at Shannon's equation for channel capacity.

$$H = B \log_2(1 + C/N) \tag{6-9}$$

where H, the "entropy" of the transmitted information, is measured in bits/s.

$$= \log_2(1 + C/N_0 B)^B$$

$$= C/N_0 \log_2[1 + (C/BN_0)]^{BN_0/C}$$

$$\boxed{\lim_{B \to \infty} H = C/N_0 \log_2 e}$$

from the definition of e, the base of natural logarithms.

Channel capacity is defined as the maximum rate at which information can be transmitted without error. It is an upper bound and says nothing about how to achieve it.

C/N_0 (measured in hertz) is proportional to the maximum information rate transmittable with a given carrier power regardless of bandwidth. C/T is sometimes used in the professional literature with the same objective—that is, to separate the RF and baseband parts of the calculation. It however does not have the intuitive interpretation given above nor the convenient units of hertz.

For digital transmission, a parameter closely related to C/N_0 is often used. It is E_b/N_0, the ratio of energy per transmitted information bit to noise density, which is C/N divided by the bit transmission rate. In the limiting case where the transmission rate is equal to the channel capacity H we see that

$$H = \frac{E_b H}{N_0} \log_2 e \tag{6-10}$$

$$\frac{E_b}{N_0} = -1.6 \text{ dB}$$

This is Shannon's limit for the bit energy to noise density ratio. For values below this, it is not possible to devise error-free coding systems. It is a "target" against which the efficacy of various error-correcting codes can be measured.

6.3 THREE SPECIAL TYPES OF LIMITS ON LINK PERFORMANCE

It is instructive to take the basic equation for C/N_0 and write it in an appropriate form for three different design scenarios.

6.3.1 Fixed Antenna Sizes at Both Ends

We first consider a link in which the physical size of the antennas is likely to be limited at both ends for reasons of cost or convenience—this is typical of terrestrial line-of-sight microwave relays.

Using Eq. (6-8) and the universal antenna formula, it is easy to show that

$$C/N_0 = \frac{\eta^2}{kT_s\lambda^2} \frac{A_R A_T}{R^2} P_T \qquad (6\text{-}11)$$

where A_R and A_T are the *physical* areas of the receiving and transmitting antennas and η is the antenna efficiency, here taken as the same for both antennas.

Note that the link performance as measured by C/N_0 is better for higher transmitter power and lower receiver system temperature. This is *always* true. In addition, for this case having fixed A_R and A_T, shorter distances R and shorter wavelengths λ also improve the performance. We shall see that this is not necessarily so as a general rule.

6.3.2 Fixed Antenna Gains at Both Ends

In some applications, both the antenna beamwidths have specified values. In that case we look at that form of the equation with antenna gain specified at both ends since the antenna gains are inversely proportional to the product of these beamwidths. Equation (6-8) is simply rewritten as

$$C/N_0 = \frac{G_T G_R \lambda^2}{(4\pi)^2 kT_s R^2} P_T \qquad (6\text{-}12)$$

Note that in this case C/N_0 is again improved with increased P_T and decreased T and R^2, but that now it is also better at *longer* wavelengths.

This case could be of interest in satellite-to-satellite links. In such links, the accuracy of the attitude- and position-control systems of the spacecraft could determine the minimum usable beamwidths and thus the maximum usable antenna gains.

6.3.3 Fixed Antenna Gain at One End and Antenna Size at the Other

This extremely important case applies directly to the downlink of a satellite–earth system in which the satellite transmitter antenna gain is fixed by the beamwidth implications of the earth coverage requirement, whereas the receiver antenna size is as large as possible—considering convenience and cost. We start from Eq. (6-7), use the universal antenna gain–area relation and in addition the basic geometric relation among solid angle Ω, distance R, and area S on a concentric spherical surface.

$$\Omega = \frac{S}{R^2} \qquad (6\text{-}13)$$

From the definition, $\Omega = 4\pi$ for a complete sphere.

If we assume that the energy is concentrated in the main beam, the antenna gain G_T of the transmit antennas is inversely proportional to the beam's solid angle.

$$G_T = \frac{K_1}{\Omega} = \frac{K_1 R^2}{S} \qquad (6\text{-}14)$$

Substituting in Eq. (6-8), using the universal antenna formula, we have

Downlink:

$$C/N_0 = \frac{\eta K_1 A_R}{4\pi k T_s S} P_T \qquad (6\text{-}15)$$

This is a particularly interesting and important result. Note that the link performance still depends on the transmitted power P_T and inversely on the receiver system temperature T_s but, assuming that the surface area S to be covered is given, the performance is no longer dependent on either the wavelength or distance. The height in orbit of the satellite is unimportant and so is the carrier frequency if only first-order effects are considered. Clearly, device performance, atmospheric losses, and interference with other systems will be significant factors in the choice of carrier frequency, but the first-order result is still basic. Note that it is the physical size of the antenna that counts and not its gain. The widespread use of G/T as a figure of merit, although convenient, must not be allowed to obscure this result. G/T can only be used to compare different receiving systems at the same frequency.

Equation (6-15) is even more interesting when it is rewritten to apply to the uplink under the restriction that the satellite antenna coverage of the earth, now for reception, is specified.

$$G_R = \frac{K_1}{\Omega} = \frac{K_1 R^2}{S} \quad \text{and} \quad G_T = \frac{\eta 4\pi A_T}{\lambda^2}$$

Again substituting in Eq. (6-8), we arrive at:

Uplink:

$$C/N_0 = \frac{\eta K_1 A_T}{4\pi k T_s S} P_T \qquad (6\text{-}16)$$

This equation is identical to the previous expression for the downlink except that the antenna area involved is that of the *transmit* antenna. The transmitter power P_T now refers to that of the earth station, the system temperature T_s to that of the satellite receiver, *but* the physical antenna size involved is that of the earth station antenna—as it was in the downlink case.

In other words, the performance of any satellite communcation system in which fixed terrestrial coverage must be provided is dependent on the earth station antenna physical size for both its uplinks and downlinks.

6.4 SATELLITE LINKS (UP AND DOWN)

6.4.1 General

It is convenient to write Eq. (6-8) in decibels by taking the common logarithm of both sides and multiplying by 10. Thus

$$[C/N_0] = [\text{e.i.r.p.}] - [L_s] + [G_R/T_s] - [k] \qquad (6\text{-}17)$$

The brackets remind us that the "dB" form of the factor is to be used.

This equation is correct for either the uplink or downlink, with the caution that the operating values of e.i.r.p. and G/T must be used. When modified by atmospheric and other incidental losses, it is applicable to any line-of-sight communications link, either terrestrial or in space.

The units are important and must be consistent.

$[C/N_0]$ in dBHz

$[\text{e.i.r.p.}]$ in dBW

$[L_s]$ in dB

$[G/T]$ in dB/K

$[k]$ = 228.6 dBJ/K (since $k = 1.38 \times 10^{-23}$ J/K)

$[B]$ in dBHz

6.4.2 Uplink

The uplink is often handled by introducing an intermediate parameter, ψ, the flux density required to produce the maximum or saturated transponder output, P_T, for a single carrier. It is a satellite parameter and its use conven-

Sec. 6.4 Satellite Links (Up and Down)

iently separates the required satellite receive level from the rest of the link. The actual value ϕ of the flux density received is found from

$$\phi = \frac{\text{e.i.r.p.(earth station)}}{4\pi R^2}$$
$$= \frac{\text{e.i.r.p.}}{L_s} \frac{4\pi}{\lambda^2} \qquad (6\text{-}18)$$

In decibels:

$$[\phi] = [\text{e.i.r.p.}] - [L_s] + \left[\frac{4\pi}{\lambda^2}\right] \quad \text{dBW/m}^2 \qquad (6\text{-}19)$$

If $\phi = \psi$, this equation can be used to calculate the value of e.i.r.p. required at the earth station to provide the specified value of saturated flux density ψ at the satellite. Using Eq. (6-17), we also have

$$[C/N_0]_U = [\phi] - \left[\frac{4\pi}{\lambda^2}\right] + [G/T]_U - [k] \qquad (6\text{-}20)$$

6.4.3 Downlink and "Back-Off"

For the downlink, straightforwardly:

$$[C/N_0]_D = [\text{e.i.r.p.}]_D - [L_s]_D + [G/T_s]_D - [k] \qquad (6\text{-}21)$$

All amplifiers, regardless of whether they are electron tubes or solid-state devices, have a maximum or saturated power output. This is largely a reflection of the nonlinear instantaneous transfer characteristic of the amplifier. It can also be influenced by limitation of the amount of DC power available to energize the amplifier. Figure 6-2 shows the characteristic in a very general sort of way. It is easy to show (Panter, 1972) that when more than one frequency is transmitted through any nonlinear device, a spectrum of spurious frequencies is generated. Since any useful signal itself comprises a band of frequencies, the resultant spurious frequencies—the result of what is loosely called *intermodulation*—must be reckoned with in any real amplifier. In general, the relative level of these spurious components increases with the output level to which the amplifier is driven. Thus, it is common to operate amplifiers at less than their saturated output by an amount called *back-off* that depends on the specific nonlinearity and input spectrum. This topic is covered in detail in Chapter 9.

If the satellite output power is reduced an amount BO_o (the output back-off) from the saturated value, P_T, the downlink expression becomes

$$[C/N_0]_D = [\text{e.i.r.p.}]_D - [BO]_o - [L_s]_D + [G/T_s]_D - [k] \qquad (6\text{-}22)$$

If we assume fixed values of such parameters as $[G/T_s]_U$ and $[L_s]_U$, the back-off is achieved either by reducing the uplink transmitter power or by reducing the satellite transponder amplification. If the uplink e.i.r.p. is reduced an amount BO_i (the input back-off, related nonlinearly to BO_o by the amplifier power transfer characteristic), the final expression for the uplink becomes

$$[C/N_0]_U = [\text{e.i.r.p.}]_U - [L_s]_U + [G/T_s]_U - [k] \tag{6-23}$$

where e.i.r.p._U is the reduced value used on the uplink, or

$$[C/N_0]_U = [\psi] - \left[\frac{4\pi}{\lambda^2}\right] + [G/T_s]_U - [k] - [BO_i] \tag{6-24}$$

To use Eq. (6-24), the required e.i.r.p. as calculated from Eq. (6-19) must not be greater than that available at the earth station. In using Eq. (6-19) to check that sufficient e.i.r.p. is available, remember that $[\phi] = [\psi] - [BO_i]$.

If the transponder amplification is reduced to achieve the back-off, ψ becomes a new value ψ' and

$$[C/N_0]_U = [\psi'] - \left[\frac{4\pi}{\lambda^2}\right] + [G/T_s]_U - [k] - [BO_i] \tag{6-25}$$

or

$$[C/N_0]_U = [\psi] - \left[\frac{4\pi}{\lambda^2}\right] + [G/T_s]_U - [k] \tag{6-26}$$

since

$$[\psi'] = [\psi] - [BO_i]$$

Note the important difference between Eqs. (6-24) and (6-26). If the amplification is reduced, this increases the value of ψ and makes C/N_0 independent of the back-off. It is thus desirable to be able to adjust the satellite back-off by a transponder gain control rather than only by reducing the earth station transmitter power. Adjustable gain controls in a satellite transponder are critical to the provision of optimum performance with various-size earth stations. In the presence of interference from earth stations working with adjacent satellites, gain controls are mandatory.

6.5 COMPOSITE PERFORMANCE

The overall performance of the total satellite connection depends on that of the uplink, the downlink, the aforementioned nonlinear effects in the transponder, and interference. An overall noise "schematic" is given in Figure 6-1.

The nonlinear effects in the high-power amplifier generate intermodulation products treated as noise. To a good approximation, this noise can be considered as adding on a power basis to the thermal noise. We further assume that the transmitted power P_s is the amplified received power and there is no reduction attributable to the transponder's retransmitting the interference. This is a good assumption except for extremely high levels of interference, such as those encountered with deliberate jamming. Such situations are primarily military and are not considered here.

We note that the uplink noise—and any interfering signal—is amplified by the transponder, as is the uplink signal. Downlink interference from adjacent satellites would add to the thermal noise in the earth station receiver. Taking the important case of only intermodulation and thermal noise N_i, we can write immediately:

$$(C/N)_T = \frac{TP_s}{N_D + \alpha T N_U + N_i T} \tag{6-27}$$

where

$$(C/N)_D = \frac{TP_s}{N_D} \quad (C/N)_U = \frac{P_s/\alpha}{N_U}$$

and

$$(C/N)_i = \frac{P_s}{N_i}$$

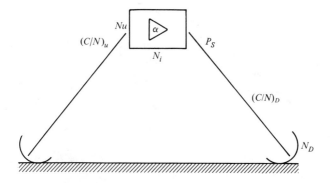

T = downlink transmission ratio, $T < 1$
α = transponder amplification, $\alpha > 1$
N_T = total noise at downlink receiver
N_U = uplink noise at satellite receiver
N_D = downlink noise only at earth station
N_i = intermodulation noise generated in transponder
P_S = satellite transmitter power

Figure 6-1 Overall noise schematic.

With routine algebra:

$$(C/N)_T^{-1} = (C/N)_D^{-1} + (C/N)_U^{-1} + (C/N)_i^{-1} \qquad (6\text{-}28)$$

Note that this has exactly the form of the formula for the resistance of resistors in parallel.

This is a result of considerable utility and significance. It is straightforward to generalize it further to show that an interfering signal I on either the uplink or downlink can also be considered, using

$$(C/N)_T^{-1} = (C/N)_D^{-1} + (C/N)_U^{-1} + (C/N)_i^{-1} + (C/I)^{-1} \qquad (6\text{-}29)$$

Similar terms can be added for other sources of interference. The form of the equation is a result of the important assumptions that the thermal noise, spurious intermodulation frequencies, and interference are all additive and that none of the interference is high enough to steal appreciable amounts of

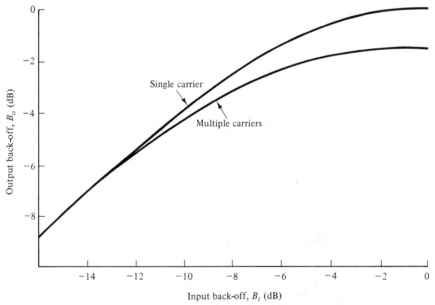

Figure 6-2 TWT amplitude transfer characteristics. (Reprinted by permission of Communications Satellite Corp., COMSAT Technical Memorandum, CL-12-71, by R. McClure, 1971.)

Sec. 6.5   Composite Performance   157

transmitter power. Intermodulation can also have the latter effect. It is considered in more detail in Chapter 9. Note that in the transfer characteristic of Figure 6-2 there is a lower output for multiple carrier operation. This can be attributed to the power lost in the intermodulation products.

The calculation of the ratio $(C/N)_i$ is an entire subject in itself (Westcott, 1967; Panter, 1965; Shimbo, 1971). Briefly, it is a function of the number of carriers, their modulation characteristics, and the amplitude and phase characteristics of the transponder high-power amplifier. It must be reemphasized that all high-power amplifiers saturate and are thus severely nonlinear when operated near their rated outputs. This is qualitatively true, regardless of whether the HPA is a tube or solid-state amplifier. Solid-state amplifiers tend to remain more nearly linear as they approach saturation, and then to turn over sharply. Many kinds of amplifiers also have overall phase shifts that are a function of signal level—giving rise to AM-to-PM conversion and the further generation of spurious frequencies. This phenomenon should not be confused with the nonlinear phase versus frequency characteristic that is a property of *linear* circuits. It is a nice coincidence that both nonlinear intermodulation and AM-to-PM conversion produce the same family and distribution of spu-

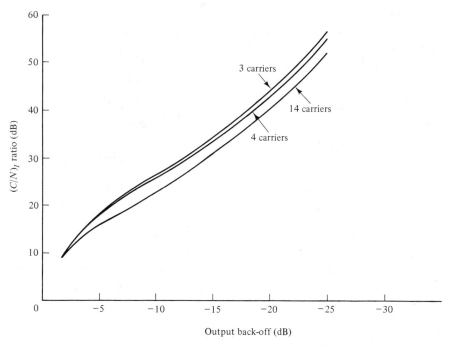

Figure 6-3   Carrier-to-intermodulation noise versus back-off.

rious frequencies (Westcott, 1967). They are usually combined in a composite value of $(C/N)_i$, as measured or calculated. The transmission impairments, both linear and nonlinear, are discussed in more detail in Chapter 9. In this section we group all the nonlinear effects together and characterize them by a composite ratio $(C/N)_i$, either measured or calculated. Characteristics are usually available for an amplifier in the form of input–output power transfer characteristics, as shown in Figure 6-2. Note that this is not the same as the instantaneous transfer characteristic which actually determines intermodulation performance. That curve can be inferred from the power transfer characteristic as described in Chapter 9. Intermodulation data, as given by amplifier manufacturers, is usually sparse and limited to two carrier measurements. They are not very useful as presented and must be extended to multiple-carrier operation.

Typical curves of composite carrier-to-intermodulation ratio as a function of back-off, and with the number of carriers as parameter, are shown in Figure 6-3. Such families can be calculated from various mathematical modules or measured using satellite transponder simulators.

6.6 OPTIMIZATION OF THE RF LINK

The carrier-to-noise ratio due to intermodulation improves as the HPA power level is reduced and, at the same time, the downlink carrier-to-noise density ratio, as calculated from Eq. (6-17), deteriorates. The procedure for determining the operating level to achieve the highest overall value of C/N_0 is basic and straightforward. Figure 6-4 shows the uplink and downlink thermal carrier-to-noise density ratios plotted versus input back-off. In this example it is assumed that back-off is achieved by reduced uplink transmitter power. The overall carrier-to-noise density ratio $(C/N_0)_T$ is plotted on the same curve and shows a conspicuous maximum.

The flow diagram of Figure 6-5 should be an aid in organizing this calculation. (See Table 6-1 for the notation.) Since all factors are expressed in "dB" form (except where we see "10 $\log_{10}$"), the brackets indicating the "dB" form will be omitted from here on. An input back-off value BO_i is assumed, and the overall carrier-to-noise density $(C/N_0)_T$ for a single RF carrier is calculated following the routine on the chart. This is repeated over a wide range of back-offs until the maximum point is clear. It is assumed that transfer characteristics and data on carrier-to-intermodulation versus back-off are available—either as curves or formulas suitable for numerical calculation.

If the number of carriers n is specified—as is usually the case in an FDMA system where operational requirements dictate the number of carriers—then the optimization can be carried out either with $(C/N_0)_t$ for an individual carrier or $(C/N_0)_T$ for all of them—or, for that matter, with (C/N). They differ among each other only by constants:

Sec. 6.6 Optimization of the RF Link

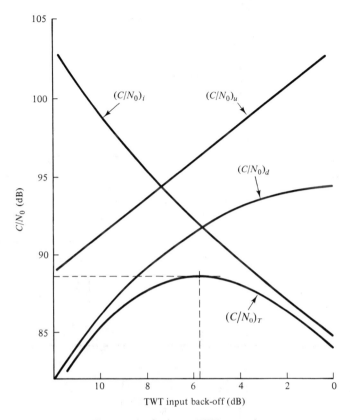

Figure 6-4 Optimum TWT operation.

$$(C/N_0)_T = (C/N_0)_t + 10 \log n \tag{6-30}$$

$$C/N = (C/N_0) - B \tag{6-31}$$

If only a single carrier is involved, the flow diagram is still applicable. A value of back-off may be prescribed, based on such consideration as spillover into adjacent channels and other nonlinear effects discussed in Chapter 7; in some cases, no back-off at all may be used.

Single-channel-per-carrier systems are more complicated. They usually have "voice activation"; that is, when a carrier is not needed—because of a

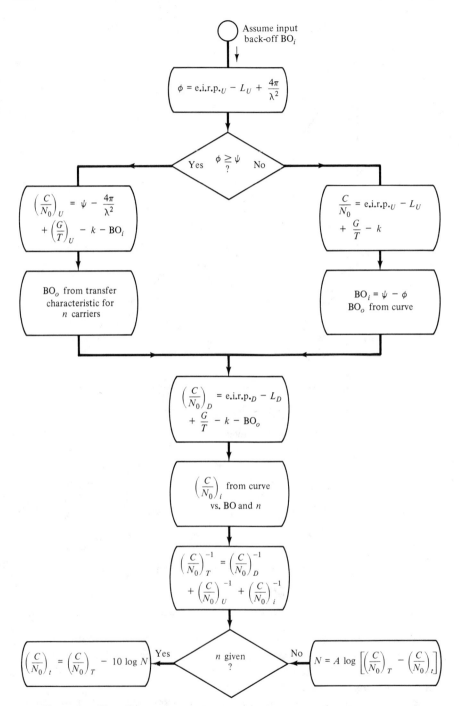

Figure 6-5 Flow diagram for RF link: the calculation of carrier-to-noise density $(C/N_0)_t$ for a single RF carrier or the calculation of the number of possible RF carriers given that value.

Sec. 6.6   Optimization of the RF Link   161

TABLE 6-1   NOTATION FOR RF LINK FLOWCHART

e.i.r.p.$_u$	Available uplink equivalent isotropic radiated power (dBW)
e.i.r.p.$_D$	Available downlink equivalent isotropic radiated power (dBW)
$(G/T)_U$	Figure of merit—uplink receiving system (dB/K)
$(G/T)_D$	Figure of merit—downlink receiving system (dB/K)
L_u	Total uplink loss (free space L_{SU} + atmospheric L_{RU} + margin M_u) (dB)
L_D	Total downlink loss (free space L_{SD} + atmospheric L_{RD} + margin M_D) (dB)
λ	Transmit wavelength (m)
ψ	Single-carrier saturation flux density (dBW/m^2)
ϕ	Operating single-carrier flux density (dBW/m^2)
BO$_i$	Input back-off (dB)
BO$_o$	Output back-off (dB)
$(C/N_0)_D$	Downlink carrier-to-noise density (dBHz)
$(C/N_0)_U$	Uplink carrier-to-noise density (dBHz)
$(C/N_0)_i$	Intermodulation carrier-to-noise density (dBHz)
$(C/N_0)_T$	Total carrier-to-noise density at receive earth station (dBHz)
N	Number of RF carriers
$(C/N_0)_t$	Carrier-to-noise density for a single carrier (dBHz)

speech pause—it is turned off. This produces two good effects: satellite transmitter power is conserved and intermodulation is reduced. This is seen in Figure 6-6. $(C/N)_i$ is plotted versus back-off for a large number of carriers with activity a as a parameter. Note that there is something like a 3.5-dB increase in this ratio when the carrier activity drops to 40%—a typical value with SCPC telephone channels since a channel, as half of a two-way circuit, is busy slightly less than half the time. The 40% activity factor would also produce a 4-dB gain in available transmitter power per carrier.

The activity factor can be considerably higher for one-way data channels with gains in available power and carrier-to-intermodulation ratio noticeably less.

Optimization of the back-off is carried out in the same manner, but in this case it is necessary to work with $(C/N_0)_T$ since the number of channels is unknown. The $(C/N)_i$ increased in accordance with the curves of Figure 6-6. An optimum is found as before, and the number of channels is calculated from the optimum value of $(C/N_0)_T$, the given value of $(C/N_0)_t$, which depends on the modulation system and Eq. (6-30). The same total carrier power actively permits several decibels higher power for each individual carrier, because of the aforementioned statistical effect. This "credit" is taken in the calculation of the value of $(C/N_0)_t$ for an individual channel and is discussed further in Chapter 8.

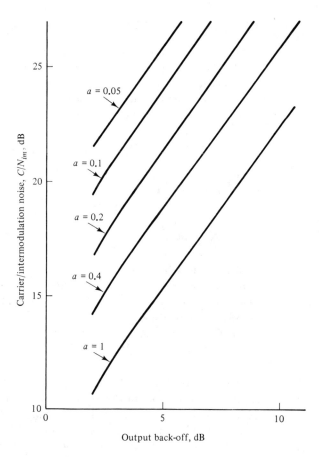

Figure 6-6 Satellite intermodulation noise model—many carriers (a = activity).

6.7 NOISE TEMPERATURE

6.7.1 General

The computation of RF link performance is usually carried out in terms of an equivalent system noise temperature T_s. This temperature is that to which a resistance at the input of an internal-noise-free receiver with the same gain as our actual receiver would have to be heated to produce a noise level at the receiver output equivalent to that observed. It is a composite measure of receiver system performance and comprises link thermal noise, radio noise from the atmosphere and outer space, and device noise. It is the temperature used in the terminal figure of merit (G/T_s).

Individual points in the system are characterized by a noise temperature T_N. This temperature is that of a passive resistor producing an available noise

Sec. 6.7 Noise Temperature

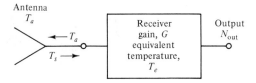

Figure 6-7 Basic receive system.

power per unit bandwidth equal to that available at the point in question at the specified frequency. Antenna temperature T_a is an important special case. It is the temperature of a resistor having the same available noise output as that measured at the antenna terminals. It depends on the radiation pattern, the physical temperature of the surroundings with which the antenna exchanges energy, and the noise received from space. It can change with elevation angle, rain loss, and time of day. It will be discussed in more detail later.

We need one further concept to facilitate the computation of T_s, given the characteristics of individual stages. The individual stages are characterized by an excess temperature T_E. This temperature, sometimes called *effective* or *equivalent input noise temperature*, is the *difference* between the system temperature T_s and the noise temperature T_N at the output of the stage, where T_s is observed with respect to the output of the stage.

Thus in Figure 6-7 we show an elemental receiving system comprising an antenna and receiver. By definition:

$$N_{\text{out}} = kGT_sB \qquad (6\text{-}32)$$

Also by definition:

$$T_s = T_a + T_e \qquad (6\text{-}33)$$

Note well the difference between antenna temperature and system temperature. Noise temperature T_N, of which T_a is a special case, is a "looking-backward" temperature. It represents the noise accepted, generated, and passed by the devices up to that point in the chain. System temperature is a "looking-forward" temperature which assumes only noiseless devices after the point in question. T_a is defined so as to represent the antenna by replacing it with a resistor at that temperature, whereas T_s is defined to represent the noise of the entire receiver system as if it all arose at the point in question. The reference point for T_s is often, but not necessarily, the antenna terminals, which can lead to confusion. T_s differs between two points in the chain by only gain or loss between those points. Thus G/T at the antenna terminals is the same as G/T at the receive antenna input, even with intervening losses.

For two networks in tandem (Figure 6-8) with excess temperatures T_{e1} and T_{e2}, their composite noise temperature T can be shown to be given

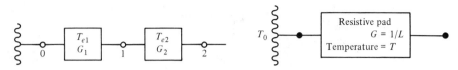

Figure 6-8 Two networks in tandem.   **Figure 6-9** The hot pad.

(Mumford and Scheibe, 1968) by

$$T_e = T_{e1} + \frac{T_{e2}}{G_1} \qquad (6\text{-}34)$$

This is a particularly useful result that can be used iteratively to ascertain the excess noise temperatures of a chain of elements of any length.

6.7.2 Some Important Special Cases

Several important results of the definitions and equations above can be determined for some special cases. We consider a purely resistive matched attenuator at temperature T when connected to an input at T_0 (Figure 6-9).

It can be shown (Mumford and Scheibe, 1968) that this *hot pad* has an excess temperature[1]:

$$T_E = T(L - 1) \qquad (6\text{-}35)$$

and the corresponding noise temperature:

$$T_N = \frac{T_0 + (L - 1)T}{L} \qquad (6\text{-}36)$$

where L is the loss of the attenuator, defined as the reciprocal of its "gain."

These expressions are very useful. They can be used to calculate the noise effects of a pure loss. Note that there is an "excess" noise equal to $(L - 1)T_0$ even for a room-temperature passive attenuator. The same relationships can also be used to assess the effects of clear sky and rain losses on the apparent antenna temperature.

6.7.3 Noise Figures

The classical concept of *noise figure* (or noise factor) was based on the idea that the deterioration in signal (or carrier)-to-noise ratio through a device

---

[1] A tedious but straightforward and informative proof is to calculate the noise output from a resistive network, comprising the source resistance and a T pad attenuator whose resistance values are chosen to yield a loss L between matched loads. These resistances are at a temperature T and generate noise voltage $\sqrt{4kTBR}$. Kirchhoff's laws are used to calculate the output voltage from each of the four resistors (three attenuators and one source) and the outputs add on a power basis. Equations (6-32) and (6-33) are then used to derive Eq. (6-34).

Sec. 6.7   Noise Temperature

was a measure of the device's noisiness. Thus

$$\frac{C_{in}}{N_{in}} = F\frac{C_{out}}{N_{out}} \quad (6\text{-}37)$$

where F is the noise figure.

If the input and output carrier levels are related by the available power gain G and if the input noise level is $kT_{in}B$, then

$$N_{out} = FGkT_{in}B \quad (6\text{-}38)$$

This can be written as

$$N_{out} = (F - 1)G\,kT_{in}B + GkT_{in}B \quad (6\text{-}39)$$

The second term is identifiable as available input noise increased by the device gain and the first term as the "excess" noise generated by the device. A defect in the definition is the dependence of F on the input noise level. If that level is increased, the deterioration in C/N will seem to be less.

To avoid a noise figure defined so that its value is a function of the input temperature, the IEEE has standardized the definition so that T is always taken at an input temperature of $T_0 = 290$ K. Note that this input temperature is only for a standard definition and measurement. If the operating input temperature, usually T_a in satellite problems, is different from T_0, the noise output becomes

$$N_{out} = GkT_aB + (F - 1)GkT_0B \quad (6\text{-}40)$$

From the definition of system temperature in terms of N_{out}, we can set the two output noise values equal and write

$$kGT_sB = GkT_aB + (F - 1)GkT_0B \quad (6\text{-}41)$$
$$T_s = T_a + (F - 1)T_0 \qquad T_0 = 290\text{ K}$$

and from the definition of excess temperature T_e:

$$\boxed{T_e = (F - 1)T_0} \qquad T_0 = 290\text{ K} \quad (6\text{-}42)$$

The latter equation is particularly useful in changing the characterization of a device from noise figure to equivalent (excess) temperature, and vice versa.

Using Eqs. (6-32) and (6-42), one can show that the noise figure of two networks in tandem is given by

$$F_{12} = F_1 + \frac{F_2 - 1}{G_1} \quad (6\text{-}43)$$

6.8 ANTENNA TEMPERATURES

6.8.1 Composite Antenna Temperature

The antenna receives energy from and radiates energy to the sky, the ground, and (at certain times) the sun.

In principle, T_a should be evaluated using an integration over the complete solid angle of the antenna (4π steradians) in accordance with

$$T_a = \int_{\Omega_1} G_1 T_{\text{sky}} \, d\Omega_1 + \int_{\Omega_2} G_2[(1 - \rho^2)T_g + \rho^2 T_{\text{sky}}] \, d\Omega_2 \quad (6\text{-}44)$$

where G_1 = antenna gain in sky directions
 G_2 = antenna gain in ground directions
 Ω_1 = solid angle region in sky directions
 Ω_2 = solid angle region in ground directions
 T_0 = assumed rain or tropospheric temperature, normally 290 K
 T_g = ground temperature, normally 290 K
 ρ = voltage reflection factor of earth

The sky temperature, T_{sky}, is found from T', the "clear" sky temperature due to galactic noise, the microwave background, and O_2 and H_2O vapor losses in accordance with the curves of Figures 6-10 and 6-11. If we have a rain loss L, the sky temperature is calculated from T' using Eq. (6-36). Thus

$$T_{\text{sky}} = \frac{T' + (L_R - 1)T_0}{L_R} \quad T_0 = 290 \text{ K} \quad (6\text{-}45)$$

This integration is a complicated undertaking and normally not necessary. An adequate approximation can be found using a simplified approach:

$$T_a = a_1 T_{\text{sky}} + a_2 T_g + a_3 T_{\text{sun}} \quad (6\text{-}46)$$

where a_1 = $\Omega_{\text{sky}}/4\pi$
 a_2 = $(\Omega_g/4\pi)(1 - \rho^2)$
 a_3 = $(\Omega_s/4\pi)(G_s/L_R)$

Sec. 6.8   Antenna Temperatures

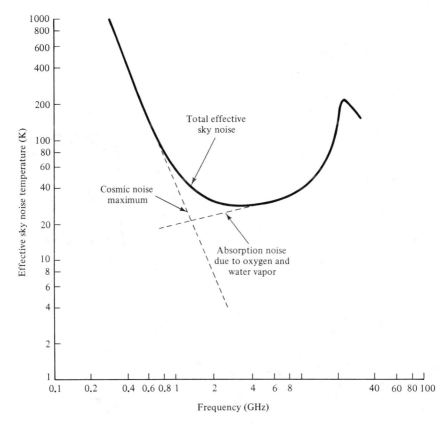

Figure 6-10 Effective sky noise temperature at 5° elevation due to atmospheric absorption and cosmic noise.

Ω_{sky} and Ω_g are the total solid angles of the antenna pattern intercepting the sky and ground, respectively. The term in a_3 applies only when the sun falls within the antenna's view, either on a main or side lobe. It assumes the antenna beamwidth is much larger than Ω_s, the solid angle occupied by the sun (about ½° squared). G_s is the antenna gain in the direction of the sun. If the antenna gain is high enough so that its beamwidth is less than half a degree or so, the antenna temperature simply becomes equal to T_{sun}—normally high enough to raise the system noise level to an inoperably high value called a sun outage. Note that rain loss L reduces the apparent sun temperature just as it does in the visible part of the spectrum. If the earth station beamwidth is wide compared to the sun (say 2° or so), the increase in noise level may be tolerable. At about a 2° beamwidth at K-band, the clear-weather increase in solar noise would cause the same deterioration as the rain loss. Since they would not occur together, a design margin suitable for one would also be adequate for the other.

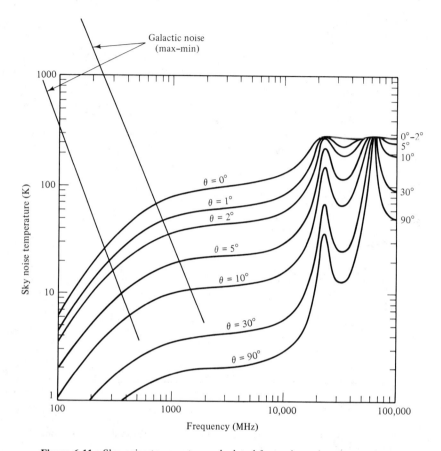

Figure 6-11 Sky noise temperature calculated for various elevation angles.

6.8.2 Sun Temperature

The sun temperature T_{sun} can be found from the curves of Figure 6-12 for the received noise density of an antenna interception of the entire sun exactly. Occasionally, one sees solar noise data in the form of a solar flux density or as an equivalent temperature. These parameters are all related through some basic ideas. The Planck expression for blackbody radiation, applied to the sun at relatively high temperatures and low frequencies with which we are concerned, becomes

$$W_f = \frac{2\pi h}{e^2} \frac{f^3}{e^{hf/kT} - 1} \simeq \frac{2\pi}{\lambda^2} k T_{\text{sun}} \quad \text{W/m}^2 \cdot \text{Hz} \qquad (6\text{-}47)$$

If we integrate this density over the total surface of the sun $4\pi R_s^2$ and diminish it by the loss due to the inverse-square law at a distance D_s (equal

Sec. 6.8 Antenna Temperatures

to the sun–earth distance), the received flux density ϕ_s is

$$S = \frac{2\pi}{\lambda^2 k T_{\text{sun}}} \left(\frac{R_s}{D_s}\right)^2 \quad \text{W/m}^2 \cdot \text{Hz} \tag{6-48}$$

We assume that the receive antenna gain is just that required to intercept all the available solar energy ($G = 4\pi/\Omega_s$) and that only half the energy can be received because of the randomly polarized solar radiation. Multiplying the solar flux density by the equivalent area of the antenna, we can write

$$N_0 = \frac{S}{2} \frac{G\lambda^2}{4\pi}$$

and (6-49)

$$G = \frac{4\pi}{\Omega_s} = \frac{4\pi D_s^2}{\pi R_s^2}$$

Therefore,

$$N_0 = k T_{\text{sun}} \tag{6-50}$$

This equation is intuitively satisfying. It can be used with the curve of Figure 6-12 to determine the sun temperature—either quiet or active—and this temperature T_{sun} is used in Eq. (6-46) for the antenna temperature. Note

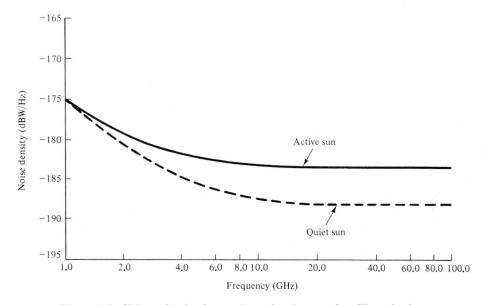

Figure 6-12 Values of noise from quiet and active sun. Sun fills entire beam (Perlman et al., 1960). (From *NASA Propagation Effects Handbook for Satellite System Design*, ORI TR 1679.)

that the temperature of the quiet sun seems to be approaching a value of 11,500 K, asymptotically.

6.8.3 Clear-Sky Temperature

As stated before, the clear-sky temperature T' can be found from the curves of Figures 6-10 and 6-11. They can also be calculated from the attenuation of the atmosphere as given in Figure 6-13, and using the hot-pad formula of Eq. (6-36). Figure 6-14 is a useful curve at 4 GHz as a function of elevation angle.

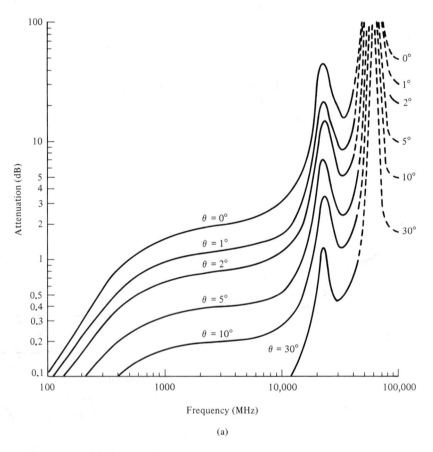

Figure 6-13 (a) One-way signal attenuation through troposphere at various elevation angles; (b) oxygen and water vapor absorption at zenith.

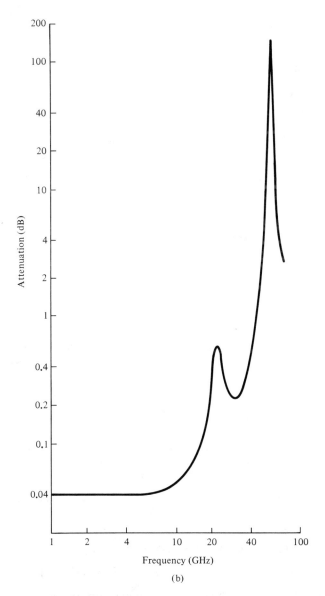

Figure 6-13 (Continued)

6.9 OVERALL SYSTEM TEMPERATURE

Figure 6-15 shows a typical microwave receiver chain, either in the satellite or at the earth station. If T_0 is the ambient temperature of the loss L, T_a is the antenna temperature, T_R is the effective receiver temperature, and F is

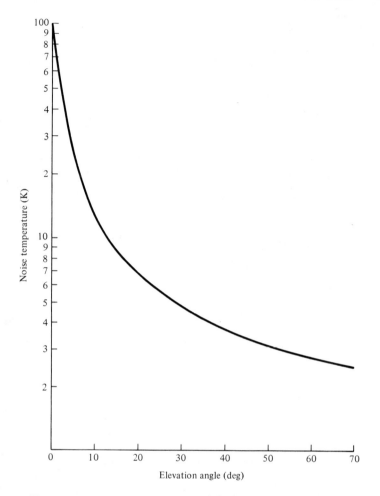

Figure 6-14 Sky noise temperature at 4 GHz versus elevation angle.

the noise figure of the down-converter, then

$$T_1 = (L - 1)T_0 \quad T_2 = T_R \quad T_3 = (F - 1)T_0 \quad (6\text{-}51)$$

Using Eq. (6-34) for the effective temperature of two networks in tandem, the definition of system temperature, and some routine algebra:

$$T_s = T_a + (L - 1)T_0 + LT_R + L(F - 1)\frac{T_0}{G_R} \quad (6\text{-}52)$$

Sec. 6.10 Rain Attenuation Model

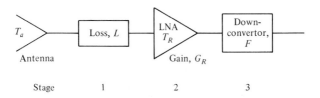

Figure 6-15 Typical microwave receiver chain.

Here the system temperature is taken at the antenna terminals. It is occasionally specified at the receiver input terminals, in which case it is simply reduced by L. Note again that the figure of merit G/T_s is independent of the choice of reference point since G must be specified at the same point and is thus reduced by the same factor.

6.10 RAIN ATTENUATION MODEL

The calculation of rain attenuation can be divided into two steps (Ippolito, 1983). The first step is to estimate the point rain rate in mm/h as a function of the cumulative probability of occurrence. This probability helps determine the grade of service to be provided and thus the values of rain attenuation that will be required as "margins." The second step is to calculate the attenuation resulting from those rain rates, given the angles of elevation and the earth station latitudes.

The first step is done using the maps of Figures 6-16 through 6-18. The appropriate regions are identified on the map, and the letter designations used to find values of rain rate for each percentage of outage from Table 6-2.

The second half of the problem is more complex. The path-averaged rain rate can vary considerably from the surface-point rain rate for the same period of time. We can only present here some of the best available (1983) empirical models for handling this problem.

Numerous theoretical and practical studies have shown that the rain attenuation A can be modeled adequately by the expression

$$\alpha = aR_p^b$$

and (6-53)

$$L_R = \alpha L$$

where α is the specific attenuation in dB/km and R_p is the point rain rate in mm/h. We may take a and b from Table 6-3.

The length of path L must be calculated from the freezing height H_0, sometimes called the *zero-degree isotherm*, and the elevation angle θ.

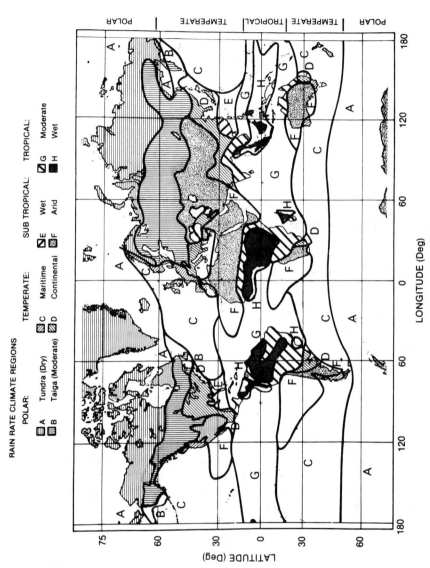

Figure 6-16 Global rain rate climate regions, including the ocean areas. (From *NASA Propagation Effects Handbook for Satellite System Design*, ORI TR 1679.)

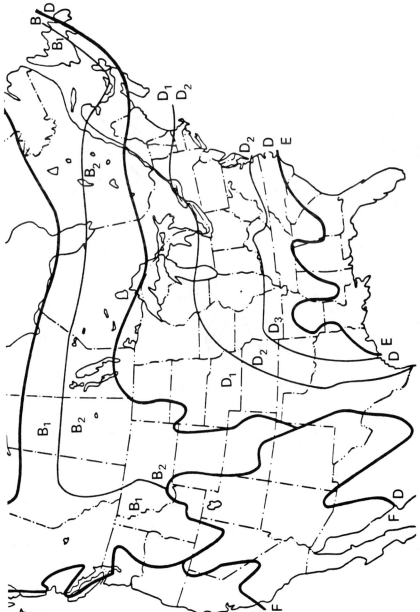

Figure 6-17 Rain rate climate regions for the continental United States showing the subdivision of region D. (From *NASA Propagation Effects Handbook for Satellite System Design*, ORI TR 1679.)

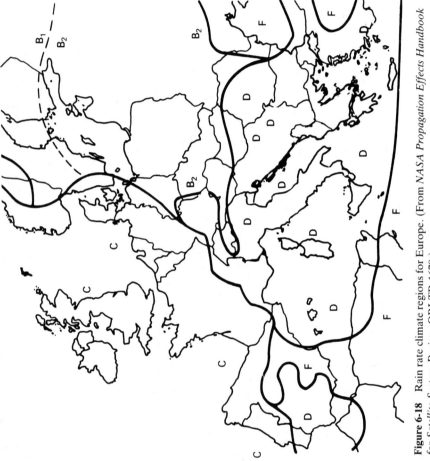

Figure 6-18 Rain rate climate regions for Europe. (From *NASA Propagation Effects Handbook for Satellite System Design*, ORI TR 1679.)

176

TABLE 6-2 Point-Rain-Rate Distribution Values (mm/h) Versus Percent-of-Year Rain Rate Is Exceeded

Percent of Year	Rain Climate Region											Minutes per Year	Hours per Year
	A	B	C	D_1	D_2	D_3	E	F	G	H			
0.001	28	54	80	90	102	127	164	66	129	251	5.3	0.09
0.002	24	40	62	72	86	107	144	51	109	220	10.5	0.18
0.005	19	26	41	50	64	81	117	34	85	178	26	0.44
0.01	15	19	28	37	49	63	98	23	67	147	53	0.88
0.02	12	14	18	27	35	48	77	14	51	115	105	1.75
0.05	8	9.5	11	16	22	31	52	8.0	33	77	263	4.38
0.1	6.5	6.8	7.2	11	15	22	35	5.5	22	51	526	8.77
0.2	4.0	4.8	4.8	7.5	9.5	14	21	3.8	14	31	1,052	17.5
0.5	2.5	2.7	2.8	4.0	5.2	7.0	8.5	2.4	7.0	13	2,630	43.8
1.0	1.7	1.8	1.9	2.2	3.0	4.0	4.0	1.7	3.7	6.4	5,260	87.66
2.0	1.1	1.2	1.2	1.3	1.8	2.5	2.0	1.1	1.6	2.8	10,520	175.3

Source: NASA Propagation Effects Handbook for Satellite System Design, ORI TR 1679.

TABLE 6-3 Average Annual Rainfall

Place	Average Annual Rainfall (cm)
San Diego, Calif.	28.0
Santiago, Chile	35.8
Madrid	42.0
Paris	56.7
London	76
Seattle, Washington	85.0
Buenos Aires	95.4
New York	110.0
Rio de Janeiro	118.0
Jacksonville, Florida	148.0
San Juan, Puerto Rico	150.0
New Orleans, Lousiana	170.0
Ho Chi Minh City	198.0
Manila	208.0
Singapore	240.0

The simplest, and for most cases, completely adequate method for calculating the path length L is first to calculate the geometric length L_0[2]:

$$L_0 = \frac{H_0 - H_g}{\sin \theta} \quad (6\text{-}54)$$

H_g is the earth station altitude and H_0 is the freezing height. In terms of latitude ϕ:

$$\begin{aligned} H_0 &= 7.8 - 0.1\phi & \phi \geq 30° \\ H_0 &= 4.8 \text{ km} & \phi < 30° \end{aligned} \quad (6\text{-}55)$$

Lin (1979) has given a rather good empirical formula for L, the corrected path length, to allow for observed variations in density and other factors:

$$L = \frac{L_0}{1 + \dfrac{L_0(R - 6.2)}{2636}} \quad (6\text{-}56)$$

---

[2] A more accurate, but rarely needed, expression for the geometric path length considers the earth's curvature and is

$$L_o = (R_E + h) \frac{\cos(\theta + \zeta)}{\cos \theta} \quad \text{where } \zeta = \sin^{-1}\left[\frac{(R_E + hg)\cos\theta}{R_E + h}\right]$$

It should not be used in the rain attenuation analysis shown here since the empirical correction factors have assumed the use of Eq. (6-50).

Sec. 6.10   Rain Attenuation Model   179

In most cases, especially in the K-band, this method is adequate for system planning. The CCIR has a variation of the NASA model used for international planning and leading to similar results.

A still more complicated model developed by NASA can be used for more accurate path-length correction. It starts from the same basic formula for α and the tabulated values of a and b, but uses a different approach for the corrected path length.

The basic variable used in the correction for the path distribution of rain intensity is D_h, the horizontal distance:

$$D_2 = \frac{H_0 - H_g}{\tan \theta} \qquad (6\text{-}57)$$

$$\theta > 10°$$

where H_g is the earth station elevation. H is now taken from the curves of Figure 6-19, using the earth station latitude and the same probability of occurrence used in finding the point rain rate for a particular climate.

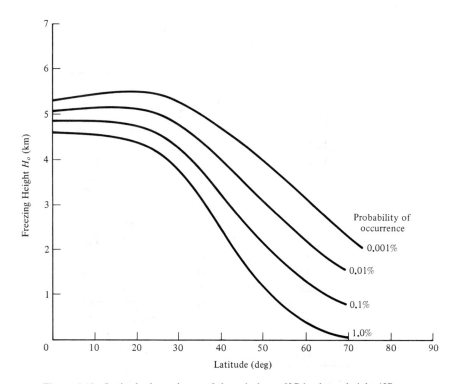

Figure 6-19 Latitude dependence of the rain layer 0°C isotherm height (H) as a function of probability of occurrence. (From *NASA Propagation Effects Handbook for Satellite System Design*, ORI TR 1679.)

It is then necessary to compute the empirical constants $X, Y, Z,$ and U for each rain rate R_p, as follows:

$$X = 2.3 R_p^{-0.17} \qquad (6\text{-}58)$$

$$Y = 0.026 - 0.03 \ln R_p \qquad (6\text{-}59)$$

$$Z = 3.8 - 0.6 \ln R_p \qquad (6\text{-}60)$$

$$U = \frac{\ln(Xe^{YZ})}{Z} \qquad (6\text{-}61)$$

The total attenuation is calculated from

$$A = \frac{aR_p^b}{\cos\theta}\left(\frac{e^{UZb}-1}{Ub} - \frac{X^b e^{YZb}}{Yb} + \frac{X^b e^{YDb}}{Yb}\right); \ \theta \geq 10° \qquad (6\text{-}62)$$

for $D_h \geq Z$. If $D_h < Z$, then

$$A = \frac{aR_p^b}{\cos\theta}\frac{e^{UDb}}{Ub} \qquad (6\text{-}63)$$

TABLE 6-4 REGRESSION CALCULATIONS FOR a AND b IN aR^b (dB/km) AS A FUNCTION OF FREQUENCY[a]

Frequency (GHz)	a		b	
	Low	High	Low	High
10	1.17×10^{-2}	1.14×10^{-2}	1.178	1.189
11	1.50×10^{-2}	1.52×10^{-2}	1.171	1.167
12	1.86×10^{-2}	1.96×10^{-2}	1.162	1.150
15	3.21×10^{-2}	3.47×10^{-2}	1.142	1.119
20	6.26×10^{-2}	7.09×10^{-2}	1.119	1.083
25	0.105	0.132	1.094	1.029
30	0.162	0.226	1.061	0.964
35	0.232	0.345	1.022	0.907
40	0.313	0.467	0.981	0.864
50	0.489	0.669	0.907	0.815
60	0.658	0.796	0.850	0.794
70	0.801	0.869	0.809	0.784
80	0.924	0.913	0.778	0.780
90	1.02	0.945	0.756	0.776
100	1.08	0.966	0.742	0.774

[a] Low: $R \leq 30$ mm/h; High: $R > 30$ mm/h.

Source: Adapted from *NASA Propagation Effects Handbook for Satellite System Design*, ORI TR 1679.

If looking directly at the zenith, $\theta = 90°$, $D = 0$, and

$$A = (H_0 - H_g)aR_p^b \qquad (6\text{-}64)$$

The calculation is repeated for each percentage of the time, using values of R_p and H for the appropriate climate and latitude from Table 6-2 and Figure 6-16.

For very low values of inclination, frequencies not tabulated in Table 6-4, and extreme cases in general, the referenced report should be consulted.

If D is greater than 22.5 km (very low angles), the probability of occurrence is adjusted according to

$$\text{true probability} = \text{assumed probability} \times \frac{D}{22.5} \qquad (6\text{-}65)$$

REFERENCES

BOUSQUET, M., G. MARAL, and J. PARES, *Les Systèmes de Télécommunications par Satellites*, Masson et Cie, Éditeurs, Paris, 1982.

GAGLIARDI, R. M., *Satellite Communications*, Lifetime Learning Publications, Belmont, Calif., 1984.

IPPOLITO, L. J., R. D. KAUL, and R. G. WALLACE, *Propagation Effects Handbook for Satellite Systems Design*, NASA Reference Publication 1082(03), 3rd ed., June 1983.

JASIK, J., *Antenna Engineering Handbook*, McGraw-Hill Book Company, New York, 1961.

KRAUS, J. D., *Antennas*, McGraw-Hill Book Company, New York, 1950.

LIN, S. H., "Empirical Rain Attenuation Model for Earth–Satellite Path," *IEEE Transactions on Communications* Vol. COM-27 No. 5, May 1979, pp. 812–817.

MUMFORD, W. W., and E. H. SCHEIBE, *Noise Performance Factors in Communication Systems*, Horizon House–Microwave, Inc., Dedham, Mass., 1968.

PANTER, P. F., *Communications Systems Design*, McGraw-Hill Book Company, New York, 1972, pp. 181–219.

PIERCE, J. R., "The General Sources of Noise in Vacuum Tubes," *IRE Trans. on Electron Devices*, Vol. ED-1, No. 4, Dec. 1954, pp. 134–167.

SHIMBO, O., "Efforts of Intermodulation, AM-PM Conversion and Additive Noise in Multicarrier TWT Systems," *Proc. IEEE*, Vol. 59, 1971, pp. 230–238.

WESTCOTT, J., "Investigation of Multiple f.m./f.d.m. Carriers through a Satellite t.w.t. near to Saturation," *Proc. IEEE*, Vol. 144, No. 6, June 1967.

7

Modulation and Multiplexing

7.0 INTRODUCTION

In this chapter the signal processing techniques used to encode, modulate, combine, and transmit signals in satellite communications applications are described. Figure 7-1 is a simplified view of the processing techniques used on voice, data, and video (imagery) signals between a user location and a satellite earth terminal. The techniques have been divided into four levels. The first level is the source coding, or single-channel modulation level, wherein an individual voice, data, or video signal is converted into a form suitable for transmission or further processing. The next level is multiplexing, wherein several voice, data, or video channels are combined to form a single, higher-speed, composite signal. The third level is modulation, wherein a baseband signal containing one or more channels is modulated on a sinusoidal carrier in a manner suitable for transmission over a radio-frequency link. The first three levels of signal processing are the subjects of Chapter 7. Chapter 8 will deal exclusively with the fourth level, multiple access.

The material in this chapter is not intended to be a comprehensive treatment of any of the individual subjects. Many papers and textbooks (Bell Laboratories, 1982; Bennett, 1970; Bennett and Davey, 1965; Bylanski, 1976; Carlson, 1975) are available, covering subjects such as pulse-code modulation (PCM), time-division multiplexing (TDM), frequency-division multiplexing (FDM), frequency modulation (FM), and phase-shift keying (PSK). It is not our intent to develop rigorous theory to support the fundamental results. Instead, we concentrate on a description of the systems as they are actually

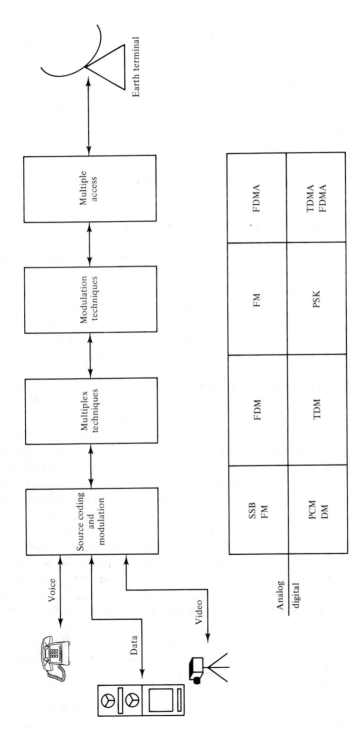

Figure 7-1 Signal processing between user location and satellite earth station.

used in satellite communications and describe their functions and characteristics. This description is complemented with a presentation of the key mathematical relationships that govern performance. The depth of treatment is designed to provide only that which is necessary to perform systems engineering calculations.

The chapter will deal with both analog and digital transmission systems. The systems described will be limited to those which are in actual use today, and those which will come into practice in the years ahead. The intent is to provide a basic understanding of the mechanics of the techniques through plausibility arguments combined with the lightest mathematical treatment.

7.1 SOURCE SIGNALS: VOICE, DATA, AND VIDEO

Before beginning a discussion of signal processing techniques a brief discussion of the characteristics of the three most common classifications of signals transmitted over satellite channels is required. These are: the telephone speech signal, data signals of various types, and video signals, both broadcast quality and business teleconferencing quality. To understand the performance of various coding and modulation schemes, it is important to know the quantitative and qualitative service requirements for each type of signal.

7.1.1 The Telephone Speech Signal

The telephone speech signal, as distinct from the general class of audio signals occupying bandwidths up to about 20 kHz, is generated by converting acoustic energy into an electrical signal using a telephone set as the transducer. The time-varying waveform associated with this speech signal is not easy to characterize. As summarized in Table 7-1, the telephone speech signal is typically band-limited (by a combination of the telephone instrument and the transmission network), to a frequency range of about 300 to 3400 Hz.

The quality of a received analog voice signal is usually specified by several parameters. The most useful figure of merit is the signal-to-noise ratio, although in telephone practice it is often not known by that term. As specified by the CCITT, the worst-case voiceband signal-to-noise ratio acceptable for long-haul transmission (considering *signal* to be the standard-level "test-tone") is about 50 dB. This corresponds to a maximum allowable noise of 10,000 pW in the voiceband.

Analog speech signals have a high peak-to-average ratio. The distribution of talker volume is broad, requiring a dynamic range of operation of 40 to 50 dB. Also, interference levels from various sources must be kept at a level 65 dB below the nominal test-tone level.

Another important characteristic is that talkers tend to pause between phrases and sentences. This results in active speech energy being concentrated

Sec. 7.1 Source Signals: Voice, Data, and Video

TABLE 7-1 Illustrative Characteristics of the Telephone Speech Signal

Bandwidth occupied	300–3400 Hz
Nominal frequency spacing per channel	4 kHz
Signal (test-tone)-to-noise ratio	50 dB
Dynamic range required	40–50 dB
Interference levels (below test-tone level)	60–65 dB
Speech activity (duty cycle)	35–40%

in talk spurts of about 1 s average duration, separated by gaps (quiet intervals) of a second or so. Thus the speech signal consists of randomly spaced bursts of energy of random duration. The activity or duty cycle in the average speech signal is about 35 to 40% active and thus 60 to 65% idle time.

If we employ digital transmission, two additional parameters must be specified to determine the ultimate quality of the reconstructed analog speech signal. These are the transmission rate in bits per second and the bit error rate. Typical digital transmission rates for commercial speech telephony, using today's most common encoding scheme, range from 32 to 64 kb/s. The higher rate represents public dial-up network quality, and the lower range private business network quality. The bit error rate (BER) required to support speech telephony is normally considered to have a threshold of about 10^{-4}. This means that on the average, 1 error in 10,000 is the threshold of acceptable speech quality. If the BER exceeds 10^{-4}, the speech quality is judged to be subjectively unacceptable. Therefore, an error rate of 10^{-4} is typically used as the design threshold for digital speech telephony systems.

7.1.2 Data Signals

Data signals can be broadly classified into three ranges: narrowband data (≤300 b/s); voiceband data (300 b/s to 16 kb/s); and wideband data (>16 kb/s). Classifying data applications into these three categories, by speed, approximately matches the transmission facilities used to support them. Narrowband data begins at telegraphy rates and includes a wide range of communications applications with terminals and teleprinters usually implemented over wire facilities. Data of many types, such as facsimile and transactional services are supported at rates up to 16 kb/s using data modems operating within the voiceband (300 to 3400 Hz). Wideband data applications such as electronic mail, high-speed file transfer, and computer-aided design utilize the efficient high-speed transmission capabilities offered by satellite, fiber optics, and digital radio channels.

The methods of data transmission, the protocols used to control computer communications, the operation of voiceband modems and multiplexers, and the techniques used in terrestrial data communications are generally outside the scope of this book. Treatments on all these subjects may be found in many available texts (Bennett and Davey, 1965; Clark, 1977; Glasgal, 1976; Lucky et al., 1968; McGlynn, 1978).

Although there is a large menu of services, rates, applications, and interfaces used in data communications, the transmission error rate required to support satisfactory performance depends critically on the application. The error rate requirements for data services are typically much more severe than that for voice. This tends to complicate system designs where both voice and data applications must share the same links. This and related issues are covered in more detail in Section 7.3.6.

7.1.3 Video Signals

There are generally two types of video signals transmitted via satellite circuits. The first is broadcast-quality commercial television, and the second is television used for business teleconferencing. The commercial broadcast-quality signals are high resolution, high-quality signals, and thus require large analog bandwidths or high data rates. The business video signal employs typically much lower data rates (≤ 1.544 Mb/s), and a great deal of signal processing is usually required to reduce the data rate and needed bandwidth. (Kaneko and Ishiguro, 1980; Tescher, 1980).

Television signals contain information in electrical form from which a picture can be recreated. To translate a complete picture into an electrical signal, the picture is scanned in a systematic manner. A video signal contains a series of scans in nearly horizontal lines from left to right, starting at the top of the image. When the bottom of the image is reached, the process is started again with alternate scanning fields interlaced. The frame rate is rapid

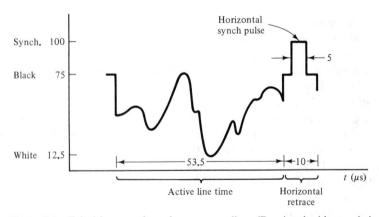

Figure 7-2 Television waveform for one scan line. (Reprinted with permission from *Communication Systems*, McGraw-Hill, Inc., 1975.)

TABLE 7-2 Comparison of U.S. (NTSC-M) and European (PAL-B) Television Standards

	NTSC-M	PAL-B
Aspect ratio (width to height)	4:3	4:3
Total lines per frame	525	625
Line rate[a]	15.75 kHz	15.625 kHz
Line time[a]	63.5 µs	64 µs
Frame rate	30 Hz	25 Hz
Field rate[a]	60 Hz	50 Hz
Video bandwidth	4.2 MHz	5.0 MHz
Assigned channel bandwidth	6.0 MHz	7 MHz
Audio carrier frequency (above video carrier)	4.5 MHz above	5.5 MHz
Audio FM deviation	25 kHz	50 kHz

[a] Nominal values; for NTSC-M, actual line rate is 15.634 kHz and field rate is 59.558 Hz.

enough (25 to 30 per second) to create the illusion of continuous motion, while the field rate (which is twice the frame rate) makes the flickering imperceptible to the human eye.

Color information (chrominance), where used, is conveyed on a subcarrier which travels along with the basic color-independent description (luminance) of the scanned image.

Two adjustments are made to the video signal after the scanning process. First, blanking pulses are inserted during the retrace intervals to blank out retraced lines on the receiving picture tube. Second, synchronizing pulses are added on top of the blanking pulses to synchronize the receiver's horizontal and vertical sweep circuits. In addition, for color transmission, a reference "burst" of the chrominance subcarrier is sent during the synchronizing pulse for each line. This is illustrated in Figure 7-2, which shows a waveform for one scan line, with amplitude levels and durations, corresponding to the U.S. TV standards. Other systems used in other parts of the world tend to be slightly higher in resolution and in performance than the U.S. system. Table 7-2 compares the characteristics of the U.S. (NTSC) system and an important European system (PAL).

7.2 ANALOG TRANSMISSION SYSTEMS

In a wide sense, there are two types of signal transmission used on satellites: analog and digital. In this section we describe those analog transmission systems that are used to transmit signals via satellite. We will focus specifically on the transmission of telephony signals, not only because voice accounts for the bulk of the traffic carried by satellites, but also because data and video signals use essentially the same techniques.

Analog transmissions systems can be classified into two distinct types. The first are multiple channel per carrier (MCPC) systems, employing carriers modulated by a multiplexed signal representing multiple channels, and the second are single channel per carrier (SCPC) techniques, wherein a single voice channel is assigned its own individual carrier. Analog MCPC systems utilize amplitude modulation (AM) of the individual voice channel, frequency-division multiplex (FDM) to combine channels, and frequency modulation (FM) on the radio-frequency carrier. Analog SCPC systems employ FM modulation to transmit a single VF channel on its own carrier frequency. AM single sideband suppressed carrier (AM SSB-SC) is also coming into use in SCPC systems. Each of these techniques is discussed in the following paragraphs, and the applications in satellite transmission are described.

7.2.1 Amplitude Modulation

In its simplest form, amplitude modulation (AM) is generated using a product modulator. Conventionally, the source signal is multiplied by a sinusoidal carrier as illustrated in Figure 7-3a. The amplitude of the modulated carrier follows the amplitude of the source signal. In the frequency domain, the spectrum of the modulated signal contains two sidebands located symmetrically about the carrier frequency as shown in Figure 7-3b. Conventional AM is not an efficient modulation method because a considerable amount of transmitter power is utilized in sending the non-information-bearing carrier component. A trivial method used to improve efficiency is to attenuate the carrier component prior to transmission. As long as the carrier is small compared to the information components, but strong enough to be recovered by a narrowband filter, coherent detection can still be achieved.

Another method is simply not to send the carrier component of the AM signal at all. This is referred to as AM *double sideband suppressed carrier* (DSB-SC).

If the highest frequency in the baseband signal is f_m, the bandwidth of the modulated signal $B = 2f_m$. This AM signal may be demodulated after passing through a noisy channel by using envelope detection (perhaps as simple as a series diode followed by a low-pass filter).

Throughout this chapter, we will be interested in signal-to-noise ratios both in the transmission channel itself and in the signal as recovered by the receiver. To examine these ratios in the case of amplitude modulation, we must first consider the formation of an AM signal and the resulting power relationships.

Assume that the modulating signal can be expressed as $a_s(t)$. Then, for a carrier signal with frequency f_c (angular frequency $\omega_c = 2\pi f_c$) and (unmod-

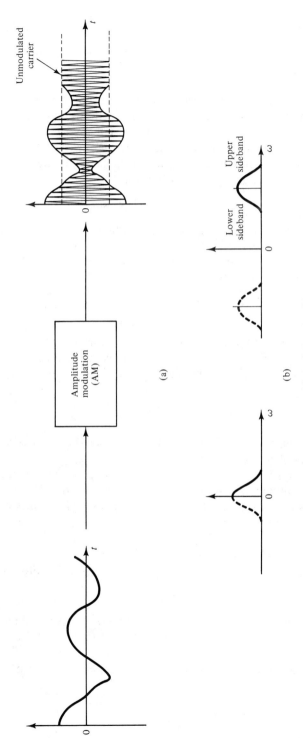

Figure 7-3 AM modulation: (a) time domain; (b) frequency domain.

ulated) amplitude A_c, the modulated signal may be expressed as

$$a_M(t) = (ka_s(t) + 1) A_c \sin(\omega_c t + \theta) \tag{7-1}$$

where θ is an arbitrary phase angle of the carrier and k is a constant representing the input sensitivity (scaling factor) of the modulating circuit. For normal AM operation, especially if envelope detection is contemplated, the term $(ka_s(t) + 1)$ cannot be allowed to go below zero.

For the important special case where the modulating signal is a sine wave with angular frequency ω_s, this becomes

$$a_M(t) = (m \sin \omega_s t + 1) A_c \sin \omega_c t \tag{7-2}$$

ignoring the arbitrary phase angles of the modulating signal and carrier. The parameter m is called the *index of modulation*, and for ordinary AM can range from 0 to 1. It is commonly stated in percent, ranging then from 0 to 100 (called "100% modulation").

With m at its greatest permissible value, 1, the effective amplitude of the modulated carrier varies, over the cycle of the modulating signal, from 0 to $2A_c$.

It can be shown that, for modulation by a sinusoidal signal with an index m of 1, the power in the two sidebands jointly is one half the power C at the carrier frequency. Thus, the total signal power is given by

$$P = \frac{3}{2} C \tag{7-3}$$

The sideband power is evenly divided between the two sidebands, giving them each a power of $C/4$.

Assume now that the modulated signal appears at the receiving demodulator accompanied by noise which, across the bandwidth of the receiver, $B = 2f_m$, has a uniform power density N_0, measured in watts per hertz of spectrum. The total noise power in the receiver bandwidth therefore is

$$N = 2f_m N_0 \tag{7-4}$$

This noise appears at the output of the demodulator as a noise power

$$N_b = kN_M \tag{7-5}$$

where k represents, arbitrarily, the "transfer gain" or scaling factor of the demodulator, and the subscript b is indicative of the "baseband," or demodulated, signal.

The signal power in each of the two sidebands (upper and lower) has been shown to be

$$S_U = S_L = \frac{C}{4} \tag{7-6}$$

Sec. 7.2   Analog Transmission Systems   191

Each of these produces demodulated signal power of

$$S_{bU} = S_{bL} = k\frac{C}{4} \qquad (7\text{-}7)$$

However, both sidebands are "images" of the identical modulating signal, and their recovered versions are thus identical. Accordingly, S_{bU} and S_{bL} add *coherently* in the demodulator, producing an output signal power

$$S_b = 2k(S_{bU} + S_{bL}) = kC \qquad (7\text{-}8)$$

We can thus relate the signal-to-noise ratio of the demodulated baseband signal to the carrier-to-noise ratio in the modulated signal domain (that is, at the input to the demodulator, reflecting noise added in the transmission channel or by the receiver itself) thus

$$\frac{S_b}{N_b} = \frac{C}{N} = \frac{C}{N_0 B} \qquad (7\text{-}9)$$

We must emphasize that this relationship assumes 100% modulation ($m = 1$) by a sinusoidal modulating signal, the conventional reference case but one which does not necessarily represent actual operation. For sinusoidal modulation at a general index m the relationship is

$$\frac{S_b}{N_b} = m^2 \frac{C}{N} \qquad (7\text{-}10)$$

As mentioned above, the carrier-frequency component of the transmitted signal in AM is wholly redundant from an information standpoint. The duplicated sidebands are also redundant as consumers of spectrum space. An important variant of AM, called single-sideband suppressed carrier AM (SSB-SC), eliminates the carrier component and one of the sidebands. This mode is used in frequency-division multiplex (FDM) and is also coming into use on the satellite radio channel itself, especially in SCPC operation.

An SSB-SC signal can be generated in various ways. A popular one uses a *balanced modulator* whose operation eliminates the " +1" term in Equation (7-1) (the source of the carrier-frequency component) and then eliminates the unwanted sideband with a filter.

The receiver no longer can use envelope detection, but must coherently demodulate the received signal with the use of a local replica of the original carrier. There is no longer any concept of modulation index. The power in the sideband is proportional to the modulating signal power, and can attain any value consistent with the power capacity of the transmitter and the transmission channel. The required transmission bandwidth is just the bandwidth of the modulating signal b, which for a voice signal can be approximated by its maximum frequency f_m.

Signal-to-noise ratio comparisons are more straightforward than with conventional AM. Consider a received signal with sideband power S, accompanied by noise in the receiver bandwidth of

$$N = N_0 B = 2N_0 f_m \qquad (7\text{-}11)$$

The recovered signal power is just

$$S_b = kS \qquad (7\text{-}12)$$

where k, as before, is the demodulator gain, and the recovered noise is

$$N_B = kN = kN_0 B = kN_0 f_m \qquad (7\text{-}13)$$

The baseband signal to noise ratio is then

$$\frac{S_b}{N_b} = \frac{S}{N} = \frac{S}{N_0 f_m} \qquad (7\text{-}14)$$

This relationship is independent of the waveform of the modulating signal.

We also note that the total power is just

$$P = S \qquad (7\text{-}15)$$

It is instructive to compare the total transmitted power requirement for equal signal-to-noise ratio of the recovered signal between conventional AM transmission and SSB-SC AM transmission under the following reference conditions:

Equal value of N_0 in both cases
Sinusoidal modulating signal (AM)
100% modulation ($m = 1$) (AM)

From Eqs. 7-3, 7-10, 7-14, and 7-15 we can determine that a total transmitted power three times greater is required for the conventional AM case than for the SSB-SC case. If the modulation index cannot be held at 1, or if the waveform of the modulating signal has a greater peak-to-average ratio than that of a sine wave, the difference is even greater.

This result suggests a major motivation for the adoption of the SSB-SC mode for most AM radio transmission today with the notable exception of the radio broadcast service. There, the issues of receiver simplicity (envelope detector vs. coherent detection) and compatibility with existing receivers have to date outweighed the advantages in required transmitted power and spectrum conservation.

It is also interesting to consider the equivalent comparison between double sideband suppressed carrier (DSB-SC) and SSB-SC modes. Assuming again a fixed value of the noise density N_0, we find that equal signal power is required for equal signal-to-noise ratios in the demodulated signals. This may seem paradoxical, since the wider bandwidth of the DSB-SC mode admits twice the noise power, N. However, the advantage of coherent detection of the two sidebands, between which the transmitted power is divided, exactly compensates for this disadvantage. This analysis gives an early insight into the properties of a general concept known as *spread-spectrum communication*, to which we will return in Chapter 8.

7.2.2 Frequency-Division Multiplex

Frequency-division multiplex (FDM) is used to combine multiple analog voiceband channels into higher-level composite signals. As illustrated in Figure 7-4, FDM signals (assemblies) are generated by first modulating each individual voiceband signal onto a sinusoidal carrier, using a balanced AM modulator. One input to the modulator is the voiceband signal, which typically occupies a bandwidth of 300 to 3400 Hz. The other input to the modulator is a sinusoidal carrier.

Adjacent carrier frequencies are separated by 4 kHz. The AM-DSB signal at the output of each modulator is passed through a filter to eliminate one of the sidebands. At the output of the filter, an AM SSB signal appears, with a bandwidth nominally equal to 4 kHz. These AM SSB signals are then combined, using a summing device to produce an FDM *baseband assembly*.

Long before the advent of satellite communications, FDM techniques were used in telephony on analog radio and cable systems. Consequently, a

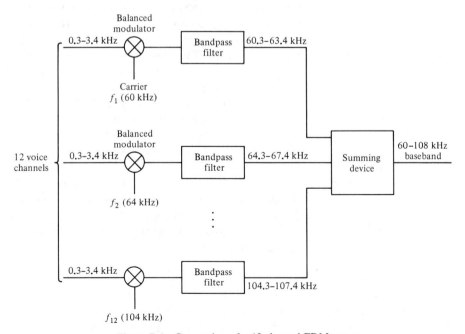

Figure 7-4 Generation of a 12-channel FDM group.

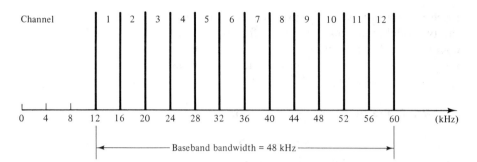

Figure 7-5 Spectrum of a 12-channel FDM group.

set of conventions called the *FDM hierarchy* was developed to organize the combination of individual voice channels in a structured way. This hierarchy is specified in CCIR, CCITT, and U.S. telephone industry (formerly Bell System) standards (Bell Laboratories, 1982; International Telecommunications Union, 1976).

The first level in the FDM hierarchy is called a *group*, consisting of 12 channels, whose baseband spectrum is depicted in Figure 7-5. The 12-channel group is the smallest FDM assembly. Several groups may be combined to form higher-level composite signals as illustrated in Figure 7-6. For example, combining five groups of 12 channels results in a *supergroup* consisting of 60 channels. Higher levels in the hierarchy include *master groups* (600 channels), *super-master groups*, and *jumbo groups*, consisting of as many as 6000 channels in a single composite signal. (Terminology for the higher-level assemblies varies between U.S. and CCITT standards.) Pilot tones are placed within the composite signals to achieve the spectrum-centering frequency control required. This FDM hierarchy formed the basis for early satellite transmission systems and is still in common use on satellite links. Such a link can be established simply by using the FDM technology to form the basebands and modulating the composite FDM signals onto radio-frequency FM carriers.

The combination of analog voice-frequency (VF) signals using FDM requires a fairly large investment in per-channel equipment (e.g., modulators, oscillators, amplifiers, filters, and summing devices). The concentration of per-channel equipment in FDM is to be compared to the time sharing of common equipment typical in time-division multiplex techniques.

Note that the FDM process actually combines the first two levels of our system model, *source coding* and *multiplexing*.

7.2.3 Frequency Modulation

Frequency modulation (FM) has been used in commercial satellite communications for both telephony and video transmission. It is in use in both SCPC and MCPC configurations. FM is one of the best covered subjects in

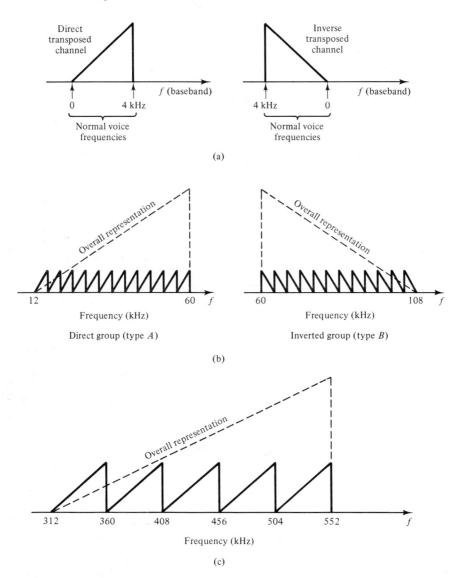

Figure 7-6 FDM hierarchy: (a) direct and inverse telephone channels; (b) direct and inverted groups; (c) supergroup.

the technical communications literature. Its applications in all forms of radio communications are truly prolific (Bell Laboratories, 1982; Bennett, 1970; Carlson, 1975; Lundquist, 1978; Panter, 1965; Taub and Schilling, 1971).

FM is created, as illustrated in Figure 7-7, by varying the frequency of a sinusoidal carrier with the amplitude of the message signal. This voltage-to-frequency conversion process results in a bandwidth expansion in the RF

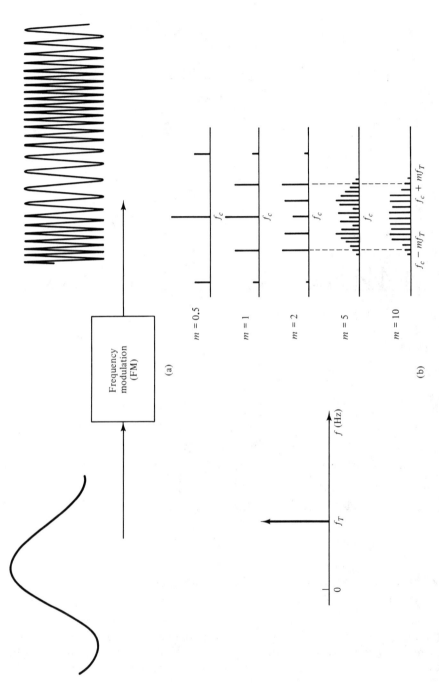

Figure 7-7 (a) FM in the time domain; (b) FM in the frequency domain.

channel which can be traded for signal to noise ratio improvement in the baseband signal, another example, although rarely so-called, of the spread-spectrum concept. To characterize the performance of FM, we need to define several parameters:

N_0 = the noise power spectral density in the RF channel.
C = the power level of the sinusoidal RF carrier (and thus the constant total transmitted power).
f = the instantaneous frequency of the modulated signal.
b = the modulating signal (baseband) bandwidth.
Δf = the peak deviation frequency.
f_m = the highest frequency in the baseband modulating signal.
$m = \Delta f/f_m$ = the FM modulation index.

Figure 7-7b shows the spectrum of an FM signal as a function of modulation index, m, for a modulating signal consisting of a single tone f_T. The FM spectrum consists of a carrier plus an infinite number of sidebands whose amplitudes are described by various-order Bessel functions of m. Therefore, the magnitude of each sideband is a function of both the amplitude and the frequency of the modulating signal.

The bandwidth of an FM modulated signal is thus theoretically infinite. However, in practical systems an FM signal must be contained within a finite bandwidth. The practical bandwidth for an FM signal depends on the modulation index, $m = \Delta f/f_m$. The peak deviation frequency Δf is the maximum departure from the nominal carrier frequency, which occurs at the peak of the input signal. The modulation index is an indicator of the RF bandwidth expansion compared to the baseband bandwidth. It is also a measure of the performance enhancement *vis-a-vis* AM available using FM.

J. R. Carson of Bell Laboratories first suggested an empirical rule for determining the practical bandwidth of FM in an unpublished memorandum in 1939. Carson's rule states that this practical bandwidth may be taken as

$$B = 2(\Delta f + f_m) \qquad (7\text{-}16)$$

Note that the bandwidth can be rewritten in terms of the modulation index showing that the bandwidth expansion is directly proportional to the modulation index.

$$B = 2f_m(m + 1) \qquad (7\text{-}17)$$

As with AM, FM performance is usually described in terms of the signal-to-noise ratio at the demodulator output S_b/N_b as a function of the carrier-to-noise ratio (C/N) in the RF channel. For a single modulating signal containing a highest-frequency f_m, FM system performance is described by

$$\frac{S_b}{N_b} = 3m^2 \frac{B}{2f_m} \frac{C}{N} \quad (7\text{-}18)$$

where $N = N_0 B$

Compare the performance of conventional AM given by equation 7-10 to the result of equation 7-18. The ratio of the FM signal-to-noise ratio to that for AM is $3m^2$, where m is the FM modulation index. (The AM modulation index must be 1 for equation 7-10 to be valid.) This factor is called the *FM improvement*. As long as $m > 0.6$, FM delivers noise performance superior to AM for equal power and equal noise density.

The power spectrum of the noise at the baseband output of an FM receiver is proportional to f^2. As shown in Figure 7-8, this "parabolic" noise spectrum of an FM signal causes higher frequencies in the baseband signal to be disturbed by a higher level of noise compared to lower frequencies. Therefore, if the signal power is uniform at all frequencies, the signal-to-noise ratios in the higher-frequency part of the spectrum will be poor compared to that for the lower-frequency components. To equalize the performance over the full range of baseband frequencies, a technique called preemphasis and deemphasis is normally used with FM signals. Preemphasis processing amounts to applying a frequency-sensitive network characteristic to the message signal prior to modulation. This characteristic tends to amplify the higher-frequency signals relative to the lower-frequency signals in a manner corresponding to the shape of the FM noise. This preemphasized signal is then modulated on an FM carrier. After demodulation of the message signal, a deemphasis network characteristic is applied which is the complement of the preemphasis network. Such networks have been standardized by the CCIR and result in an improvement of about 4 dB in the (subjective) signal-to-noise ratio in the demodulated baseband signal. Since the pre- and deemphasis characteristics are complementary, the effect on the signal is nil.

The general performance of FM is depicted in Figure 7-9. This figure

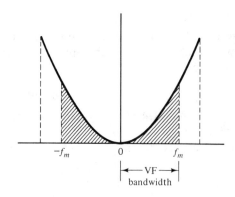

Figure 7-8 FM noise power spectrum at receiver output.

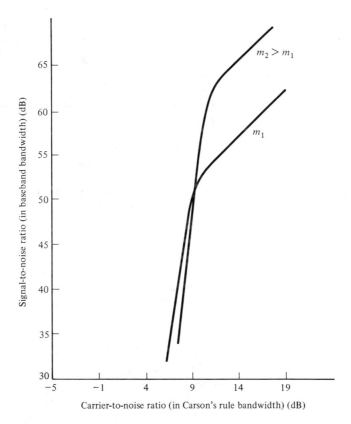

Figure 7-9 FM performance characteristic.

is a plot of the demodulated baseband signal-to-noise ratios (S/N), versus the carrier-to-noise ratio (C/N), in the RF channel. Note that the performance curve has a distinct knee, or threshold. Below this threshold, the performance of FM is nonlinear, and is extremely complicated from an analytical viewpoint. Practical FM systems are usually operated well above the threshold point. This simplifies both the description of FM performance and the required calculations. For our purposes it is always assumed that the FM system is operating above the threshold point so that the formulas cited are valid.

The performance of FM also depends on the way it is utilized. In satellite telephony transmission, FM is used in both SCPC and MCPC arrangements. Two practical telephony transmission systems in which FM is utilized will be covered. The first is an MCPC system, wherein multichannel FDM baseband assemblies are modulated on FM carriers. The characteristics of FDM signals containing multiplexed voice channels were studied in the late 1930s by Holbrook and Dixon (1939). They determined that the equivalent signal power for n voice channels in a multichannel FDM composite signal is substantially lower than the sum of the powers of the individual voice channels when each carries a standard-level test tone. The ratio, called the *multichannel loading factor*, reflects the fact that the average power of a voice waveform is less than for a sine wave of equal peak power, and also for the reduced "duty cycle" of actual speech. Empirical formulas for the factor were standardized by the CCIR in a manner that allows direct calculation of equivalent signal power of an FDM signal based on the number of channels (n) on the baseband. The CCIR multichannel loading factors are given by

$$S = -15 + 10 \log n \text{ (dB) for } n > 240$$
$$S = -1 + 4 \log n \text{ (dB) for } n \leq 240 \quad (7\text{-}19)$$

These power ratios can be converted to amplitude ratios using the formula

$$g = \text{antilog}\left(\frac{S}{20}\right) \quad (7\text{-}20)$$

It is usual practice in FDM/FM system engineering calculations to express performance in terms of an equivalent root-mean-square (rms) test-tone deviation (f_r) in a single baseband channel, rather than peak deviation Δf in the entire FM assembly. Δf and f_r are related according to

$$\Delta f = \rho g f_r \quad (7\text{-}21)$$

where ρ is the peak-to-rms factor for FDM multiplex channels. A typical design value for ρ is about 3.16 (10 dB). If Eq. (7-21) is substituted into Eq. (7-16), we can derive an expression for the equivalent single channel f_r in terms of bandwidth, the highest modulating frequency, the multichannel loading factor, and the peak-to-rms ratio. Of course, f_m and Δf depend on the number of channels in the baseband assembly.

The performance of FDM-modulated FM systems expressed in terms of signal-to-noise ratio in the voiceband channel, versus the carrier-to-noise ratio in the RF channel, can be written as

$$S/N = \left(\frac{f_r}{f_m}\right)^2 \frac{B}{b}\left(\frac{C}{N}\right)_t I_p W \quad (7\text{-}22)$$

where
S/N = signal (test-tone)-to-noise ratio in the voiceband
f_r = equivalent single-channel rms test-tone deviation
f_m = highest modulating frequency
B = RF bandwidth of the modulated signal
b = voice signal bandwidth
$\left(\frac{C}{N}\right)_t$ = carrier-to-noise ratio in the RF channel required to exceed the FM threshold
I_p = preemphasis improvement (4 dB)
W = psophometric weighting improvement (2.5 dB)

Sec. 7.2 Analog Transmission Systems

The psophometric weighting improvement results from recognition of the differing impact of noise at different frequencies on the human ear, using a standardized telephone instrument as the acoustic receiver. A psophometric weighting curve is used to quantify this variation for noise assessment purposes. Figure 7-10 shows the *C Message frequency weighting characteristic*, which is the current standard in North America. The Western Europe standard uses a similar characteristic called the *psophometric weighting curve*. Both curves reflect the fact that noise frequencies above and below about 900 Hz are of lesser impact than mid-band frequencies. There is, therefore, an apparent signal-to-noise ratio improvement due to recognition of this weighting. In the case of psophometric weighting, that improvement is 2.5 dB. The C Message weighting improvement is 3 dB.

A second way in which FM is utilized in telephony via satellite is in SCPC applications. In SCPC systems, an individual carrier is assigned to each voice channel. Therefore, an individual voiceband signal modulates a sinusoidal carrier. In this case the same basic analysis applies, but the overall performance measured in the voiceband is described by a slightly different function, given by

$$S/N = \frac{3}{2}\left(\frac{\Delta f}{f_m}\right)^2 \frac{B}{2f_m} \left(\frac{C}{N}\right)_t I_c I_p W \qquad (7\text{-}23)$$

where all of the variables except I_c (syllabic compandor improvement) are as defined previously.

Often in FM SCPC systems, voice-operated devices called syllabic compandors are used to improve the speech-to-noise ratio prior to modulation (Carlson, 1975). The word *compandor* is a contraction of *com*pressor and

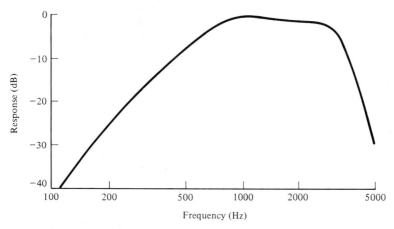

Figure 7-10 C message frequency weighing characteristic. (From *Transmission Systems for Communications*, Bell Telephone Laboratories, Inc. fifth ed. 1982.)

ex*pandor*. A compressor is associated with the transmit side of the terminal, and a complementary expandor is associated with the receiving side of the terminal. The compressor raises the weaker speech signals and thus reduces the volume range before the signals are applied to the transmission system. In the case of FM, this tends to keep up the peak deviation (which maximizes the signal-to-noise ratio). Another way to look at it is that lower-level signals tend to get a greater proportionate deviation compared to higher-level signals. Typically, the output from the compressor changes 1 dB for every 2-dB change in input signal. The expandor provides the opposite function, inserting loss which increases with decreasing speech volume. The gain of the compressor, and the loss of the expandor, are dependent on the average power of the talker. During quiet periods, the loss of the expandor is constant and very high. Thus noise during quiet periods is heavily attenuated, providing the major part of the noise advantage realized by companding.

Note the slight differences between Eqs. (7-22) and (7-23). For SCPC applications, the FM equation is preceded by a factor of 3, while for FDM modulating signals, the constant 3 disappears. The reason for the difference is illustrated in Figure 7-11, which shows that the integration of the noise power is dependent on whether there is a single channel or many channels

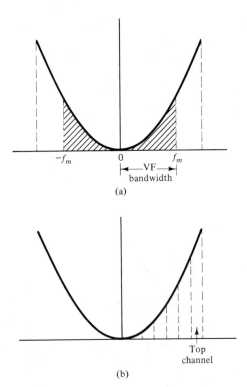

Figure 7-11 FM noise spectra for single (a) and multiple (b) channels.

in the demodulated signal. The 3 is simply the integration constant, resulting in the SCPC calculation. The factor $\frac{1}{2}$ in Equation 7-23 reflects the use of a peak deviation, Δf, compared to the rms deviation, f_r, seen in Equation 7-22. The multichannel loading factor does not appear in Equation 7-22 because that equation reflects performance as seen from the perspective of an individual voice channel. Note for FDM modulating signals the noise increases with increasing channel number. Typically, the CCIR has established recommendations for the performance of FDM/FM sytstems, and has specified that the top channel in an FDM assembly will have a signal-(test-tone)-to-noise ratio of no less than 50 dB. The standard test-tone level is 0 dBm at the zero transmission level point (0dBm0). Therefore, the maximum allowable noise in the worst-case FDM channel can be no greater than 10,000 pW0p, where the second p in pW0p indicates psophometric weighting and the 0 again implies measurement corrected to the zero transmission level point.

7.3 DIGITAL TRANSMISSION SYSTEMS

A virtual explosion in the application of digital technology in telecommunications began in the mid-1960s. The merging of computer and communications technologies has been so strong that it has caused dramatic shifts in the methods of transmission from analog to digital. Some of the reasons that digital techniques have gained wide acceptance are:

Ruggedness: Digital signals tend to be less susceptible to waveform distortions, such as crosstalk, nonlinearities, and noise, compared to analog signal transmission.

Power/bandwidth trade-off: As is the case with FM, digital transmission systems may be tailored to the application by trading off quality and transmission rate. In general, as the transmission rate increases in a digital system, the quality improves, and vice versa.

Voice/data/video integration: Converting signals into digital form provides an opportunity to combine and integrate various types of voice, data, and video signals, thus providing a common language on the communications link.

Compatibility with switching machines: The overwhelming increase in digital transmission technology has helped increase the use of digital switching technology. Digital switching technology, encouraged by reduced cost and improved efficiencies, has, in turn, influenced the further development of digital transmission applications.

Security: Signals in digital form are much easier to secure. Compared to analog signals, the processes of enciphering and deciphering digital signals are much more efficient and effective.

Economics: The dramatic improvements in integrated circuit technology and microprocessing capabilities continue to drive the economics of increasingly complex digital communication into the affordable range.

7.3.1 System Types

Digital transmission systems are in use on satellites in both SCPC and MCPC applications. The growth of both digital FDMA and TDMA systems is truly remarkable.

A digital SCPC system is implemented by first converting the analog voice-frequency (VF) signal into digital form using one of several coding techniques. These include pulse-code modulation (PCM), delta modulation (DM), or adaptive systems such as nearly instantaneous companding (NIC) or adaptive differential PCM (ADPCM). The digital signal representing the voice (or voiceband data) signal is then modulated on an individual radio frequency carrier using phase-shift keying (PSK). In MCPC systems, multiple digital voice signals, after analog-to-digital conversion, are combined using time-division multiplexing (TDM). The composite digital signal containing multiple digital voiceband signals is then modulated on a wideband radio frequency carrier using PSK. In TDMA applications, multiple digital channels are also combined using TDM and then modulated on a wideband carrier, using PSK at high speed in burst mode.

In the following sections we describe various aspects of source coding, multiplexing, and modulation that are used in these systems. Chapter 8 will deal more specifically with multiple access characteristics.

7.3.2 Source Coding

Digital source coding, for both voice and image signals, has been the subject of an enormous amount of research and development. In the paragraphs that follow, we will concentrate specifically on the coding of speech from the analog to the digital domain, using conventional waveform coding methods. A large body of literature is available describing digital source coding methods for image signals (Kaneko and Ishiguro, 1980; Tescher, 1980). The discussion that follows concentrates on speech coding, but much of the material is applicable to other signal forms.

Methods used to encode the human voice signal cover a range of data rate from about 2.4 to 400 kb/s and depend specifically on the application. Broadcast-quality audio program applications cover the range from 64 to 400 kb/s. Local haul telephone network quality may be accommodated between 32 and 64 kb/s. Speech quality suitable for some private networks may be obtained between 4.8 and 32 kb/s, and synthetic-quality voice may be obtained at rates between 2.4 and 4.8 kb/s. A good deal of literature is available covering various types of source coding systems for each of these applications

Sec. 7.3 Digital Transmission Systems

(Bell Laboratories, 1982; Bylanski and Ingram, 1976; Carlson, 1975; Cuccia, 1979; Jayant, 1975; O'Neal, 1980; Taub and Schilling, 1971).

Our efforts will focus on the telephone network quality voice-coding applications, covering the range from 32 to 64 kb/s. In this range of applications, there are several techniques to be considered. Clearly, the most widely accepted of these is pulse-code modulation (PCM). First introduced in the early 1960s, PCM has grown dramatically and at 64 kb/s has become the international standard for digital voice transmission. Other contenders for this application, particularly at the more economical rate of 32 kb/s, include an adaptive PCM system using nearly instantaneous companding (NIC), and two differential coding systems called adaptive delta modulation, sometimes known as continuously variable slope delta modulation (CVSD), and adaptive differential pulse-code modulation (ADPCM). In the paragraphs that follow, each of these systems will be briefly described and their relative performance discussed.

Pulse-code modulation (PCM). PCM is a conventional waveform coding technique, which directly converts the analog speech waveform into a sequence of multidigit (bit) binary numbers. No attempt is made to take advantage of the redundancy in the source signal. Figure 7-12 is a block diagram of a PCM coder/decoder (codec), designed to convert the electrical analog speech waveform into bits and reconstruct the analog signal on the receiving side from the received bit stream. At the input to the coder, a voiceband signal nominally occupying a 4 kHz bandwidth is first applied to a bandpass filter with a bandwidth of 300 to 3400 Hz. The band-limited output of the filter is usually adjusted in gain, so that for a standard level test-tone the maximum value of the waveform will fall about 3 dB below the maximum levels accommodated by the coder. This filtered and gain-adjusted signal is then applied to a sampler. The purpose of the sampler is to determine the instantaneous amplitude of the voice signal at regular time intervals. The upper frequency of the signal input to the sampler determines the minimum rate at which the signal can be sampled without loss of information content. According to the Nyquist sampling theorem, as long as the signal is sampled at a rate greater than twice the highest frequency in the band-limited analog signal, $2f_m$, the original signal may be recovered without distortion, for example by passing the samples through a low-pass filter with an upper cutoff frequency, f_m. Sampling at any rate less than $2f_m$ (undersampling) results in an irremovable distortion called *aliasing* in the reconstructed signal.

Referring again to Figure 7-12, a voiceband signal is sampled at the slightly higher than Nyquist rate of 8 kHz. This is the internationally standardized sampling rate for telephone bandwidth digital voice. The output of the sampler is a pulse amplitude modulation (PAM) waveform with pulse amplitude samples occurring every 125 μs. The amplitude of a pulse equals the instantaneous amplitude of the analog waveform at the sampling instant.

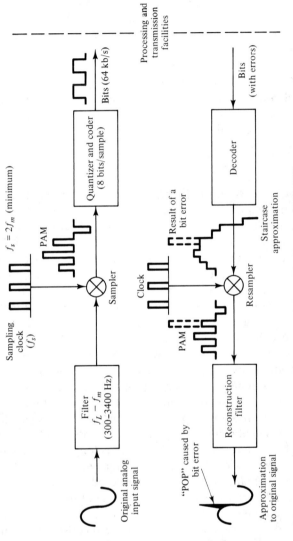

Figure 7-12 PCM coder/decoder (CODEC).

Sec. 7.3   Digital Transmission Systems   207

The PAM signal is then converted into a PCM signal by applying it to a quantizing encoder as illustrated in Figure 7-13. The range of amplitude occupied by the signal from minimum to maximum, called the *dynamic range*, is divided into many quantizing levels. In the systems used for telephony, approximately 256 levels are used. Each PAM sample amplitude is compared to these levels, and sample by sample, each pulse amplitude is assigned the level that is closest to the sample amplitude. Each of these levels is identified by a number, represented by an 8-bit binary code, as illustrated in Figure 7-13.

The code words for successive samples are transmitted in bit-serial form to the distant end. (The transmission format is determined by the multiplexing system applied at the next system level.)

At the receiving end, each successive 8-bit word is converted into a pulse, whose amplitude is specified by the value of the word. The output of the decoder is then a PAM pulse sequence (or perhaps a staircase signal), which closely follows the shape of the original analog signal waveform. This signal is typically resampled with a pulse narrow enough to minimize the zero-

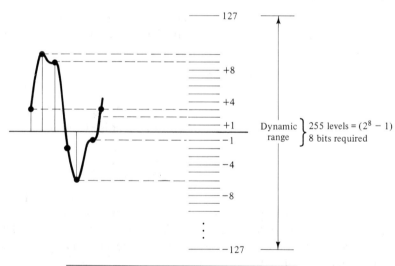

Sample	Decimal level	Binary code word
1	+3	0 0000011
2	+10	0 0001010
3	+9	0 0001001
4	−2	1 0000010
5	−6	1 0000110
6	−1	1 0000001
7	+3	0 0000011

Figure 7-13   PCM quantization.

order-hold effect, which tends to cause spectral droop (Carlson, 1975). The output of the resampler is then passed on to a low-pass reconstruction filter, which produces a replica of the original signal.

The process of quantized encoding cannot be accomplished without error, since in the general case the quantized amplitude represented by a sample's binary word is not precisely the amplitude of the sample. Since the receiver reconstructs a signal based on the coded samples, there is a discrepancy between the original and reconstructed signals. This discrepancy is said to be caused by *quantizing distortion*. It can be considered a type of noise and is thus called *quantizing noise*. This quantizing noise is signal dependent but in general is proportional to the square of the distance between adjacent quantum levels. Therefore, the more levels (and hence the larger number of bits that must be transmitted per sample), the lower the quantizing noise. Conversely, reducing the number of levels (or reducing the number of bits required per second) increases the quantizing noise. If the range of signal amplitudes to be encoded is divided into $L = 2^B$ levels, B bits are required per sample. (Sometimes only $2^B - 1$ levels are used, but this is ignored in the analysis to follow.)

If we consider a sinusoidal signal whose amplitude extends over the full range of the L levels, the ratio of signal power S_{max} to quantizing noise N_Q is

$$\frac{S_{max}}{N_Q} = \frac{3}{2} L^2 = \frac{3}{2} \cdot 4^B \qquad (7\text{-}24)$$

In telephony applications, where a test-tone falls 3 dB below such a maximum amplitude signal, the test-tone signal to quantizing noise ratio is

$$\frac{S_0}{N_Q} = \frac{3}{4} L^2 = \frac{3}{4} \cdot 4^B \qquad (7\text{-}25)$$

Since quantization noise depends on the size of the step between levels, a uniform quantizer using equal steps causes the quantization noise to be the same level for low-amplitude signals as it is for high-amplitude signals. Therefore, the S/N_Q in such a system is poorer for low-level signals than it is for high-level signals (such as the test-tone).

For a fixed number of levels (steps), corresponding to a certain bit rate requirement and a certain dynamic range of signal to be accommodated, better subjective performance can be achieved by having steps which vary in size. The steps are smallest (finer resolution) at low signal amplitudes and increase in size (coarser resolution) for increasing amplitudes. The effect of this is illustrated in Figure 7-14. In practice, this *nonlinear quantization* may be achieved by passing the signal samples through a circuit with a nonlinear tranfer characteristic before presenting it to a quantizing encoder with uniform

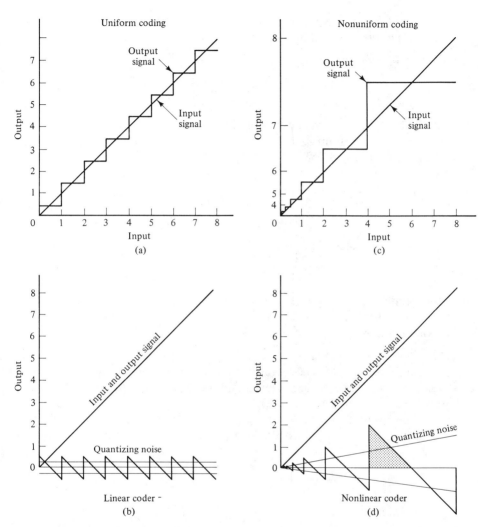

Figure 7-14 Comparison of the quantization noise for uniform and nonuniform step size. (Reprinted with permission from *PCM and Digital Transmissions* by G.H. Bennett, Marconi Instruments, 1976.)

steps. That circuit is sometimes called an *instantaneous compressor*. At the receiving end, samples reconstructed in accordance with the values of their binary code words can then be passed through a complementary *instantaneous expander*.

Whether or not the nonuniform encoding is actually done with a compressor, the variation of the encoding steps can be described in terms of the equation of the effective "compression" curve. These two equations are widely used in the telephone industry:

North American ("mu law")

$$y = \frac{\ln(1 + \mu x)}{\ln(1 + \mu)} \qquad (7\text{-}26)$$

(Magnitude only; apply sign of x as sign of y)
ln = natural logarithm function
Common value for μ: 255
Known as "μ-law" (mu-law) or "μ = 255" (for example)

European ("A law")

$$y = \frac{Ax}{1 + \ln(A)} \qquad 4c\ 0 \leq x \leq \frac{1}{A}$$
$$y = \frac{1 + \ln(Ax)}{1 + \ln(A)} \qquad 4c\ \frac{1}{A} \leq x \leq 1 \qquad (7\text{-}27)$$

(Magnitude only; apply sign of x as sign of y)
ln = natural logarithm function
Common value for A: 87.6
Known as "A law"

In these equations, x represents the signal amplitude input to the compressor (normalized to maximum amplitude), and y represents the output presented to a "uniform step" encoder.

The result of instantaneous companding (non-linear encoding) is illustrated in Figure 7-15, which shows a plot of the S/N_Q as a function of the

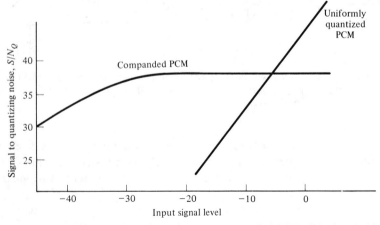

Figure 7-15 S/N_Q versus input level for linear companded PCM. (Reprinted with permission from *Transmissions Systems for Communication* Bell Telephone Laboratories, 4th ed, 1970.)

Sec. 7.3 Digital Transmission Systems

input signal level. Notice that without companding, the S/N_Q decreases decibel for decibel with decreasing signal level. However, the S/N_Q of the companded PCM system is relatively constant over a wide range of input signal level.

The effect of a bit error in the received signal is to reconstruct one or more of the pulse-amplitude samples at an incorrect amplitude level. After passing through the reconstruction filter, the impulse corresponding to the level error, looking like a damped sinusoid, is added to the desired signal and is perceived aurally as a click or a pop.

The quantitative effect of such bit errors depends on which of the bits in the code word are in error. An error in the most significant bit (MSB) will create the largest impulse noise component on the output signal, while an error in the least significant bit (LSB) generates the smallest noise component. The "clicks" and "pops" caused by bit errors are not bothersome during a two-way conversation as long as they do not occur too often. For PCM speech transmission it is generally agreed that as long as the bit error rate (BER) does not exceed 10^{-4}, the speech quality is acceptable. This is the threshold BER above which PCM voice systems degrade from a subjective viewpoint.

A bit error that occurs in the LSB of a PCM code word corresponds to an incorrect determination by amount L (one quantum level) in the reconstructed signal. A bit error in the next higher significant bit causes an error of 2L, in the next, 4L; and so on. As shown in Taub and Schilling (1971), a signal-to-noise ratio due to bit errors caused by thermal noise can be calculated and is given by

$$(S/N)_E = \frac{1}{4P_b} \tag{7-28}$$

where P_b is the probability of a single bit error (i.e., BER). We can combine both the quantizing noise and the noise due to bit errors to express an overall signal-to-noise ratio given by

$$(S/N)_T = \frac{C4^B}{1 + 4P_b C4^B} \tag{7-29}$$

where C is a constant accounting for the use of instantaneous companding. Figure 7-16 is a family of curves showing $(S/N)_T$ in the baseband as a function of C/N in the transmission channel for PCM systems. As shown in Section 7.3.5, the value of P_b in Eq. (7-29) is directly related to C/N. Therefore, for a given value of C/N and B (number of bits/sample), Eq. (7-29) can be used to plot the curves of Figure 7-16. Note that for high C/N (low P_b), the performance is dominated by quantizing noise and that increasing C/N results in no corresponding improvement in $(S/N)_T$. As C/N decreases, performance is dominated by the effects of bit errors. Note also that as the quantization resolution increases (B increases) the S/N_Q improves and the "knee" of the curve moves to the right.

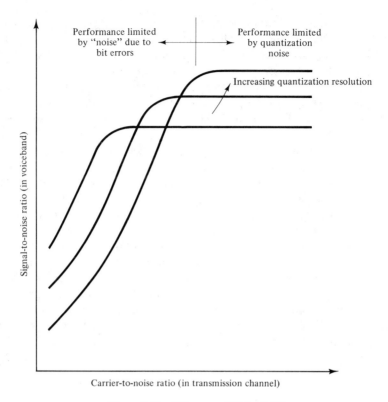

Figure 7-16 S/N versus C/N for PCM.

7.3.3 Other Practical Voice Coding Systems

In addition to conventional PCM, several other digital speech coding systems provide satisfactory performance for many applications at 32 kb/s. These include nearly instantaneous companding (NIC), an adaptive form of PCM, adaptive delta modulation (ADM), also known as continuously variable slope delta modulation (CVSD), and adaptive differential PCM (ADPCM).

Nearly instantaneous companding (NIC). The NIC coding system achieves bit-rate reduction by taking advantage of the short-term redundancies in the speech signal. That is, over a time interval of a few milliseconds the envelope of the speech signal is relatively constant. From interval to interval, the dynamic range occupied by the signal varies. Therefore, a reduction in required coding rate can be obtained by tracking the dynamic range dynamically. That is, for each interval, the available quantizing levels are always spread over the full dynamic range occupied by the signal during that interval. Since the speech signal will tend to occupy a far smaller dynamic range than that allotted by PCM operation, the required bit rate can be reduced with an acceptably small quality degradation. By transmitting approximately 4 bits per sample (representing 16 levels), together with a small amount of overhead information to communicate the changes in the dynamic range assignment from interval to interval, we can approach a data rate of approximately half that required for PCM.

Adaptive delta modulation (ADM). Another way to reduce the coding rate is to take advantage of the redundancies in the speech signal from sample to sample. Such systems use differential encoding. Delta modulation is a bit-oriented differential encoding scheme which samples the speech signal at a much higher rate than the Nyquist rate and uses a single bit (one quantizing level) to track the differences in amplitude from sample to sample. Systems employing a fixed step size are called *linear delta modulation* systems. Systems of the companding type using a variable step size are called *adaptive delta modulation* (ADM) techniques, sometimes called *continuously variable slope delta modulation* (CVSD). The basic process is illustrated in Figure 7-17.

The ADM systems are also capable of providing acceptable voice coding quality at a data rate of approximately 32 kb/s. The ADM coding techniques include both digital and analog implementations using the feedback loop to control effectively the step size and match it to the short-term variations in the speech signal.

Adaptive differential PCM (ADPCM). ADPCM also uses differential encoding techniques to reduce the effective bit rate without seriously affecting quality. This system exploits the fact that the mean-square error between

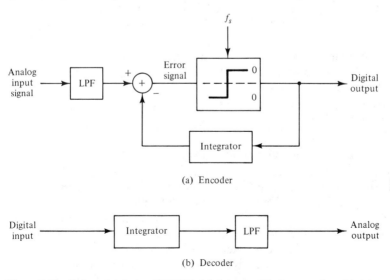

Figure 7-17 Delta modulation CODEC: (a) delta modulation encoder; (b) delta modulation decoder.

adjacent PCM samples is much smaller than the mean-square value of all the PCM samples taken together. Therefore, the difference, or error signal, can be quantized and encoded using fewer bits per sample than that required using PCM. Of course, the sample-to-sample correlation in the speech waveform is nonstationary. Therefore, the coder must have an adaptive quantizer and coder in order to track the variations in sample-to-sample correlation. An adaptive filter is required as the predictor for such a device. ADPCM has some advantages over the other techniques in that it is easily constructed from PCM samples, making it compatible with PCM switching machines. A standard 32 kb/s ADPCM system is now undergoing standardization both in the U.S. and Europe.

Relative performance of speech coding systems. All digital source coding schemes, including those described above, are governed by quantizing noise and noise due to bit errors. However, because of the difference in bit rate between PCM at 64 kb/s and the 32-kb/s systems, it is not easy to compare these techniques directly using only objective signal-to-noise ratio comparisons. For example, Figure 7-18a shows a plot of signal-to-quantizing noise ratio (S/N_Q) versus the input signal level for each of the four systems (GTE Lenkurt *Demodulator*, 1982).

Comparing the 32-kb/s systems, ADPCM consistently outperforms both ADM and NIC over a wide dynamic range of input signal level. Figure 7-18b shows the overall signal-to-noise ratio (S/N) as a function of bit error rate for each of the four 32-kb/s systems. Again, the ADPCM system seems to be best as long as the error rate is less than 10^{-3}.

On a subjective basis ADPCM also seems to be the leader and tends to outperform PCM at the higher bit error rates, as indicated in Figure 7-18c.

Another important consideration in evaluating the effectiveness of a reduced bit rate system is its ability to accommodate the transmission of voiceband-modulated data through the channel. Several practical 32-kb/s systems have been shown to be essentially transparent to such data transmission at data rates up to at least 4800 kb/s.

7.3.4 Time-Division Multiplexing

Time-division multiplexing (TDM) is used to combine multiple digitally-encoded signals into a higher-level composite signal at a bit rate greater than or equal to the sum of the input rates. The multiplexing can be either bit interleaved or character interleaved, depending on the type of source coding used. The digital inputs could represent a combination of digital voice, data, or video signals. If each of the input signals to the multiplexer has been generated from either the same or phase-coherent clock sources, the multiplexing is synchronous. *Pulse stuffing* techniques are also used to combine

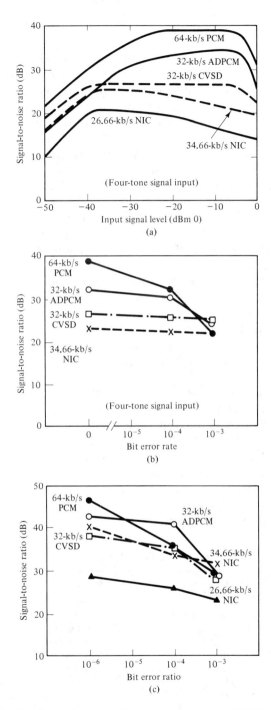

Figure 7-18 Relative performance of speech coding systems: (a) S/N_Q versus input signal level; (b) overall objective S/N versus bit error rate; (c) overall subjective S/N versus bit error rate. (GTE Lenkurt *Demodulator*, Nov./Dec. 1982, Vol. 31, No. 6. Reprinted with permission of GTE Communication Systems Corporation.)

nonsynchronous digital signals (Bell Laboratories, 1971; Bylanski and Ingram, 1976). In addition to combining the outputs of digital coders, multiplexers must also add framing information into the composite output bit stream. This usually consists of a known pattern for which the demultiplexer is designed to search within the receved bit stream. The framing pattern establishes the beginning and end of each frame in time. A frame consists of the sequential samples from each of the n inputs, plus the frame synchronization information.

As in the case with FDM, TDM techniques used in commercial applications are organized into a well-structured hierarchy. There are two basic TDM hierarchies in worldwide use. These are the so-called T-carrier hierarchy, used principally in North America, and the European (CEPT) hierarchy, used in Europe and South America. Similar systems are used in other parts of the world. For example, in Japan, a system based on the North American arrangement (but with certain special rates) is used. Table 7-3 summarizes the data rates and the channel organization in each level of the two major multiplexing hierarchies. Each level in the hierarchy corresponds to a primary cross-connect point from which signals are multiplexed and demultiplexed in each direction. The levels in the North American hierarchy are designated with DS- (*digital signal*) numbers. The systems used to transmit the various levels of the North American hierarchy are identified as "T-carrier" systems.

The first level in the North American digital hierarchy is called the DS1 level, employing a bit rate of 1.544 Mb/s. As shown in Figure 7-19, this signal can accommodate up to 24 voice and/or data signals, each consuming a nominal data rate of 64 kb/s. The DS1 frame uses 24 8-bit time slots, each reserved for a single PCM encoded voice or digital data signal. Therefore, a frame contains 192 bits (24 × 8) of information, plus a 193rd bit, used for framing. In normal PCM telephony, the eighth bit of each time slot is borrowed every sixth frame to transmit signaling information. Note that the frame time is 125 µs, consistent with the standardized 8000-Hz sampling rate.

TABLE 7-3 NORTH AMERICAN AND WESTERN EUROPEAN DIGITAL HIERARCHIES

North American Hierarchy				Western European Hierarchy			
Level	Number of Voice Circuits	Equivalent Build-up	Bit Rate (Mb/s)	Level[a]	Number of Voice Circuits	Equivalent Build-up	Bit Rate (Mb/s)
DS1	24	24 × voice	1.544	1	30	30 × voice	2.048
DS1C	48	2 × DS1	3.152	2	120	4 × level 1	8.448
DS2	96	4 × DS1	6.312	3	480	4 × level 2	34.368
DS3	672	7 × DS2	44.736	4	1920	4 × level 3	139.264
DS4	4032	6 × DS3	274.176	5	7680	4 × level 4	565.148

[a] Called in text E1, etc. to differentiate from DS levels.

Sec. 7.3 Digital Transmission Systems

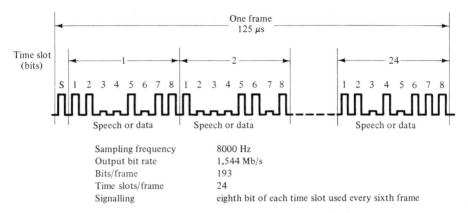

Figure 7-19 T1 frame.

Figure 7-20 shows typical interconnections among the levels of the North American digital hierarchy. The boxes labeled with M- designations are higher level multiplexers. Note that rates for all higher levels are slightly larger than the sum of the input data rates. For example, 4 × 1.544 = 6.176 Mb/s. The additional bit rate required at the DS2 level (6.312 Mb/s) represents the overhead required to accommodate asynchronous pulse stuffing multiplexing functions.

The first level in the European hierarchy uses a data rate of 2.048 Mb/s and will be referred to here as the E1 level. Figure 7-21 depicts the standard E1 PCM frame. Note that there are 32 8-bit time slots in each 125-μs frame. Only 30 of these 8-bit time slots are used for the transmission of voice and/or data signals in essentially the same way as the DS1 frame uses them. The E1 frame uses a dedicated 8-bit channel (time slot) for synchronization, and a second 8-bit channel (time slot) for the communication of the signaling information for all 30 voice channels. Again, pulse stuffing synchronization accounts for the overhead in each of the successively higher-level data rates.

7.3.5 Digital Modulation Techniques

The field of digital modulation has also been the subject of many theoretical and practical studies, and the literature abounds with treatments of the subject (Angello et al., 1978; Bennett and Davey, 1965; Bylanski and Ingram, 1976; Carlson, 1975; Galko and Pasupathy, 1981; Glave and Rosenbaum, 1975; Gronemeyer and McBride, 1976; Huang and Feher, 1979; Lundquist, 1978; Lundquist et al., 1974; Morais and Feher, 1979, 1980). Most of these treatments are highly theoretical and require a good deal of study. A comprehensive treatment will not be presented here. Rather we will concentrate on the basic notions and definitions and summarize the key results.

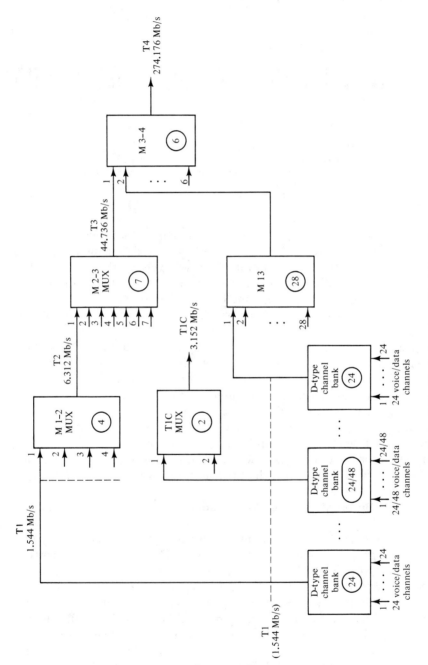

Figure 7-20 North American digital hierarchy.

Sec. 7.3 Digital Transmission Systems

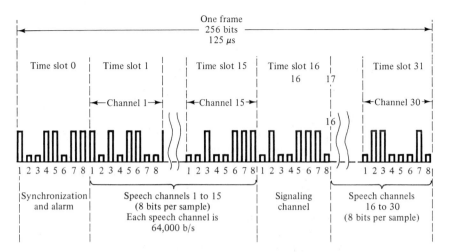

Figure 7-21 Frame organization of first level in Western European digital hierarchy.

The function of a digital modulator is to accept the digital bit stream to be sent and modulate this information on a sinusoidal carrier in a manner suitable for transmission over a radio-frequency channel. Typically, the amplitude, phase, or frequency of the carrier is modulated by the binary (or M-ary) values of the data (or the M-ary values of a recoded form of the data). The modulated carrier is then transmitted through the RF channel, where it may be corrupted by noise, resulting in bit errors in the demodulated data. The function of the demodulator is to accept the modulated carrier, and make binary (or M-ary) decisions to reconstruct the original bit stream.

There are two basic approaches to the demodulation process, which is the key to the effectiveness of the modulation technique. The first is coherent detection, which requires knowledge of the carrier phase within the received carrier signal. The second is noncoherent detection, which makes decisions without knowledge of the carrier phase of the received signal. To visualize the difference between coherent and noncoherent detection, consider the vector diagram shown in Figure 7-22 for an amplitude modulation system. The carrier, represented by the vector **C**, is perturbed in the channel by a noise vector **N**. If the demodulator can track the apparent phase changes of the receive carrier, and develop a phase coherent reference carrier, coherent detection can be achieved. Notice that the total noise component can be resolved into in-phase (N_i) and the quadrature (N_q) components. If the receiver is locked to the phase of the incoming carrier, only the relative amplitude of the carrier is unknown, and hence only the in-phase noise (N_i) can perturb the signal and cause bit errors. The quadrature component (N_q) simply modifies the apparent phase of the carrier, a change which the receiver ignores. In noncoherent detection the receiver does not develop knowledge of the phase of the signal, and hence the total noise vector **N** affects and modifies

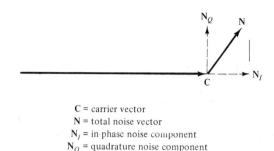

C = carrier vector
N = total noise vector
N_I = in-phase noise component
N_Q = quadrature noise component

Figure 7-22 Carrier and noise components of an amplitude modulation system.

the apparent amplitude of the carrier as it is passed through the channel. Therefore, both the in-phase and quadrature noise components affect the outcome of the decision process in noncoherent detection.

Digital modulation of a sinusoidal carrier can be accomplished in three basic ways: on/off keying (OOK) of the amplitude, frequency-shift keying (FSK), in which the frequency of the carrier is shifted, and phase-shift keying (PSK), in which the phase of the signal is changed in accordance with the transitions in the data. There are also many variations and combinations of these techniques which are utilized in the transmission of digital signals on paired and coaxial cable and radio systems. For satellite applications, the most efficient of these techniques is PSK with coherent detection. This technique has the desirable characteristic of transmitting a constant envelope signal with the information in carrier phase transitions, lending itself readily to coherent detection. Many references cover the operation of other digital modulation techniques (Bell Laboratories, 1971; Carlson, 1975; Cuccia, 1979; Lundquist, 1978; Taub and Schilling, 1971; Van Trees, 1979). In this chapter we concentrate on PSK because it is the predominant technique used in digital satellite communications.

Phase-shift keying (PSK). The simplest form of PSK is binary PSK (BPSK), wherein the digital data modulates a sinusoidal carrier, as illustrated in Figure 7-23a. The modulated output may be envisioned as assuming one of two possible phase states (say, 0 and π radians), during each bit time interval (T_b), representing either a binary 0 or a binary 1. This form of modulation is actually identical to amplitude modulation with suppressed carrier and a modulating signal having positive and negative values (for binary 0 and 1). In such modulation the carrier amplitude becomes *negative* during the negative excursions of the signal. This corresponds to the π radian phase of the carrier as described in BPSK terminology. In the time domain, the modulated carrier appears as a constant envelope sinusoid, with rapid phase changes occurring at a rate called the *keying rate*, dependent on the digital

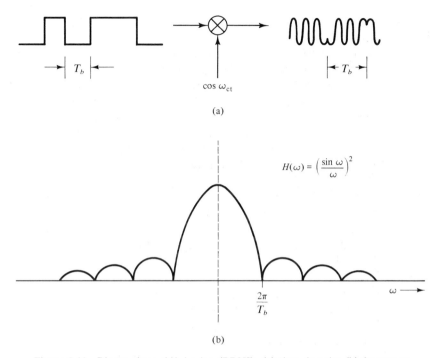

Figure 7-23 Binary phase-shift keying (BPSK): (a) time domain; (b) frequency domain.

data rate. In the frequency domain, the power spectral density of the modulated carrier varies in accordance with

$$H(\omega) = \left|\frac{\sin \omega}{\omega}\right|^2 \qquad (7\text{-}30)$$

As illustrated in Figure 7-23b, most of the energy in the modulated signal is contained in the major lobe, the width of which depends on the bit time or keying rate. The bandwidth of the modulated signal is considered to be that portion of the spectrum contained within the major lobe. The minor lobes of the spectrum repeat indefinitely, at lower and lower amplitudes. Therefore, as with FM, the spectrum of a PSK modulated signal is theoretically infinite. In BPSK, by limiting the bandwidth to be approximately equal to the bit rate, the energy in the sidelobes is lost, although with a resulting impact on performance. In practice, more sophisticated modulation techniques are used to squeeze more of the modulated carrier's energy into the major lobe and reduce the power in the side lobes. This is done by processing the digital signal in such a way as to cause the data transitions to be much less abrupt, resulting in more smooth transitions in phase. This tends to limit the essential bandwidth of the signal, and provide more efficient operation. The technique most often used for this purpose is called *minimum-shift keying* (MSK) (Angello et al., 1978).

Consider the BPSK signal in the phase domain as illustrated in Figure 7-24. Note that the phase can assume one of two possible states, representing binary 1 and a binary 0. The transmission of the modulated PSK carrier through a channel disturbed by noise can cause errors in the demodulated output if the noise disturbs the carrier phase sufficiently. Figure 7-24 illustrates the effect of the noise signal on the phase of the carrier. If we assume that the demodulator has acquired a coherent reference, the quadrature component of that noise cannot disturb the receiver's perception of the phase of the signal, since we know that the true phase must lie on the horizontal axis. Therefore, only the in-phase noise can cause the phase to shift from say π, along the horizontal axis, toward 2π. If the noise is sufficient to cause the carrier to be perceived as on the opposite side of the origin from the phase state transmitted, an error occurs. To specify the effect of the errors, we must determine the relationship between the probability of error and some figure of merit relating to the carrier-to-noise ratio in the RF channel.

Since digital demodulators make decisions at each bit time interval, independently from adjacent bit time intervals, it can be shown that the error rate is a function of the type of signal-to-noise ratio known as the bit energy-to-noise density ratio, written E_b/N_0.

A review of the relationships between the various types of signal-to-noise ratios is appropriate. First, thermal noise is a function of noise temperature and bandwidth, that is,

$$N = KTB \tag{7-31}$$

where K is Boltzmann's constant, T is the noise temperature in kelvin, and B is the measurement bandwidth in hertz. This is often written as N_0B, where N_0 is called the noise density, which is the noise power normalized to 1 Hz of bandwidth. In practice, we can only measure the carrier-to-noise ratio, which is the ratio of carrier power to noise power, measured in the entire

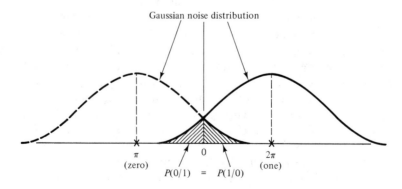

Figure 7-24 Binary PSK in the phase domain.

Sec. 7.3 Digital Transmission Systems   223

bandwidth B. We also use the term carrier-to-noise density ratio, C/N_0, which is related to the carrier to noise ratio, C/N, by the formula

$$\frac{C}{N} = \frac{C}{N_0 B} \tag{7-32}$$

The bit energy-to-noise density ratio, E_b/N_0, is related to these quantities in accordance with

$$\frac{C}{N} = \frac{C}{N_0 B} = \frac{E_b}{T_b N_0 B} = \frac{E_b R_b}{N_0 B} \tag{7-33}$$

where E_b is the energy in the carrier signal devoted to a single bit time interval. T_b is the bit time interval, and R_b is the bit rate, with $T_b = 1/R_b$. Notice that if the bandwidth of the measurement is equal to the bit rate (as suggested by the essential bandwidth concept), then

$$\frac{C}{N} = \frac{E_b}{N_0} \tag{7-34}$$

In practical systems, we first measure C/N and convert it to E_b/N_0 using Eq. (7-33). It is also usual practice to add implementation margin to this equation to account for imperfections in the equipment compared to theoretical performance.

Referring again to Figure 7-24, we can develop a relationship between the probability of a bit error, P_b, and E_b/N_0 by integrating a form of the Gaussian density over the limits represented by the crosshatched area. This integration yields the relationship

$$P_b = \tfrac{1}{2} \operatorname{erfc} \sqrt{\frac{E_b}{N_0}} \tag{7-35}$$

where the error function, erfc, is tabulated in all standard books of tables [REF] and represents the integral over the Gaussian density as illustrated. Figure 7-25 shows a plot of P_b versus E_b/N_0 for the theoretical performance of coherent binary PSK. The typical bandwidth occupied by the modulated BPSK signal is approximately 1.1 to 1.2 times the bit rate. Below error rates of 10^{-7}, the rule of thumb is to reduce P_b by a factor of 10 for each 1 dB increase in E_b/N_0.

A more efficient utilization of radio-frequency bandwidth can be achieved with no degradation in error-rate by employing four-phase or quaternary PSK (QPSK), as illustrated in Figure 7-26. A QPSK modulated signal is constructed by operating two BPSK modulators in quadrature. Odd-numbered bits of the input signal are routed to the i-(in-phase) channel, and even-numbered bits are sent to the q (quadrature) channel. The carrier frequency is fed directly into the i-channel modulator but is shifted in phase by 90° prior to entering the quadrature channel modulator. The ouputs of the two channels are then

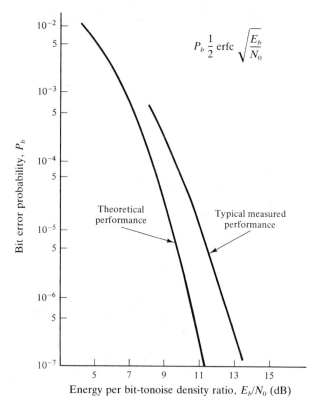

Figure 7-25 Theoretical and practical performance of coherent BPSK and QPSK.

added together to form the QPSK signal. Note that the phase state of the output of the summing device depends on both an i bit and a q bit. Therefore, the output state for each interval of signal (called a *symbol*) is dependent on a pair of bits. In the frequency domain, the power spectrum again takes the shape $[(\sin \omega)/\omega]^2$. However, the essential bandwidth of the QPSK signal is exactly half that of the BPSK signal for the same bit rate. This reduction in bandwidth is a result of the fact that the keying rate at the output of the modulator (the *symbol rate*) has been reduced by a factor of 2 in the QPSK case compared to the BPSK case.

Figure 7-27 shows the four phase states that may be assumed by a QPSK signal. Note that each phase state is dependent on a pair of bits. Viewed vectorially, i channel bits operate on the horizontal axis, at phase states 0 and π radians, while q channel bits operate on the vertical axis at phase states $\pi/2$ and $3\pi/2$. The vector sum of an i-channel phase and a q-channel phase produces one of the four states shown in the phase diagram. Since the i and q channels are orthogonal to each other, the advantage of coherent detection can be fully realized in each channel independently. This means that on the

Sec. 7.3   Digital Transmission Systems   225

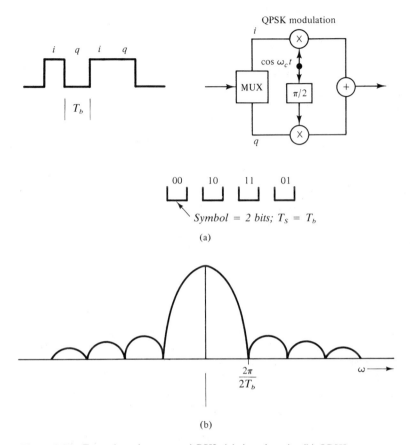

Figure 7-26 Four-phase (quaternary) PSK: (a) time domain; (b) QPSK spectrum.

i channel, only the in-phase component of the noise can cause a bit error, while in the q channel, only the quadrature component of the noise can cause an error. Therefore, the probability of bit error in either channel is identical to that realized in BPSK operation at a corresponding *symbol* rate. Much confusion has been created when comparing the symbol error rates of binary and QPSK. The important issue is that on a probability-of-bit-error basis, both BPSK and QPSK perform identically in accordance with Eq. (7-35). Therefore, the use of QPSK provides a significant advantage compared to BPSK, since QPSK can obtain the same error rate performance in a given noise environment (fixed E_B/N_0), while utilizing only half the bandwidth required by BPSK.

Extending this idea further to an eight-phase system is illustrated in Figure 7-28. In this case we can achieve a further reduction in keying rate, and hence, occupied bandwidth. This can be achieved since each symbol state now depends on a group of 3 bits. Therefore, the bandwidth required is a

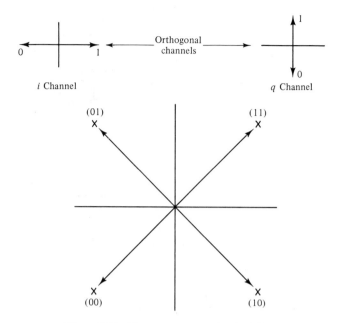

Figure 7-27 Phase-state generation in QPSK.

factor of 3 less than that of BPSK. However, as is obvious in Figure 7-28, we can no longer maintain the orthogonality characteristic of QPSK. As a result, additional power is required to maintain the same bit error rate. In fact, compared to QPSK, eight-phase PSK results in a bandwidth reduction of a factor of $\frac{3}{2}$ but requires more than doubling the required carrier power. The relationship of Eq. 7-35 must be modified for PSK systems having three or more bits per symbol.

The implementation of PSK modulators and demodulators requires careful design techniques and a comprehensive knowledge of the theoretical principles of filtering and phase-locked techniques. A general implementation is illustrated in Figure 7-29. Input data are applied to a specially designed filter preceding the modulator. This filter shapes the pulse waveforms to minimize the bandwidth of the modulated signal. The PSK signal is created by simply multiplying the output of the filtered data by the sinusoidal carrier. At the receiver, the PSK signal is first passed through a bandpass filter with a bandwidth of approximately 1.1 to 1.2 times the keying rate. A coherent reference carrier is derived from the received signal through a carrier recovery phase-locked loop. The phase coherent reference is then multiplied by the modulated signal, producing a signal containing the transmitted symbols. From the transitions between these symbols a data clock, derived from a second phase-locked loop, is then used to sample the signal, usually at mid symbol to reconstruct the original data stream.

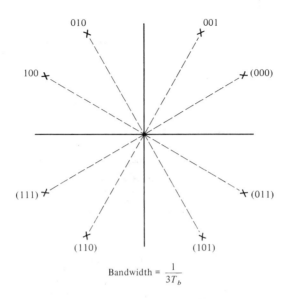

Figure 7-28 Eight-phase PSK.

7.3.6 Figures of Merit for Digital Transmission

Although the principles underlying the transmission of digital data rests firmly on probability theory, in practice, we must make measurements on actual data in limited time intervals. In this section we outline the figures of merit used to describe the quality of digital communications, which rely on analyses of digital errors in the signal.

In actual systems, there are several potential sources of error, including:

Thermal noise (in the RF channel)
Hardware errors (in the equipment)
Software errors (in programs driving the equipment)
Interference (both human-made and environmental)

Theoretical treatments deal only with the effect of thermal noise and its relation to bit errors. However, in practical systems we must also analyze errors caused by intermittent failures in hardware and software, as well as the effects of interference caused by human-made and environmental sources. These errors are not easy to predict.

In practice, several figures of merit are utilized to quantify the performance of digital communications systems. The first of these is the bit error rate (BER). Theoretical performance is usually specified in terms of the probability of bit error. In practice, we can measure BER only over a finite time

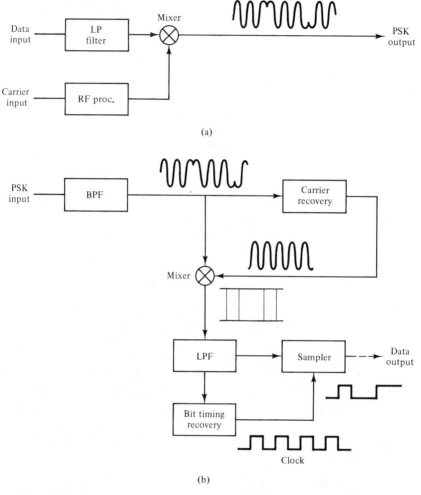

Figure 7-29 Implementation of a PSK modem: (a) modulator; (b) demodulator.

interval, and compare it to the bit error probability predicted by theory. The BER is calculated according to

$$\text{BER} = \frac{E}{RT_M} \tag{7-36}$$

where E is the number of bit errors observed, R the data rate in bits per second, and T_M the measurement interval in seconds. This is, of course, equivalent to the ratio of the number of errors observed divided by the number of bits over which the measurement is made. Clearly, because of the probablistic nature of the error generation process, the significance of the bit error rate measurement in projecting system performance is dependent on both the interval of time over which the measurement is made, and the number of errors observed during that interval. The longer the measurement interval and the more errors actually observed, the more certainly will the measurement reflect the actual error rate of the system. Figure 7-30 shows the dependency of the confidence of the measurement on both the error rate and the measurement interval.

If all errors were caused by thermal noise, the BER by itself would be a sufficient measure of digital transmission quality. However, because of hardware-, software-, and interference-caused errors, practical systems require a block error rate measurement as well. Thermal-noise-induced errors tend to be spread uniformly in time, while hardware-, software-, and interference-generated errors tend to occur in bursts. Therefore, the concept of a block-error-rate measurement, coupled with a bit-error-rate measurement, can more precisely define the nature of the errors.

Information systems typically transmit information in blocks, with each block containing error detection bits that are used to accept or reject each block. Rejected blocks are typically retransmitted. A general block-error-rate measurement is made by computing the ratio of the number of blocks in error to the number of blocks observed during the measurement interval. A block error is defined as a block in which at least one error has occurred. Therefore, in a block of 1000 bits, only one block error has occurred, whether 1 error, 10 errors, or 100 errors occurred in the block.

Another type of block error rate used by communications carriers to specify the quality of data communications circuits is known as *percent error-free seconds* measurement. Percent error-free seconds (%EFS) is computed according to

$$\%\text{EFS} = \frac{T_M - S_E}{T_M} \times 100 \ (\%) \tag{7-37}$$

where S_E is the number of seconds during which at least one error occurs, and T_M is the measurement interval in seconds. Notice that the EFS measurement is identical to a block-error-rate measurement where the block size is 1 s. Since a carrier will lease a subscriber a circuit based on the number of bits per second per month utilized, error-free seconds is a convenient way to specify the quality of that circuit. Typical high-quality data circuits are expected to perform at a rate of 95% to 99% error-free seconds.

Figure 7-31 shows the sensitivity of BER and block error rate to the distribution of errors. Suppose that 1000 bits are transmitted in 10 blocks, each consisting of 100 bits. If 10 errors occur in the 1000 bits, the bit error rate is 10^{-2} independent of the distribution of those errors. If the errors are distributed as in Figure 7-31a, the bit-error-rate is 10^{-2} and the block-error-rate is 0.9. Contrast this with the distribution shown in Figure 7-31b, where

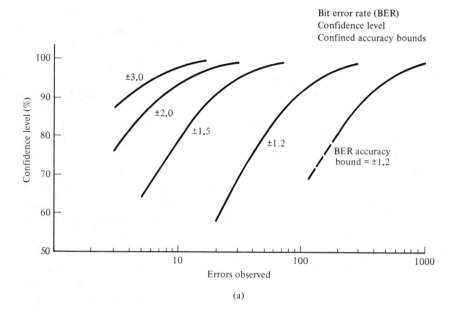

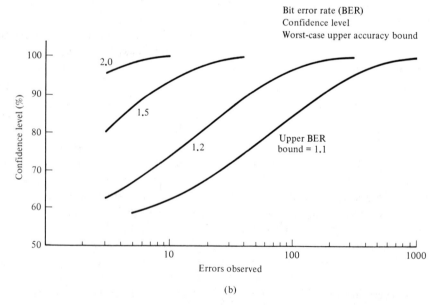

Figure 7-30 BER accuracy dependence on measurement interval: (a) confidence level related to number of errors observed during a test interval; (b) confidence level on worst-case value of BER. (Reprinted from *Telecommunications*, No. 11–12, Dec. 1977, a publication of Horizon House.)

Sec. 7.4   Television Transmission

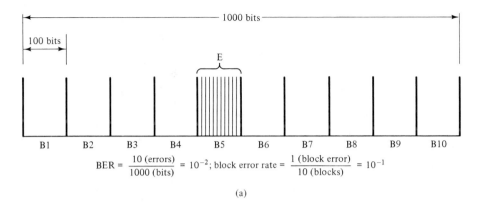

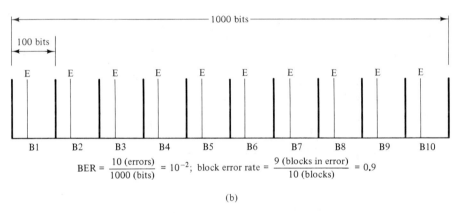

Figure 7-31 Sensitivity of error rate measurements to the distribution of errors: (a) burst-error distribution; (b) uniformly distributed errors.

the bit-error-rate is still 10^{-2} and the block-error-rate is 10^{-1}. Therefore, in typical systems, both block-error-rate and bit-error-rate measurements are required simultaneously to specify the quality of the data communications circuit.

Table 7-4 summarizes the requirements for both digital voice and various types of data services in terms of the BER performance required. Notice there is a wide range of application and performance requirements.

7.4 TELEVISION TRANSMISSION

7.4.1 Signal-to-Noise Ratios

The long distance transmission of broadcast quality television signals was one of the first commercial applications of satellite communications. Television transmission continues to account for a major portion of the total

TABLE 7-4 Typical Performance Requirements for Digital System Applications

Application	Transmission Rate	Required BER Performance	Connectivity	Typical Connect Time
Digital voice	19.2–64 kb/s	10^{-4}	Point-to-point via switched channel	3–4 min
Business video	56 kb/s–1.544 Mb/s	10^{-5}	Point-to-point via switched channel	30 min–1 h
File transfer	56 kb/s–6.312 Mb/s	10^{-8}	Point-to-point via switched channel	2–30 min
Electronic mail and high-speed fax	4.8–56 kb/s	10^{-8}	X-25 mesh network	2–10 min
Data-base "refresh and downline loading"	9.6–56 kb/s	10^{-8}	X-25 mesh network	2–10 min
Cad/cam	56–224 kb/s	10^{-7}	Point-to-point via switched channel	1-h session intermittent use
Remote job entry station	9.6–56 kb/s	10^{-8}	Point-to-point via switched channel	10–30 min
Computer graphics	9.6–56 kb/s	10^{-8}	X-25 mesh network	30 min–1 h intermittent use

Sec. 7.4 Television Transmission

satellite transponder utilization. In this section we examine the systems engineering calculations used to design satellite TV links with FM modulation. Earlier, we introduced Equation 7-18 which relates the signal-to-noise ratio, S/N, measured in the bandwidth of the demodulated signal, to the C/N and the modulation parameters. This equation is:

$$\frac{S}{N} = \frac{3}{2} \left(\frac{\Delta f}{f_m}\right)^2 \frac{B}{f_m} \frac{C}{N} \qquad (7\text{-}38)$$

It applies for a single (nonmultiplexed) signal and assumes sinusoidal (test-tone) modulation with peak deviation Δf and operation above the FM threshold. The term S/N is the ratio of the signal power to the noise power.

According to CCIR Recommendations 421-1, 421-3, and 451, the signal-to-noise ratio for television signals should be defined as the ratio of the power equivalent of the peak allowable amplitude of the luminance signal, S_{TV}, to the actual weighted noise power. We treat the luminance signal as "one sided" (having one extreme at zero) and thus the peak allowable amplitude equals the maximum allowable peak-to-peak value. That value is 0.714 of the peak-to-peak range of the composite video signal (100 IRE units/140 IRE units).

The value of S/N in Eq. (7-38) is based on a sinusoidal test signal modulation with amplitude f and thus peak-to-peak deviation of $2\Delta f$. The television signal is applied to produce peak-to-peak deviation $2\Delta f$ by the full range composite video signal. We can then relate the equivalent video signal power, S_v, to the test-tone power S, following our usual assumption of a one-ohm impedance in converting voltage ratios to power ratios:

$$S = \frac{\Delta f^2}{2} \qquad (7\text{-}39)$$

$$S_{TV} = (0.714 \cdot 2\Delta f)^2 \qquad (7\text{-}40)$$

$$\frac{S_{TV}}{S} = \frac{2 \cdot (0.714)^2 \cdot 2^2 (\Delta f)^2}{(\Delta f)^2} = 4.08 \doteq 4 \qquad (7\text{-}41)$$

This ratio can be used to adjust Eq. (7-38) to give the equivalent signal-to-noise ratio as defined for television. Preemphasis/de-emphasis is used in TV transmission, and the effect of noise weighting must be accounted for. This calls for a further adjustment in the effective value of S/N by a combined weighting improvement factor K_w, the specific value of which depends on which of the TV standards (NTSC-M, PAL-B, SECAM-B, etc.) is in use.

Therefore, we can adjust the value of S/N given in Equation A for the TV case to be

$$\left(\frac{S}{N}\right)_{TV} = 4 \left(\frac{S}{N}\right) K_w \qquad (7\text{-}42)$$

The CCIR, in Recommendation 421-3, achieves the same result by substituting $2\Delta f$ for Δf in Eq. (7-38). Note that this substitution into the modification for weighting improvement yields

$$\left(\frac{S}{N}\right)_{TV} = 4\left[\frac{3}{2}\left(\frac{\Delta f}{f_m}\right)^2 \frac{B}{f_m} \frac{C}{N}\right] k_w \qquad (7\text{-}43)$$

which is equivalent to equation B. For the 525 line NTSC system in use in the USA and Canada, the following values are useful in making systems engineering calculations:

Highest video modulating frequency, $f_m = 4\ MHz$
Combined weighting factor, $K_w = 13.8\ dB$
Objective luminance signal to weighted noise ratio $= 56\ dB$.

The RF bandwidth required to support broadcast-quality television via satellite is approximately 30 MHz. A high quality TV signal transmission, therefore, consumes a full transponder in a satellite of the class of INTELSAT V. Table 7-5 summarizes the technical performance characteristics of both 525 line and 625 line TV systems occupying the full transponder bandwidth. Note the difference in performance depending on the transmit/receive pairing of Standard A and Standard B earth stations. With some caveats, INTELSAT also permits TV transmission occupying only half the transponder bandwidth so that two TV channels may be supported in a single transponder. Table 7-6[1] summarizes the performance characteristics and lists the disclaimers for half-transponder TV transmission.

7.4.2 Television-Associated Audio

The audio baseband associated with television transmission contains not only the program audio but coordination, cue and commentary channels as well. In broadcast format the audio program signal is carried on a separate FM-subcarrier located adjacent to the video baseband spectrum (at approximately 4.5 MHz for the NTSC system). In satellite television applications TV-associated audio may be transmitted by one of several methods. One method specified by the CCITT is to transmit the audio within a FDM telephony baseband group occupying the band from 12 to 60 KHz. Engineering service channels are transmitted in the band below 12 KHz. TV-associated audio in INTELSAT networks may also be carried on SCPC carriers by agreement between connected stations.

[1] Standard B—Performance of earth stations in the INTELSAT IV, IVA & V systems, Doc. BG-28-74E (Rev. 1), 15 Dec. 1982.

Sec. 7.4 Television Transmission

TABLE 7-5 Characteristics of Full Transponder TV

General Characteristics		
Allocated satellite bandwidth (MHz)	30	
Receiver bandwidth (MHz)	22 to 30	
Television standard	525/60	625/50
Maximum video bandwidth (MHz)	4.2	6.0
*Peak-to-peak low frequency (15 kHz) deviation of a pre-emphasized video signal (MHz)	6.8	5.1
Differential gain	10%	10%
Differential phase	±3°	±3°
Receive Characteristics		
Earth station	STD "A" or "B" to "B" (1)	"B" to "A" (2)
Television standard	525/60 or 625/50	525/60 or 625/50
Elevation angle (receive)	10°	10°
C/T total at operating point (dBW/K)	−143.9	−140.4
Video signal-to-weighted noise ratio (dB)**	47.6	51.1
C/N in IF bandwidth (dB)	11.3	13.4
Receive IF bandwidth (MHz)	22.0***	30.0
Amount of over deviation 525/60 in IF bandwidth (dB) 625/50	4.0 5.2	—

*Excluding the maximum peak-to-peak deviation of 1 MHz due to the application of an energy dispersal waveform.

**As defined in CCIR Rec. 421-3 using Systems D, K and L weighting networks for a 625/50 channel, and System M for 525/60.

***For elevation angles greater than 10°, wider bandwidth IF filters may be employed which will reduce the amount of over deviation. Higher elevation angles should result in a better S/N.

(1) Assumes Standard A or B is transmitting 85 dBW at 10° elevation.
(2) Assumes Standard B is transmitting 80.8 dBW at 10° elevation.

TABLE 7-6 Characteristics of Half Transponder TV

General Characteristics					
Allocated satellite bandwidth (MHz)	17.50				
Receiver bandwidth (MHz)	15.75				
Television standard	525/60		625/50		
Maximum video bandwidth (MHz)	4.2		6.0		
*Peak-to-peak low frequency (15 kHz) deviation of a pre-emphasized video signal (MHz)	4.75		4.22		
Differential gain	10%		10%		
Differential phase	+4°		+4°		
Receive Characteristics					
Earth station	STD "A" to "B" (1)		STD "B" to "B" (2)		"B" to "A" (3)
Television standard	525/60 or 625/50		525/60 or 625/50		525/60 or 625/50
Elevation angle (receive)	10°	55°	50°	90°	10°
C/T total at operating point (dBW/K)	−147.1	−144.6	−147.1	−145.4	−141.0
Video signal-to-weighted noise ratio (dB)**	SEE NOTE	SEE NOTE	SEE NOTE	SEE NOTE	47.3
C/N in IF bandwidth (dB)	9.5	12.0	9.5	11.2	15.5
Transmit and receive IF bandwidth (MHz)	15.75		15.75		15.75
Amount of over deviation in IF bandwidth (dB) 525/60 625/50	6.2 12.0		6.2 12.0		6.2 12.0

*Excluding the maximum peak-to-peak deviation of 1 MHz due to the application of an energy dispersal waveform.

**As defined in CCIR Rec. 421-3 using Systems D, K and L weighting networks for a 625/50 channel, and System M for 525/60.

Note: The quality achieved with 17.5 MHz TV reception at Standard B earth stations is the responsibility of the earth station owner. The calculated S/N is:
   (1) For Standard A to B at 10° and 55°; 41.0 and 43.5 dB; and
   (2) For Standard B to B at 50° and 90°; 41.0 and 42.9 dB;

however, there may be visual impairments some users would find unacceptable which are caused by the low C/N available, particularly with color television. In some cases, the quality achieved is not acceptable for broadcast.

   (1) Assumes Standard A is transmitting 88 dBW at 10° elevation.
   (2)&(3) Assumes Standard B is transmitting 85 dBW at 10° elevation.

Sec. 7.4 Television Transmission

Another interesting method employed for the transmission of TV associated-audio utilizes a digital subcarrier transmitted within a combined video/audio baseband. In this application the video baseband and the digitally modulated audio subcarrier form a single multichannel baseband signal which is, in turn, transmitted via satellite on an FM carrier. The modulated audio carrier may, therefore, be considered the "top channel" of a FDM baseband and we can utilize the familiar concepts and equations. Accordingly, the signal-to-noise ratio in the bandwidth of the digitally modulated audio carrier after FM transmission of the combined baseband signal is given by:

$$\left(\frac{S}{N}\right)_A = \left(\frac{\Delta f_A}{f_A}\right)^2 \frac{B}{2B_A} \frac{C}{N} \qquad (7\text{-}44)$$

where f_A is the peak deviation of the main carrier produced by the modulated subcarrier, f_A, and B_A is the bandwidth of the modulated subcarrier. To determine the relationship between the "quality" of the subcarrier transmission and the C/N in the channel, in which the combined signal is transmitted, we use the following rationale. We can treat $(S/N)_A$, as the "carrier-to-noise" ratio in the bandwidth in which the digital carrier is transmitted. The relation between E_B/N_0 and $(S/N)_A$ is then

$$\left(\frac{S}{N}\right)_A = \frac{E_B}{N_0} \frac{R}{B_A} M \qquad (7\text{-}45)$$

where R is the information bit rate of the modulating signal and M is the implementation margin. Eq. (7-35) defines the relationship between E_B/N_0 and the probability of bit error in the demodulated digital signal. The relationship can be developed further by recognizing that the bandwidth required for a digitally modulated carrier is proportional to the keying rate. For example, for 4-phase PSK the keying rate is R/2 and the bandwidth required to support this signal is therefore

$$B = 1.2 \ (R/2) \qquad (7\text{-}46)$$

where the factor 1.2 is used by system designers to account for implementation imperfections relative to theoretical performance. For example, it is impossible to design filters with perfectly flat passbands, infinite passband to stopband ratios, perfectly rectangular cutoff and perfect linear phase response. Since Equations (7-44) and (7-45) are equivalent, a relation between the E_B/N_0 of the digital carrier and the C/N of the FM carrier is given by

$$\frac{E_B}{N_0} = \left(\frac{\Delta f_A}{f_A}\right)^2 \frac{B}{2RM} \frac{C}{N} \qquad (7\text{-}47)$$

REFERENCES

ANGELLO, P. S., M. C. AUSTIN, M. FASHANO, and D. F. HORWOOD, "MSK and Offset Keyed QPSK through Band Limited Satellite Channels," *Proc. 4th Int. Conf. Digital Satellite Commun.*, Montreal, Oct. 1978.

BELL LABORATORIES—Technical Staff: *Transmission Systems for Communications*, Bell Telephone, 1982.

BENNETT, W.R., *Introduction to Signal Transmission*, McGraw-Hill Book Company, New York, 1970.

BENNETT, W.R., and J. R. DAVEY, *Data Transmission*, McGraw-Hill Book Company, New York, 1965.

BYLANSKI, P., and D. INGRAM, *Digital Transmission Systems*, Peter Peregrinus Ltd. (EEE), Huddersfield, England, 1976.

CARLSON, J., *Communication Systems*, 2nd ed., McGraw-Hill Book Company, New York, 1975.

CLARK, A. P., *Advanced Data-Transmission Systems*, Wiley-Halsted Press, New York, 1977.

CUCCIA, C. L., *The Handbook of Digital Communications*, EW Communications, Inc., Palo Alto, Calif., 1979.

DODDS, D. E., A. M. SENDYK, and D. B. WOHLBERG, "Error Tolerant Adaptive Algorithms for Delta-Modulation Coding," *IEEE Trans. Commun.*, Vol. COM-28, No. 3, Mar. 1980.

FRANKS, L.E., "Carrier and Bit Synchronization in Data Communication—A Tutorial Review," *IEEE Trans. Commun.*, Vol. COM-28, No. 8, Aug. 1980, pp. 1107–1121.

GALKO, P., and S. PASUPATHY, "On a Class of Generalized MSK," *Proc. 1981 IEEE Int. Commun. Conf.*, Denver, Colo., June 1981.

GLASGAL, R., *Advanced Techniques in Data Communications*, Artech House, Inc., Dedham, Mass., 1976.

GLAVE, F. E., and A. S. ROSENBAUM, "An Upper Bound Analysis for Coherent Phase-Shift Keying with Cochannel, Adjacent-Channel, and Intersymbol Interference," *IEEE Trans. Commun.*, Vol. COM-23, No. 6, June 1975.

GRONEMEYER, S., and A. MCBRIDE, "MSK and Offset QPSK Modulation," *IEEE Trans. Commun.*, Vol. COM-24, No. 8, Aug. 1976.

HOLBROOK, B. D., and DIXON J. T., "Load Rating Theory for Multichannel Amplifiers," *Bell System Technical Journal*, Vol. 18 (Oct. 1939), pp. 624–644.

HUANG, J., and K. FEHER, "Performance of QPSK, OKQPSK and MSK through Cascaded Nonlinearity and Bandlimiting," *Proc. IEEE Int. Conf. Commun.*, ICC-79, Boston, June 1979.

INTERNATIONAL TELECOMMUNICATIONS UNION, "Transmission Systems," *Economic and Technical Aspects of the Choice of Transmission Systems Gas 3 Manual*, Vol. 1, Geneva, 1976.

JAYANT, S. N., "Digital Coding of Speech Waveforms: PCM, DPCM and DM Quantizers," *Proc. IEEE*, May 1975.

KANEKO, H., and T. ISHIGURO, "Digital Television Transmission Using Bandwidth Compression Techniques," *IEEE Commun. Mag.*, July 1980.

Lenkurt Demodulator, Vol. 31, No. 6, 1982 (San Carlos, Calif.)

LINDSEY, W., and M. SIMON, *Telecommunications System Engineering*, Prentice-Hall, Inc., Englewood Cliffs, N.J., 1973.

LINUMA, K., Y. LIJIMA, T. ISHIGURO, H. KANEKO, and S. SHIGAKI, "Interframe Coding for 4 MHz Color Television Signals," *IEEE Trans. Commun.*, Vol. COM-23, No. 12, Dec. 1975.

LUCKY, R. W., J. SALZ, and J. WELDON, *Principles of Data Communication*, McGraw-Hill Book Company, New York, 1968.

LUNDQUIST, L., "Modulation Techniques for Band and Power Limited Satellite Channels," *Proc 4th Int. Conf. Digital Satellite Commun.*, Montreal, Oct. 23–25, 1978.

LUNDQUIST, L., M. LOPRIORI, and F. M. GARDNER, "Transmission of 4-Phase-Shift-Keyed Time-Division Multiple Access over Satellite Channels," *IEEE Trans. Commun.*, Vol. COM-22, No. 9, Sept. 1974, pp. 1254–1360.

MARTIN J., *Future Developments in Telecommunications*, Prentice-Hall, Inc., Englewood Cliffs, N.J., 1977.

MARTIN, J., *Communications Satellite Systems*, Prentice-Hall, Inc., Englewood Cliffs, N.J., 1978.

MCGLYNN, D. R., *Distributed Processing and Data Communications*, John Wiley & Sons, Inc., New York, 1978.

MIYA, K., *Satellite Communications Engineering*, Lattice Company, Tokyo, 1975.

MORAIS, D., and K. FEHER, "Bandwidth Efficiency and Probability of Error Performance of MSK and OKQPSK Systems," *IEEE Trans. Commun.*, Vol. COM-27, No. 12, Dec. 1979.

MORAIS, D. H., and K. FEHER, "The Effects of Filtering and Limiting on the Performance of QPSK Offset QPSK and MSK Systems," *IEEE Trans. Commun.*, Vol. COM-28, No. 12, Dec. 1980.

NBS (NATIONAL BUREAU OF STANDARDS), *Handbook of Mathematical Functions*, NBS Applied Mathematics Series 55, June 1964.

NYQUIST, H., "Certain Topics of Telegraph Transmission Theory," *Trans. AIEE*, Vol. 47, Feb. 1928.

O'NEAL, J. B., "Waveform Encoding of Voiceband Data Signals," *Proc. IEEE*, Feb. 1980.

PANTER, P. F., *Modulation, Noise and Spectral Analysis*, McGraw-Hill Book Company, New York, 1965.

PAPOULIS, A., *Probability, Random Variables and Stochastic Processes*, McGraw-Hill Book Company, New York, 1965.

RODEN, M. S., *Analog and Digital Communication Systems*, Prentice-Hall, Inc., Englewood Cliffs, N.J. 1979.

ROSENBAUM, A. S., and F. E. GLAVE, "An Error Probability Upper Bound for Coherent Phase-Shift Keying with Peak-Limited Interference," *IEEE Trans. Commun.*, Vol. 22, No.

SHANMUGAN, K. S., *Digital and Analog Communication Systems*, John Wiley & Sons, Inc., New York, 1979.

SHANNON, C. E., "A Mathematical Theory of Communications," *Bell Syst. Tech. J.*, 1948, Part 1, pp. 379–423, Part 2, pp. 623–656.

SPILKER, J. J., *Digital Communications by Satellite*, Prentice-Hall, Inc., Englewood Cliffs, N.J. 1977.

TAUB, H., and D. L. SCHILLING, *Principles of Communication Systems*, McGraw-Hill Book Company, New York, 1971.

TESCHER, A. G., "Adaptive Coding of NTSC Component Video Signals," *Proc. IEEE*, NTC, Houston, Dec. 1980.

VAN TREES, H. L., ed., *Satellite Communications*, IEEE Press Selected Reprint Series. IEEE Press, New York, 1979.

8
Multiple Access Systems

8.0 INTRODUCTION

Multiple access is defined as the technique wherein more than one pair of earth stations can simultaneously use a satellite transponder. It is the technique used to exploit the satellite's geometric advantage. It is at the core of satellite networking.

Most satellite communications applications involve a number of earth stations, communicating with each other through a satellite channel. Note that the word "channel" is used in the satellite field to describe both a baseband voice, data or video channel, and an RF channel provided through a transponder. The engineering aspects of these RF channels are covered in Chapter 9. The concept of multiple access involves systems that make it possible for multiple earth stations to interconnect their communication links through a single transponder. A transponder may be accessed by single or multiple carriers. These carriers may be modulated by single- or multiple-channel basebands which include voice, data, or video communication signals. The systems used for coding and modulation in multiple access systems are discussed in detail in Chapter 7. In this chapter we concentrate on the basic multiple access techniques used primarily in commercial communication satellite systems.

Although there are many specific implementations of multiple access systems, there are only three fundamental system types:

Frequency-division multiple access (FDMA): FDMA systems channelize a transponder using multiple carriers. The bandwidth associated with each carrier can be as small as that required for a single voice channel. FDMA can use either analog or digital transmission in either continuous or burst mode.

Time-division multiple access (TDMA): TDMA is characterized by the use of a single carrier frequency per transponder, where the bandwidth associated with the carrier is typically the full transponder bandwidth. This bandwidth is time shared among all users on a time-slot-by-time-slot basis. Although the primary advantage of TDMA is realized in the single-carrier-per-transponder arrangement, there are cases where the TDMA bandwidth may be a fraction of the transponder bandwidth. TDMA is suited only for digital transmission and operates in burst mode.

Code-division multiple access (CDMA): CDMA is a method that transforms the signal using a unique code sequence for each user. The users all transmit simultaneously in the full transponder bandwidth at the same time. CDMA is suited only for digital transmission.

Variations of all three of these basic multiple access systems are employed in commercial satellite communications applications. The original FDMA method using multiple channels per carrier (MCPC) was derived from terrestrial frequency-division multiplex systems. A typical spectral occupancy plan for a transponder using this system is shown in Figure 8-1a. Either analog or digital transmission can be employed. In the case of analog transmission, multiple channels are multiplexed to form a frequency-division multiplex (FDM) assembly, then modulated on an FM carrier. In the case of digital transmission, time-division multiplexing (TDM) is used to combine multiple digital channels. The digital baseband signal is then modulated on a digital carrier, typically using phase-shift keying (PSK). In both cases, multiple carriers are present in the same transponder, and the attendant nonlinear impairments due to the transponder characteristics (described in Chapter 9) are major concerns for the system designer. For example, if analog FDM/FM transmission is used, the AM-to-PM effects can be manifested as intelligible crosstalk from one set of channels into another. Digital transmission tends to be impervious to intelligible crosstalk because the signal is encoded in digital form. As illustrated in Figure 8-1a, the carriers can have varying bandwidths (representing different numbers of channels) per frequency slot. In all cases, a significant portion (perhaps as high as 10%) of the total bandwidth of the transponder is consumed by the guard bands required for spectral separation.

Another type of FDMA system employs a single channel per carrier (SCPC). As illustrated in Figure 8-1b, an SCPC spectrum consists of many carriers in adjacent frequency slots, occupying the transponder bandwidth. Each carrier is modulated with the information from a single voice or data

Sec. 8.0   Introduction   243

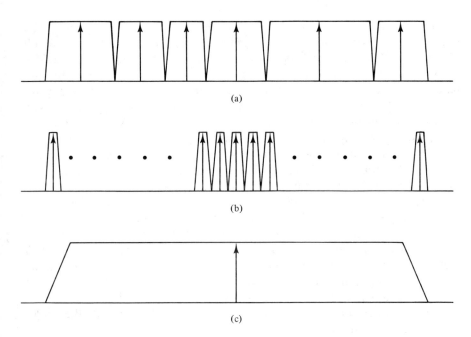

Figure 8-1 Transponder spectra: (a) MCPA/FDMA; (b) SCPC/FDMA; (c) TDMA (single carrier).

channel. The transmission can be either analog (using FM, or in some cases, AM SSB-SC) or digital (using PSK). Again, because of multiple carrier operation and the attendant impairments, the system designer must account for intermodulation and adjacent channel effects. The nonlinear impairments are controlled by operating the transponder at a point close to linear response.

Often a SCPC system is designed for burst operating mode, using voice-activated carriers. In such cases, individual carriers are turned off during the silence intervals between speech bursts in normal conversational telephony. In this case the carrier-to-interference ratio is a function of the number of individual carriers that are on at the same time in the same transponder. This type of system results in about a 4 dB overall power savings due simply to a reduction in the average power required to support only the channels that are active simultaneously. An improvement is obtained in the intermodulation distortion in addition to the power savings due to the on/off characteristics of speech. This feature is inherent when using AM SSB-SC modulation. The spectral utilization in SCPC is affected by the guard bands required by the multiple narrowband filters used to achieve channel separation.

The second basic system used in satellite multiple access is TDMA. As illustrated in Figure 8-1c, only a single carrier frequency is utilized, the bandwidth of which is usually equivalent to the full transponder bandwidth. The single bandwidth is time shared among many individual users. Using a single carrier occupying a full transponder bandwidth renders the usual impairments due to transponder nonlinearity essentially nonexistent. Therefore, a full transponder TDMA system may attain significant advantages over FDMA by operating the satellite channel at its full output power. Because the full bandwidth is occupied, no loss in spectral utilization is caused by any requirement for guard bands. Another way to get more flexibility into TDMA networks for smaller system applications is to build TDMA networks within an FDMA system operating in the same transponder bandwidth. Although the effectiveness of narrowband TDMA compared to the single carrier per transponder approach is eliminated in this case, network system requirements may sometimes favor the narrowband implementation.

As we will see in this chapter, the most important distinction among multiple access systems is that of single-carrier versus multiple carrier operation. This distinction is critical to the ultimate channel capacity of a particular multiple access method. In general, a single-carrier-per-transponder TDMA technique provides more capacity and higher flexibility compared to FDMA systems. In the material that follows, we discuss various engineering aspects of multiple access systems. This is followed by more detailed descriptions, together with the methods for calculating capacity for each of the major multiple access systems.

8.1 SYSTEMS ENGINEERING CONSIDERATIONS

In this chapter we describe the operation and application of several multiple access systems which are used in commercial satellite communications. A system designer is always faced with selecting the proper multiple access technique to meet the requirements of the communications services to be provided. In deciding which technique best fits the application, a number of factors must be considered. The factors that are normally used to evaluate the effectiveness of a multiple access technique for a particular application are:

> *Capacity:* The capacity of a multiple access system is usually defined in terms of the number of voice and/or data channels of a specified quality that can be accommodated using the power and bandwidth of a single transponder. Usually in selecting a system, the highest-capacity system is the most desirable. However, the system network requirements may lead to the choice of a system providing less total capacity, but higher cost-effectiveness.
>
> *RF power and bandwidth:* Power and bandwidth are the fundamental resources of the satellite RF link. The power and bandwidth available in a satellite communication system are directly reflected in its cost. To use the available power and bandwidth efficiently, a multiple access

system should be designed so that it is simultaneously bandwidth- and power-limited.

Interconnectivity: The network geometry for various communications services dictates the interconnectivity requirements. Simple point-to-point networks can often be served economically by other wideband transmission techniques, such as fiber optics. However, in a multinode geometry, the ability of a multiple access technique to provide interconnectivity among many users at various data rates and quality levels often makes satellites the most cost-effective method.

Adaptability to growth: Since the investment in multiple access equipment can be a significant portion of the ground system cost, the designers must consider the ability of the technique chosen to adapt to traffic growth and changes in the traffic patterns.

Accommodation of multiple services: Modern approaches to telecommunications rely heavily on digital techniques and multiservice transmission. The use of integrated services digital networks (ISDN) implies that multiple services, such as voice, data, and imagery applications, share the same transmission facilities. Multiple access systems must be designed to accommodate ISDN services.

Terrestrial interface: Interconnecting with existing terrestrial facilities that provide the "last mile" between an earth station and the user is extremely important to the overall economic and technical effectiveness of the multiple access system. As more digital interconnections become available, it becomes more attractive to employ all digital techniques.

Communication security: Although in the past most considerations of communication security have been relegated to military applications, modern commercial satellite communications systems must now face the problem of protecting confidential corporate and government data in a satellite communications environment that is vulnerable to unauthorized reception.

Cost-effectiveness: The cost per channel of implementing multiple access is an important consideration for systems engineers. Because of the dramatic development in digital techniques in recent years, their economic desirability continues to increase. However, analog techniques can be still more cost-effective in certain situations.

8.2 DEFINITIONS

There is no accepted notation for designating the various levels of signal processing within a multiple access system. To avoid confusion, we will adopt notation, using four levels of processing to specify each multiple access method (Figure 8-2). When read from left to right, this notation specifies the sequence

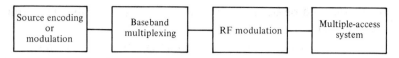

Figure 8-2 Notation chain. (Adapted from J. Puente and A. Werth, "Demand-Assigned Service for the INTELSAT Global Network," IEEE *Spectrum*, Jan. 1971.)

of processing from signal source to satellite link. The following abbreviations are used throughout this chapter to specify various processing combinations:

NBP	No baseband processing
ADC	Analog-to-digital coding (generalized)
PCM	Pulse-code modulation
ADPCM	Adaptive differential PCM
DM	Delta modulation
SSB	Single-sideband modulation
SCPC	Single-channel-per-carrier
MCPC	Multiple channels per carrier
TDM	Time-division multiplex
FDM	Frequency-division multiplex
PSK	Phase-shift keying
FM	Frequency modulation
FDMA	Frequency-division multiple access
TDMA	Time-division multiple access
CDMA	Code-division multiple access

In some cases, two or more of the levels may coalesce into a single process (such as when SSB source modulation of individual channels is the means of baseband multiplexing). In such cases, the designators of the coalesced processes will be enclosed in parentheses.

8.3 FDMA SYSTEMS

In this section we describe two generic types of FDMA systems. The first FDMA system type accommodates multiple channels per carrier (MCPC), and the second employs single-channel-per-carrier (SCPC) techniques. In each of the two classes of FDMA systems, we describe both analog and digital transmission techniques. For each case, the basic operation of the system is discussed, the methods used to calculate system capacity are outlined, and system performance is described, in both quantitative and qualitative terms.

8.3.1 (SSB/FDM)/FM/FDMA: Analog MCPC

The first multiple access technique to be employed in satellite communications was the analog MCPC system. It is designed for analog transmission and was primarily an outgrowth of the FDM hierarchy developed for terrestrial multiplex systems. Figure 8-3 shows a typical implementation of this system. Individual voiceband channels are first single-side-band modulated on frequency-division multiplex carriers to form FDM baseband assemblies as described in Chapter 7. The voiceband channels from the terrestrial telephony system are connected at a satellite earth station with FDM multiplex equipment which is used to create FDM baseband signals in accordance with a preassigned frequency plan as illustrated in the example of Figure 8-3. At each station, the FDM basebands are frequency modulated on preassigned carriers and transmitted through the satellite in a preassigned portion of the transponder bandwidth. Receiving stations demodulate each received carrier and, using FDM multiplex equipment, strip off those channels assigned to that particular station. In the example of Figure 8-3, a user located in the serving area of station A is assigned the appropriate frequency slot in the FDM baseband assembly. Since the user at location A is attempting to reach a correspondent at location F, his voiceband information is assigned to baseband group F constructed by the multiplex equipment. Other users attempting to establish a link with other locations are assigned to the appropriate groups within the 60-channel supergroup baseband. This composite baseband signal is then modulated on an FM carrier and transmitted through the transponder to all other stations in the network. At station F after demodulation only the 12-channel group F is extracted from the baseband, demultiplexed into voice channels, and interfaced with the local telephone system.

This method has for many years provided excellent voice quality and service, but tends to be inflexible in adapting to the redistribution of traffic demands. Also, because of its high per-channel hardware requirements, it does not become more cost-effective as the number of channels increases. Because it employs multiple-carrier operation, this system is subject to the penalties caused by the nonlinear performance of the transponder. Therefore, in multiple-access (multiple-carrier) operation, such a system does not always achieve the highest capacity. However, for high-density point-to-point links, it can achieve competitively high transponder capacities.

Figure 8-4 shows a block diagram of the organization of the earth station equipment used in this system. Interfacing with the terrestrial facilities requires both demultiplex and multiplex equipment to break down and reassemble the FDM basebands in accordance with the frequency plan. For each baseband an FM modulator operating at an intermediate frequency (IF) is provided, followed by up-converters to translate the frequency from IF to the RF frequency used on the uplink. Each individual RF carrier is then combined and provided to a high-power amplifier and ultimately to the an-

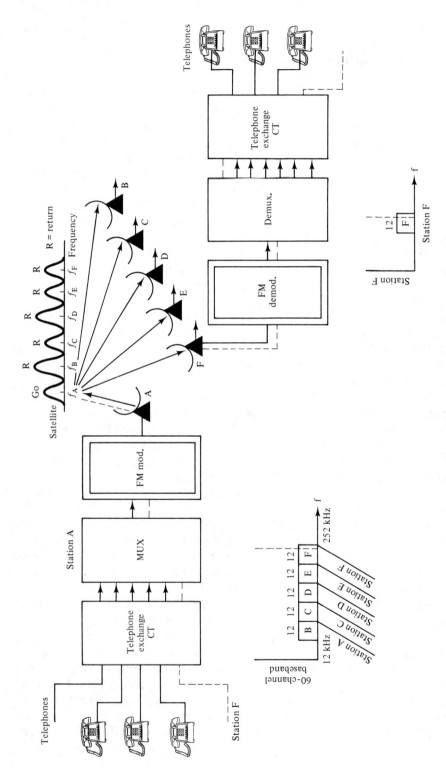

Figure 8-3 Preassigned multidestinational SSB/FDM/FDMA.

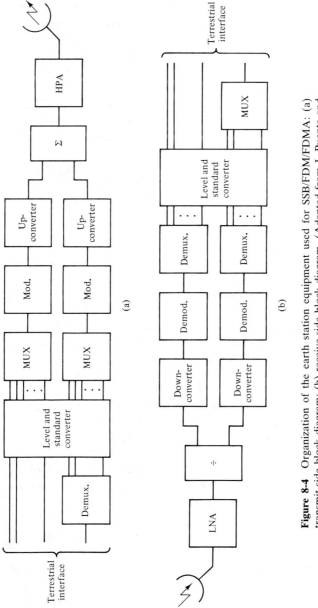

Figure 8-4 Organization of the earth station equipment used for SSB/FDM/FDMA: (a) transmit-side block diagram; (b) receive-side block diagram. (Adapted from J. Puente and A. Werth, "Demand-Assigned Service for the INTELSAT Global Network," IEEE *Spectrum*, Jan. 1971.)

tenna system. On the receiving side, downlink carriers are received, amplified, and channelized to the appropriate carrier bandwidths. Each carrier frequency is first down-converted to IF, and an FM demodulator, typically employing threshold extension techniques, is used to reconstruct the FDM baseband. The FDM baseband is then demultiplexed and reassembled to the proper configuration for interface to the terrestrial facilities.

Capacity calculations. In performing the systems engineering calculations required to determine the channel-carrying capability of multiple access systems, we must start by computing the total available carrier-to-noise density (C/N_0) in the satellite RF link. The RF calculations are performed as outlined in Chapter 6. The next step is to determine the required carrier-to-noise density ratio to achieve a specified performance or quality level in a single channel or a group of channels carrying voice, data, or video signals. The required C/N_0 is then compared with the available C/N_0 to determine channel capacity.

Specifically, in the case of (SSB/FDM)/FM/FDMA, the typical approach is to calculate capacity using an interactive procedure among four relationships. FM is the RF modulation used here, and the calculations employ the equations presented in Section 7.2.3. The basic approach is first to determine the RF carrier-to-noise density available in the RF channel for each carrier as well as the associated bandwidth of each carrier. The next step is to estimate the number of channels that can be multiplexed on that particular carrier. This is followed by calculating the resulting signal-to-noise ratio after detection using Eq. (7-22) for FDM-modulated FM carriers.

The CCIR has specified the standards for toll-quality voice transmission using FDM techniques. This universally accepted standard specifies that the worst-case noise (occurring in the highest-frequency channel in the FDM assembly) shall not exceed 10,000 pWOp. This means that the noise measured at the zero transmissions-level point (0TLP) using psophometric weighting must not exceed 10,000 pW. The test signal level used to make this measurement is a sinusoid located in the voiceband at a frequency of approximately 1 kHz at a power level of 1 mW at the 0TLP. Therefore, the test-tone signal-to-noise ratio in the worst-case FDM channel should not exceed

$$S/N = \frac{10^{-3} \text{ W}}{10{,}000 \times 10^{-12} \text{ W}} = 10^{-5} \text{ or } 50 \text{ dB} \qquad (8\text{-}1)$$

Using this signal-to-noise ratio as the standard of comparison, we can proceed to calculate the number of channels that can be accommodated with at least this quality level, using FDM/FM transmission. For each carrier, the procedure begins by estimating the number of channels. Then, using the procedure outlined in Chapter 7, we employ the CCIR multichannel loading factors to determine an equivalent rms test-tone deviation as well as a maximum baseband frequency. Using these values in Eq. (7-22) we calculate the signal-to-noise ratio expected to be achieved in the top channel. If the resulting S/N is less than 50 dB, our initial estimate of the number of channels is too high. If the calculated S/N exceeds 50 dB, the number of channels assumed is too small. The procedure, then, is, after each assumption, to calculate the S/N and then increase or decrease the estimate of the number of channels per carrier in a successive approximation fashion until we arrive at the maximum number of channels that result in a test-tone-to-noise ratio in the worst-case channel of 50 dB. This calculation must be made for each of the carriers in the transponder bandwidth. The transponder capacity is then the sum of the capacities of the individual carriers.

8.3.2 ADC/TDM/PSK/FDMA: Digital MCPC

The second type of MCPC system employed in commercial satellite communications, digital MCPC, is used for the transmission of digitally encoded baseband signals. The baseband information for each carrier typically consists of multichannel PCM-TDM bit streams. In North America these signals are constructed using the so-called T-carrier hierarchy. In Western Europe the CEPT hierarchy is employed. The first level in the North American hierarchy, DS1, assembles 24 64-kb/s channels at a bit rate of 1.544 Mb/s. The first level in the European hierarchy combines 32 64-kb/s channels at a bit rate of 2.048 Mb/s. These multiplexed signals are modulated on digital carriers, typically employing four-phase coherent PSK. Such systems are of practical interest since they are compatible with FDM/FDMA carriers sharing the same transponder. The operational requirements are similar to those used in analog FDM/FM transmission, requiring no network clock synchronization and only the rather simple frequency coordination typical of FDMA systems. The use of digital time-division-multiplex baseband permits the potential use of digital speech processing to provide a significant enhancement of voice-channel capacity by taking advantage of silence intervals in multichannel speech telephony using speech interpolation techniques. The digital baseband coding of individual channels may use one of several techniques. Although the predominant technique in use is PCM, for which well-developed international standards exist, variations of delta modulation have also been used in this application. Most recently, the use of adaptive differential PCM (ADPCM) reduces the standard voice coding rate from 64 kb/s to 32 kb/s without significant reduction of quality. The exact configuration employed depends on the traffic requirements.

Capacity calculations. To calculate the capacity of this system, we start, as usual, with the calculation of the available C/N_0 in the RF link. Knowing the available C/N_0 and the bandwidth of each carrier, we proceed to calculate the capacity of the system using the theory provided in Chapter

7. The bit rate required to support each channel depends on the analog-to-digital coding method used. For example, if PCM is employed, 64 kb/s per channel is required. If ADPCM is employed, 32 kb/s is sufficient. With PCM 24 channels can be multiplexed at a 1.544-Mb/s rate. With ADPCM twice as many channels can be multiplexed at the same bit rate. With a bit rate of 1.544 Mb/s, the keying rate of a four-phase PSK modulator will be exactly one-half of 1.544 Mb/s, or 772 kHz. The noise bandwidth required for the modulated signal is typically 1.2 times the keying rate or 926 kHz. In addition, guardband requirements between adjacent carriers consume an additional 20% of bandwidth. Therefore, using the 1.544-Mb/s example, this carrier can be accommodated in a channel spacing of 1111 kHz. To determine the power level required, the error rate at the threshold of acceptance must be specified. Typically in digital transmission, a threshold error rate of approximately 10^{-4} is assumed. As specified in Chapter 7, the bit error rate is related to the bit energy-to-noise density ratio in accordance with Eq. (7-35).

To calculate the required carrier-to-noise ratio to support each carrier, we employ Eq. (7-33), and rewrite it as a sum of terms expressed in decibels:

$$(C/N)_t = (E_b/N_0)_t - B_N + R + M_I + M_A \qquad (8\text{-}2)$$

where $(C/N)_t$ is the carrier-to-noise ratio at the threshold error rate, $(E_b/N_0)_t$ the bit energy-to-noise ratio at the threshold error rate, B_N the noise bandwidth associated with this carrier, R the data rate of the digital signal, M_I a margin associated with the implementation of the modem, and M_A a margin for adjacent channel interference.

This result is converted to carrier-to-noise density using the following relation:

$$(C/N_0)_t = (C/N)_t + 10 \log B_N \qquad (8\text{-}3)$$

$(C/N_0)_t$ must then be compared with the total available carrier-to-noise density for the entire transponder. Noting that the sum of the carrier-to-noise densities required to support each individual carrier cannot exceed the total available carrier-to-noise density, the channel capacity can be calculated using a successive-approximation procedure similar to that used in the FDM/FDMA case. This procedure will determine the power-limited capacity of the system. The final step is to determine the bandwidth-limited capacity by summing the bandwidths of all individual carriers. The true capacity of the system is the maximum number of channels at which the system is simultaneously power- and bandwidth-limited.

8.3.3 ADC/SCPC/PSK/FDMA: Digital SCPC

Another important class of FDMA systems employs SCPC techniques wherein each voice and/or data channel is modulated on a separate radio-frequency carrier. No multiplexing is involved except within the transponder bandwidth, where frequency division is used to channelize individual carriers, each supporting the information from a single channel, as illustrated in Figure 8-1. Figure 8-5 depicts a typical organization of a SCPC system. The terrestrial system connects to the earth station SCPC equipment on a channel-by-channel basis. Associated with each incoming signal is a channel unit, which contains all the equipment required to convert the voiceband or digital data signal into a PSK-modulated RF carrier for transmission over the satellite channel using only that station's assigned part of the transponder bandwidth. To establish a conversation between two locations, a pair of channel frequencies is selected, one for each direction of transmission. On the receive side, the channel unit associated with each radio-frequency carrier contains all the equipment required to demodulate the radio-frequency carrier and deliver either a voiceband signal or a digital data signal to the terrestrial end links.

The carrier frequencies in the satellite transponder may either be preassigned to individual channel units, and used exclusively by that channel unit, or they may be demand assigned. In demand assignment, neither end of a channel is permanently associated with a particular carrier frequency, and the channels are paired to form a connection as required on a demand basis. Each of the carrier frequencies within the satellite transponder bandwidth becomes part of a pool of available frequencies that may be assigned to any channel unit as required. The first fully demand-assigned SCPC system was the SPADE system, developed for use by Intelsat in the late 1960s [J.G. Puente and A.M. Werth, "Demand Assigned Service for the Intelsat Global Network," *IEEE Spectrum*, January 1971]. Outgrowths of this basic system have included many types of SCPC implementations, using preassignment and demand assignment, as well as analog and digital transmission techniques.

Voice activation. One important characteristic of SCPC systems is the ability to employ voice-activated carriers. This means that the radio-frequency carrier is turned on (and therefore consumes power) only when active speech is present in that channel. In normal conversational telephony, one speaker is talking while the other is listening. Normal hesitations and punctuating silences create a condition that has been extensively studied by speech researchers for many years [P.T. Brady, "A Statistical Analysis of On-off Patterns in Sixteen Conversations," *Bell System Technical Journal*, Vol. 47, 1968, pp. 73–92]. It has been determined that the average single talker's speech activity consumes only 40% of the total available channel time. Therefore, by turning the individual carrier off during silence intervals, an SCPC system can save approximately 4 dB of satellite power, and thereby accommodate a proportionally larger number of carriers in a single transponder. For example, a 36-MHz bandwidth transponder can support 800 SCPC channels using a channel spacing of 45 kHz. We can model this ensemble of channels as a sequence of Bernoulli trials and apply the binomial distribution to estimate the potential gain. For n independent channels, each

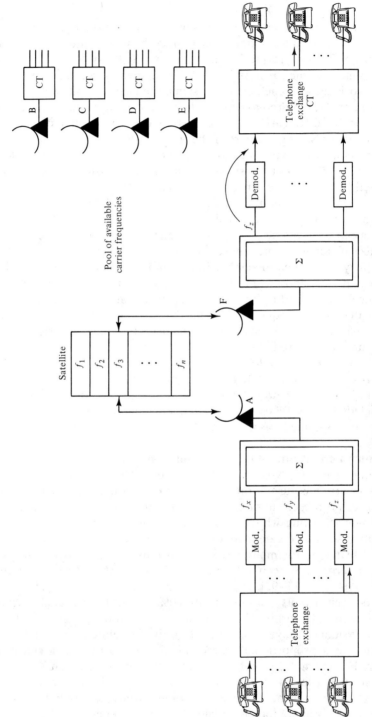

Figure 8-5 Signal flow in SCPC system.

Sec. 8.3 FDMA Systems

with a probability (or activity) equal to α, the probability that j or more are active (carriers on) at any instant is given by

$$P(n, j, \alpha) = \sum_{i=j}^{n} \binom{n}{i} \alpha^i (1 - \alpha)^{n-i} \qquad 0 \leq j \leq n \qquad (8\text{-}4)$$

With 800 independent carriers and assuming a channel activity level of 0.38, during a fully loaded condition with 400 trunks "off hook" (400 conversations underway), the probability that more than 320 of these channels will contain active speech simultaneously is less than 0.1. Therefore, we may reasonably consider the voice-activation advantage in this case to be the ratio of 800/320, or 2.5. This converts into a power saving of 4 dB. Note that the probability is less than 0.1 that *more* than 320 trunks will be active, even during the peak busy hour when all channels are in use. The improvement, of course, is available with any SCPC implementation, whether analog or digital, as long as there is a large number of channels in the transponder. Note also that the improvement applies only for two-way voice communications. If data channels become a significant portion of the total SCPC traffic, the voice-activation advantage diminishes because data channels are necessarily full-time channels without the redundancy and silence intervals typical of speech conversations. If (SSB/FDM/SSB)/FDMA or (SSB/SCPC/SSB)/FDMA operation is used, the benefits of the activity level advantage accrue without requiring actual voice actuation of carriers (since there are really no true carriers).

Digital SCPC channel unit. Figure 8-6 shows a typical implementation of a digital SCPC channel unit. The baseband interface to this channel unit can either be a voiceband signal or a digital data input. In the case of voiceband signals, the digital codec using PCM or ADPCM is typically employed. A digital speech detector is used to determine the presence of active speech and to provide a signal to turn on the RF carrier only during active speech intervals. This voice-activated carrier approach implies a burst-mode transmission that requires channel synchronization on a burst-by-burst basis. Therefore, to each speech burst, overhead information must be added to allow the modem to recover carrier, and the codec to synchronize and determine the start of the active speech message. The digitally encoded voiceband signal is provided to a PSK modem, which modulates a four-phase PSK carrier for transmission over the satellite and performs the inverse demodulation operation. In the case of digital data inputs, a proper digital interface is provided together with error correction channel encoding to improve the error rate without increasing the bit energy-to-noise density ratio. In the case of preassigned carriers, the modem is provided a fixed carrier frequency from a local oscillator. In the case of demand assignment operation, a frequency synthesizer is required to create the SCPC channel frequency used during that conversation. The digitally modulated PSK carriers are then combined

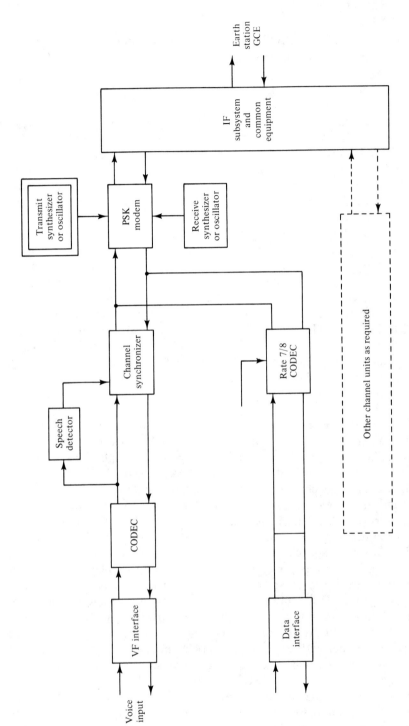

Figure 8-6 Digital SCPC channel unit.

Sec. 8.3 FDMA Systems   257

in an IF subsystem and transmitted to standard earth station up- and down-converter chains.

Capacity calculations. The calculation of the system capacity for digital SCPC follows a procedure similar to that for used digital MCPC systems. Normally, the voice coding rate and the threshold error rate are established by the system requirements. For example, if the SCPC system uses a 64-kb/s-per-channel coding rate, and a threshold error rate of 10^{-4}, we can proceed to calculate the required carrier-to-noise density ratio using Eq. (8-2). Assuming four-phase PSK modulation, the noise bandwidth will be approximately 1.2 times the symbol rate, and the channel spacing 1.2 times the noise bandwidth. Typical factors for the modem implementation margin (1.5 to 2 dB), and adjacent channel interference (0.5 dB) are also used. We then calculate the carrier-to-noise ratio and convert it to carrier-to-noise density ratio using Eq. (8-3). This required carrier-to-noise density ratio per channel is then compared with the overall available carrier-to-noise density ratio. Because the bandwidth of each carrier is identical, we determine the number of channels that can be supported in the transponder bandwidth, using the formula

$$\phi_B = \frac{B_T}{B_C} \qquad (8\text{-}5)$$

where B_T is the transponder bandwidth and B_C is the individual channel bandwidth.

Next, the voice-activation advantage can be used to increase the power-limited capacity by a factor of 2.5 (4 dB). This voice-activated power-limited capacity is then compared to the bandwidth-limited capacity, which is calculated using Eq. (8-5). The true capacity is bounded by the lesser of the bandwidth- and power-limited capacities.

8.3.4 NBP/SCPC/FM/FDMA: Analog SCPC

Analog transmissions can also be accommodated in a SCPC implementation. The basic analog SCPC system is the same as that shown in Figure 8-5. The analog SCPC channel unit, however, uses frequency modulation, as illustrated in Figure 8-7. On the transmit side, the voiceband signal is provided to an input circuit which limits the peaks of the signal to set the peak FM deviation. This is followed by a VF filter to limit the bandwidth of the baseband signal. An analog speech detector is employed to provide voice-activated carrier operation similar to that used in digital SCPC. The delay unit provides time for the speech detector to determine the difference between active speech and silence. FM/SCPC normally employs a compressor/expandor (compandor) of a syllabic type to improve the relative subjective voice quality for given value of signal-to-noise ratio. Preemphasis and deemphasis networks

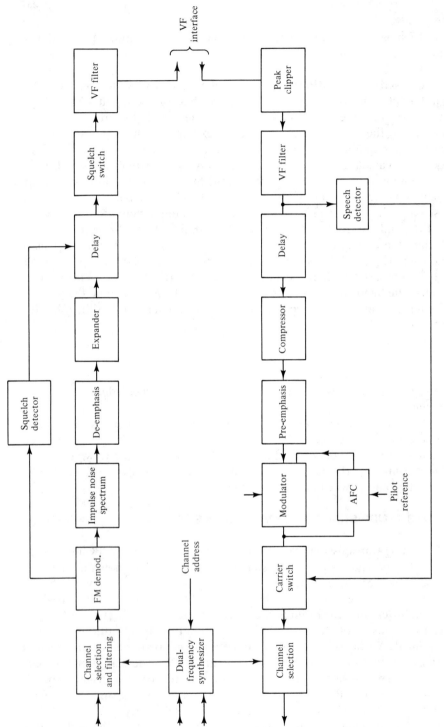

Figure 8-7 FM SCPC channel unit block program.

Sec. 8.3 FDMA Systems

are used to account for the parabolic noise spectrum of FM as described in Chapter 7.

Automatic frequency control (AFC). Another important aspect of SCPC systems is the use of automatic frequency control (AFC) to solve the spectrum centering problem and to minimize adjacent channel interference between SCPC channels. The problem is important because of the narrow bandwidth of each individual SCPC channel, compared with the total transponder bandwidth. It is compounded by the number of channels and the potential for adjacent channel interference. Therefore, an AFC system is employed by all SCPC implementations to control the spectrum centering on an individual channel-by-channel basis. AFC is typically accomplished from a system viewpoint by transmitting a pilot tone, located in the center of the transponder bandwidth. This pilot tone is transmitted by a reference station. All other stations in the network receive the pilot and slave their AFC system to it. The AFC system then controls the frequency of the individual carriers by locking the local oscillators (or frequency synthesizers in the case of demand-assignment systems) to the AFC control system.

Capacity calculations. Calculation of the channel capacity of FM/SCPC systems employs the FM performance equation (7-10). Notice that this equation employs the familiar factor of 3 in the FM improvement, since we are dealing with a single-channel modulating baseband as opposed to an FDM baseband. The first step in the capacity calculation is to ensure that the performance requirements for commercial toll quality using FM transmission are met. That is, the noise must be limited in the VF channel to 10,000 pWOp, corresponding to a 50-dB test-tone-to-noise ratio. In these cases it is always assumed that the FM system is operating above the FM threshold. This typically implies that the minimum required carrier-to-noise ratio per channel is at least 10 dB. The hardware (modem) implementation margin and adjacent channel interference margin typically add 2 to 2.5 dB to the carrier-to-noise ratio required for a single channel.

We can proceed in a manner similar to that employed in digital SCPC calculations by first comparing the required carrier-to-noise density ratio per channel to total available carrier-to-noise density, as long as the voice-frequency parameters chosen provide a signal-to-noise ratio (test-tone-to-noise ratio) equal to at least 50 dB. If the signal-to-noise ratio is in excess of 50 dB, it may be possible to achieve a higher channel capacity by reducing the FM deviation, thereby reducing the FM bandwidth, and achieve more channels in the same transponder bandwidth. Assuming that we trade power and bandwidth in the FM system in a manner that provides slightly more than a 50-dB test-tone-to-noise ratio, the power-limited capacity can be computed by comparing the carrier-to-noise density required per channel to the carrier-to-noise density available overall in the full transponder bandwidth. The next step is to increase this number of carriers by the voice-activity advantage of 2.5, thus determining the power-limited capacity. This value must be compared with the bandwidth-limited capacity computed by calculating the ratio of the transponder bandwidth to the bandwidth per channel. Again, the true capacity is the lesser of the power- or bandwidth-limited capacities.

8.4 TDMA SYSTEMS

In this section, time-division multiple access (TDMA) systems, which are in use in commercial satellite applications, are described. The first system type is the classic TDMA implementation employing a single modulated carrier occupying the full transponder bandwidth. This type of system is the most common for TDMA networks and is also the most efficient from a capacity standpoint. In this section we also describe briefly another class of TDMA system using only a fraction of the transponder bandwidth. It may be used for smaller TDMA networks which share the transponder bandwidth with other carriers in an FDMA configuration.

8.4.1 ADC/TDM/PSK/TDMA: Full Transponder TDMA

The basic concept of TDMA is illustrated in Figure 8-8. Several stations in the network use a single carrier frequency whose bandwidth occupies the full transponder. This carrier is time-shared to allow each station to transmit its information, with digital modulation, using synchronized bursts. That is, a station will receive information from a continuous source, compress it into a short time interval, and transmit it within a high-speed burst at the correct time so that bursts from all stations arrive at the satellite sequentially interleaved without interburst interference. All bursts received from all stations are then retransmitted from the satellite to all stations. Synchronization is achieved by defining a reference station whose timing information and burst position are used as a reference by all other stations in the network to time their transmissions.

To properly control the interleaving of bursts from multiple earth stations a TDMA system uses a frame organization. An example of this frame structure is illustrated in Figure 8-9. A frame usually begins with reference bursts transmitted by the primary reference station and a redundant secondary reference station used for backup. The two reference bursts are then followed by information-carrying bursts transmitted sequentially from each station in the network. The frame ends when the transmission from the last station is complete. A new frame then begins with the transmission of reference bursts followed by traffic from each station in the network. The frame time interval, T_F, is typically a few milliseconds.

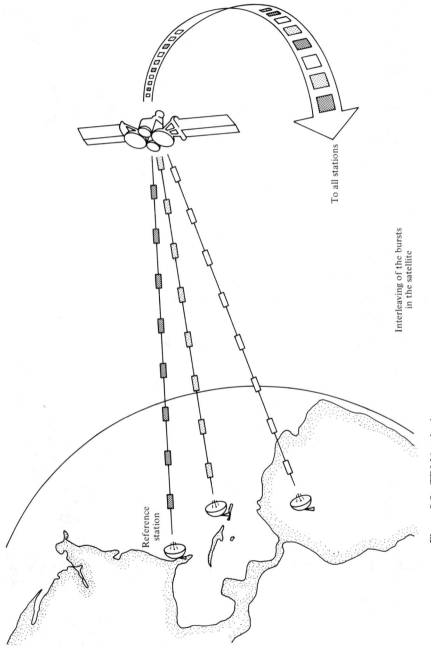

Figure 8-8 TDMA—basic concept.

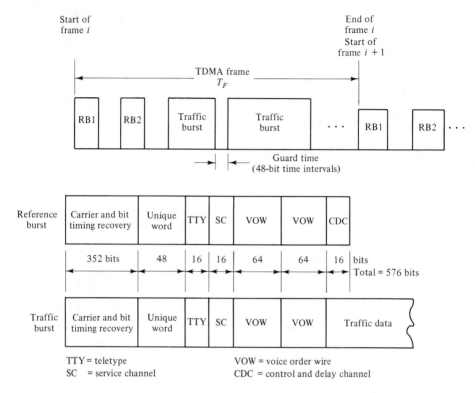

Figure 8-9 TDMA frame and burst structure: TTY, teletype; SC, service channel; VOW, voice order wire; CDC, control and delay channel.

The structure of a TDMA burst provides insight into the way the system operates. For example, as shown in Figure 8-9, each burst consists of *overhead information* and *traffic information*. The overhead data are used for system implementation and control, and the traffic data are the "useful" or revenue-generating part of the burst. The overhead portion of the burst is usually referred to as the *preamble*. A *reference burst* consists only of this preamble. The preamble begins with the transmission of a predetermined digital bit pattern used by the high-speed PSK burst demodulator to recover carrier and acquire bit timing for each burst. This sequence of bits is transmitted at the beginning of each burst. As shown in the example of Figure 8-9, 176 symbols (corresponding to 352 bits in a four-phase PSK system) are used to "train" the modem to recover carrier and bit timing even at relatively low carrier-to-noise ratios. Typically, carrier and bit timing recovery is accomplished well before the end of this sequence of bits. The next portion of the preamble is a sequence of 48 bits, constituting a unique word chosen for its correlation properties. This unique word is essentially a frame synchronization word, for which the receive side of the TDMA terminal searches as soon as the modem has achieved lock. The unique word has a high probability of correct detection and a low probability of false detection. As soon as the system recognizes this unique word, it updates its timing counters relative to the beginning of the frame and its position in the frame. The next elements of the preamble contain service-oriented information, including a teletypewriter channel for system network control and a service channel into which bit patterns are inserted for use in performing error analysis while the system is in service. A channel for a digital voice order wire is also provided in this portion of the preamble. A control information channel, called a *control and delay channel*, is also inserted within the reference burst for use by the reference station to transfer information on acquisition, synchronization, and system control and monitoring to other stations in the network. A traffic-carrying burst uses the same preamble information as a reference burst, with the exception of the control and delay channel. Following the preamble, traffic data consisting of voice, data, and perhaps video information, multiplexed in the time domain, is added to the burst and the entire burst is transmitted at the appropriate time within the frame. Between bursts a guard time interval is provided to minimize the probability of burst overlap.

A block diagram of a typical TDMA system is shown in Figure 8-10. Interfaces for various kinds of signals are provided between the TDMA terminal equipment and the terrestrial telecommunications system. Information signals typically include voice, voiceband data, direct digital data or imagery in the form of facsimile or television signals. A specific interface is provided for each type of signal. For example, in the case of voice signals, the function of the interface is to encode the incoming voice signals digitally and to multiplex a number of channels together using TDM. In the case of video information, analog-to-digital conversion is usually required. For each of the interface modules, a data compression and expansion function must be provided to create subbursts on the transmit side, and to convert subbursts into continuous data streams on the receive side. In each module, a subburst is formed using two buffer memories operated in a "ping-pong" mode. The continuous digital data stream is written into buffer memory A at the natural data rate of the signal. At the same time, data previously written into memory B are read at a high rate in a short period of time corresponding to the subburst length. Each memory is alternately read and written in a ping-pong mode. This process is reversed on the receive side, thereby creating continuous data streams from the received subbursts. Each of the interface modules is next sampled by a channel multiplexer that combines the subbursts from the outputs of each of the interface modules to form the complete traffic burst. The multiplexer then adds the appropriate preamble information to each burst and provides it to a scrambler. The scrambler is applied only to the information-bearing part of the burst and is used to prevent patterns that may occur naturally in the data stream from creating strong spectral components in the modulated signal.

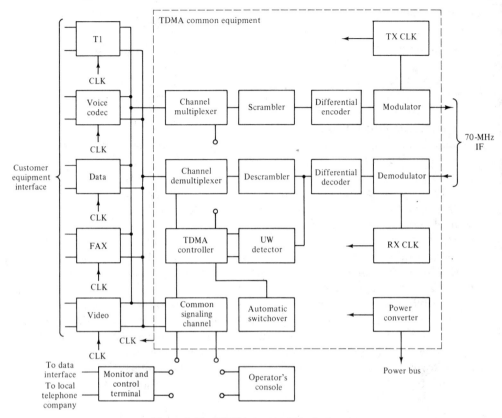

Figure 8-10 TDMA terminal block diagram.

Each burst is then provided to a differential encoder and four-phase PSK modulator, whose output is a 70-MHz IF signal providing bursts of RF energy at the proper time for transmission through the satellite transponder. All the multiplexing hierarchy, including the burst and subburst length and content, as well as the burst transmission time, are under the control of the TDMA controller, which can be accessed either via the operator's console or remotely via a separate monitor and control system. On the receive side, RF bursts are received and demodulated on a burst-by-burst basis and provided to a unique word detector which synchronizes the TDMA controller. The preamble information is stripped off and utilized by the TDMA controller for additional network control purposes. The descrambler is applied to the traffic data, which are then demultiplexed and sent to the appropriate interface modules after reconstruction from burst mode into continuous serial form.

Burst synchronization. One of the principal problems in the design of a TDMA system is synchronizing the bursts from many users in a TDMA network into an organized frame where the bursts are packed as closely as possible into the frame without colliding. There are two steps in the synchronization process. The first is the acquisition phase, which refers to the process by which a TDMA earth station enters the network. The second is the synchronization phase, which occurs after the user has entered the TDMA frame and must maintain accurate positioning of the burst within the frame during operation. As long as the TDMA system operates within a single transponder and the same antenna beam, the problem of network synchronization is simplified by each user's ability to receive the bursts from all users in the network. This means that a feedback loop can be established through the satellite back to each individual user, employing bursts received from every station in the network.

During the startup or acquisition phase, the reference burst is the first to be transmitted. Since no other bursts exist within the frame, the reference burst position is a free choice. Each additional burst from other stations enters the system by first synchronizing to the reference burst to establish a local timing reference. The next step for a new entrant is to transmit an abbreviated burst, consisting only of the preamble at a time following the reception of the reference burst, which will result in the arrival of the new burst at a time approximating its desired location within the frame. Initially, this time delay is a coarse estimate which may be determined in several possible ways. One way is to transmit a low-power-level burst, which is used to search for the proper location without significant interference with any other bursts. Another desirable method is to determine the initial value of time delay by computing it with a priori knowledge of the exact location of the earth station and the distance between the earth station and the satellite. This method may be referred to as an open-loop initial acquisition phase. The newly acquiring station observes the position of its burst within the frame during each frame, measures the error between the burst's actual location and its desired location, and refines the estimate of the time delay on each succeeding frame. When the error between the actual and desired location is sufficiently small, the acquisition phase is complete and the synchronization phase can begin with the transmission of the full-length burst, including the traffic data. This is followed by the initiation of a closed-loop synchronization process, whereby the error in the burst position is continuously measured and the burst position time delay is continuously refined.

In those cases where the TDMA system employs transponder hopping or multibeam operation, the closed loop through the satellite does not exist since the bursts from all other users are not available to each user. Other synchronization methods may be employed, including open-loop synchronization, which depend on accurate computation of the time delay through knowledge of satellite and earth station coordinate positions as well as the variations in satellite movement with time. Another method, known as *cooperative feedback*, can also be employed, where the satellite positioning information and the real-time variations in it are actually communicated to stations attempting the acquisition and synchronization process through the control and delay channel of the reference burst. These methods are more complicated because a TDMA user may only see a small number of the total bursts in the TDMA frame. These techniques are discussed in more detail in Feher (1981).

Frame efficiency. The calculation of the system capacity for a TDMA network is dependent on a figure of merit known as the *frame efficiency*. The frame efficiency is defined as the ratio of the number of bits available for carrying revenue-generating traffic to the total number of bits in the frame. It is straightforward to develop the following relationship:

$$\eta = \frac{R_T T_F - K b_p - n_r b_r - (n_r + K) b_g}{R_T T_F} \qquad (8\text{-}6)$$

where η = frame efficiency
T_F = frame time, seconds
R_T = total TDMA bit rate, b/s
b_g = number of bit positions used for guard time
b_p = number of bits in preamble of traffic burst
b_r = number of bits in reference burst
K = number of traffic earth stations in the network
n_r = number of reference earth stations

Notice that the frame efficiency is extremely sensitive to the frame time and to the number of TDMA users in the network. Early TDMA systems used rather short frame times because of the unavailability of large-capacity memory devices needed to create long frames. Modern systems use much longer frame intervals, in the neighborhood of several milliseconds, to achieve frame efficiencies on the order of 95% while servicing a typical TDMA network consisting of 15 to 20 users.

Capacity calculations. To determine the TDMA channel capacity the first step is to calculate the carrier-to-noise ratio required to achieve the threshold error rate. Again we use Eq. (8-2) to determine this value. As long as the total available carrier-to-noise ratio is somewhat higher than that required to achieve the quality or service desired, the TDMA system will operate satisfactorily at the data rate selected. If the carrier-to-noise ratio is not high enough, the TDMA bit rate must be reduced sufficiently to achieve the required carrier-to-noise ratio. Assuming that the chosen bit rate leads to a carrier-to-noise density ratio requirement which is not greater than that available, we may then proceed to calculate the TDMA capacity using the following approach. The voice-channel capacity of a TDMA system may be computed as a function of the number of accesses (or earth stations in the network) by computing a ratio of the information bit rate to the equivalent voice-channel bit rate.

Sec. 8.4 TDMA Systems

We use the following method. Let the total available TDMA bit rate, R_T, be given by

$$R_T = \frac{b_T}{T_F} \quad (8\text{-}7)$$

where b_T is the total number of bits in a TDMA frame. Let the preamble bit rate, R_p, be

$$R_p = \frac{b_p}{T_F} \quad (8\text{-}8)$$

the reference burst bit rate, R_r, be

$$R_r = \frac{b_r}{T_F} \quad (8\text{-}9)$$

and the guard time bit value be R_g. Then

$$R_g = \frac{b_g}{T_F} \quad (8\text{-}10)$$

The available bit rate for revenue-generating traffic, R_i, is given by

$$R_i = R_T - n_r(R_r + R_g) - K(R_p + R_g) \quad (8\text{-}11)$$

The equivalent voice-channel capacity is then

$$\chi = \frac{R_i}{R_c} = \frac{R_T}{R_c} - \frac{n_r(R_r + R_g)}{R_c} - \frac{K(R_p + R_g)}{R_c} \quad (8\text{-}12)$$

where R_c is the equivalent voice-channel bit rate.

8.4.2 Narrowband TDMA

TDMA signals are sometimes transmitted within a subband of the total transponder bandwidth. As illustrated in Figure 8-11, a single transponder may be employed to provide multiple services (e.g., video, SCPC, and TDMA) in an FDMA configuration. Of course, this TDMA application does not enjoy

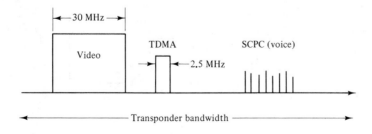

Figure 8-11 Narrowband TDMA in a multiservice transponder organization.

the usual single-carrier advantage of no intermodulation, but it can share resources with other multiple access systems. An advantage of this approach is found in networking applications that do not require the complete resources of a full transponder and thousands of channels. Using this narrowband TDMA, the requirements of smaller networks, such as those employed for corporate communications or regional services, can be met. This application still enjoys the flexibility and interconnectivity provided with TDMA, as well as its excellent suitability for digital transmission. Data rates for this type of system are typically in the range 1.544 Mb/s (DS1) to 6.312 Mb/s (DS2). This is to be contrasted with full transponder TDMA operating at 60 Mb/s in a 36-MHz transponder or 120 Mb/s in a 72-MHz transponder. A block diagram of a typical narrowband TDMA system is illustrated in Figure 8-12. This version of TDMA has the ability to accommodate simultaneously voice services from a PBX, low-speed freeze frame, video, or graphics communications, audio

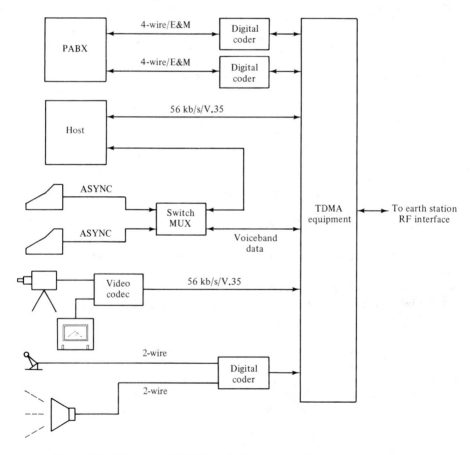

Figure 8-12 Narrowband TDMA typical user connections.

teleconferencing, and a wide range of low- to medium-speed data communications.

8.4.3 Demand Assignment

Demand assignment in a TDMA system amounts to a reassignment of capacity through reorganization of subbursts within the TDMA frame. That is, during periods of peak traffic, the total capacity of the TDMA system can be divided so that the bursts from the heavy-traffic stations are expanded, while the lighter-traffic routes are assigned shorter bursts. Such a reorganization of the burst-time plan can be implemented in several ways. The simplest uses a manual approach in which the system operator reconfigures the network plan through the operating console. This is an unsophisticated method which is typically controlled from a central location with the remote sites slaving their burst-time plans to the central location. The next higher level of sophistication uses a semiautomatic system which accomplishes demand assignment through the use of "canned" or stored burst-time plans, designed to optimize the distribution of capacity for various network conditions. Such plans are developed with a priori knowledge of the network requirements and its changing traffic conditions during normal time cycles. In such a system, the demand-assignment system can be implemented under simple operator control, or it can be implemented based on a time-of-day clock which switches between various burst plans matched to the typical traffic conditions as they change during a 24-h period. Yet a third level of sophistication consists of a fully automatic system, using complex demand-assignment algorithms which have the capacity to reconfigure the burst-time plans instantaneously, based on the instantaneous traffic distributions. Clearly, the more sophisticated the system, the more computing power that is required in the TDMA terminal. This additional sophistication affects cost, reliability, and maintenance requirements of the system. Probably the most cost-effective is the semiautomatic method, which employs a fixed set of canned plans. In most cases this approach solves the problem and approaches closely the ideal distribution of capacity at any time of day.

8.5 BEAM SWITCHING AND SATELLITE-SWITCHED TDMA

Modern communication satellites are typically designed with several spot antenna beams providing service to different regions on the earth's surface. Each beam has associated transponder receivers and transmitters, and the interconnections between receivers and transmitters are switchable. Such satellites are typically fitted with a network of RF switches that can be commanded from the ground to establish the required channel connections. Rapid electronic reconfiguration is provided by the switching system to maximize the traffic flow. In such systems it is usually possible for a station in any beam to communicate with stations in all the other beams. Either FDMA or TDMA may be used. Utilization of TDMA has a particular advantage in that it permits the use of a satellite switch which selectively connects individual up beams to individual down beams. A typical configuration of the satellite-switched TDMA network is illustrated in Figure 8-13. Notice that transmitting stations can send bursts to any station that operates through the satellite by simply tagging that burst and addressing the proper location. The on-board switching matrix is used to sort the bursts and direct them to the proper earth terminal, thereby expanding the dimensional potential of the network.

8.6 CODE-DIVISION MULTIPLE ACCESS

In code division multiple access (CDMA) operation, several stations use the same carrier frequency and associated bandwidth at the same time. This seemingly-paradoxical activity utilizes a technique which lies within the broad area of *spread-spectrum communications*. Its application is essentially limited to digital transmissions. [Pickholtz et al.; Cook and Marsh]

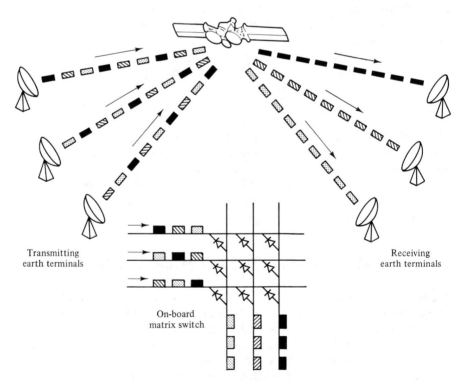

Figure 8-13 Satellite-switched TDMA.

Sec. 8.6 Code-Division Multiple Access

In CDMA, each bit of the digital "message" (for example, an ADC/PCM bit stream) is transmitted as a sequence of bits. This occurs as the original message bit stream is convolved with a predetermined code sequence at a somewhat higher bit rate. The bandwidth required for transmission is thus greater than would have been required for direct transmission of the message. However, through advance knowledge of the encoding sequence, the receiver is capable of reconstructing the message under an extremely adverse signal-to-noise ratio. If several stations transmit simultaneously in this mode, using different encoding sequences, then to any receiver all the undesired signals appear to be merely noise components within the tolerable noise power budget.

To avoid confusion in discussion between the bits of the message and the bits of the transmitted signal, the latter are described as *chips*. The rate at which they are transmitted is called the *chip rate*, or sometimes *chipping rate*.

The arrangement shown in Figure 8-14, while not generally employed in real CDMA systems, demonstrates the principle of recovery of the original message from the received "chip stream." When the chip sequence for a message bit "1" is aligned within the delay line, the summer produces an output of +5 units. When the chip sequence for a message bit "0" is aligned, the summer produces an output of −5 units. These outputs are recognized as signifying a received message bit "1" or "0", respectively. At other times, the output of the summer will be some value less in magnitude than 5. The output is ignored at such times.

This delay line arrangement is actually a special type of filter whose behavior can be more easily described in the time domain than in the customary frequency and phase domains. Because its action is complementary to the encoding function, it satisfies the classical criterion for "matched filter" reception, giving optimum performance in the presence of uniform density noise.

Because the chip energy of noise, or other signals cohabiting the same band, are combined incoherently by the summer, while the energy in the chips of the desired message is combined coherently, the simple receiving system exhibits an apparent enhancement in signal-to-noise ratio, called *processing gain*, of

$$10 \log [\text{number of chips per bit}].$$

Of course, without this benefit reception would be impossible, since with several stations active the apparent ratio of signal-to-noise-plus-other-signals would be far less than unity (negative in dB).

As a consequence of the processing gain relationship, the number of stations which can be accommodated in the same band by CDMA techniques is largely determined by the number of chips per bit. Another important factor is the degree of equality between the received power levels from the

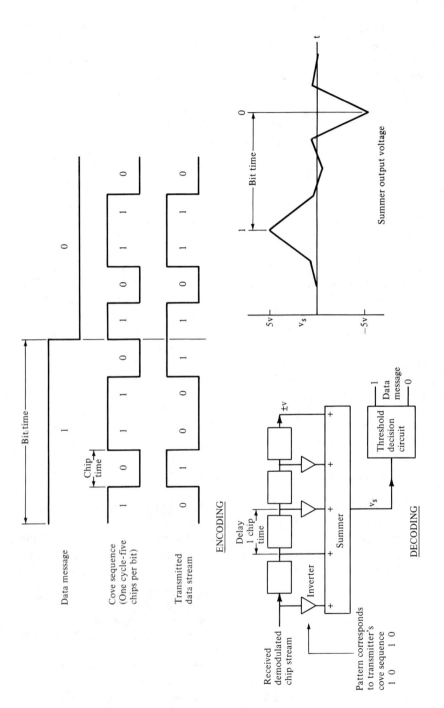

Figure 8-14 Principle of code-division multiplex.

different stations. If substantial differences exist, the ratio of the weaker signal-to-noise-plus-other-signals may be too small, even with the benefit of the processing gain, to allow extraction of the data message with adequately-small bit error rate.

In practice, the code sequences are usually far longer than five chips. In fact, the sequence may be extremely long, and each bit of the message may be encoded by just a part of it. A different decoding technique than our delay line arrangement must obviously be used in such systems.

For the system to be most effective, the code sequence should have essentially the same statistics as a random sequence of bits. The sequences are thus described as "pseudo-random." Messages encoded with such code sequences have energy spectrums which are quite uniform over the occupied bandwidth. Since such a distribution is characteristic of noise as well, coding sequences of this type are in fact called, in this context, "pseudo-noise" (PN) sequences.

In addition to the requirement that the coding sequence have apparently-random statistics, there are other criteria. In particular, to provide the greatest separation between the signals from different stations, there needs to be a low cross correlation among the assigned coding sequences. Extensive analytical work has been done to generate "catalogs" of suitable sequences. The so-called "Gold" codes are a prominent result. [Gold, 1967]

In most actual CDMA systems, a different approach is used to extract the message bits from the received chip stream. Figure 8-15 illustrates its principles. This correlator receiver, however, requires that the local replica of the code sequence be properly timed with respect to the chip stream—timing that can range over many message bits, for a long sequence. This requirement has been the target of much clever design work, and many attractive methods are now known. Conceptually, we may think of the receiver shifting the timing of its sequence replica until 1's and 0's are being found at essentially the message bit rate.

As a result of the noise-like properties of the transmitted signals, the process by which a receiver rejects noise is the same as the one by which it rejects the received signal components from all the "undesired" messages. An original motivation for the study of this technique, as well as of other forms of "spread spectrum" transmission, was to protect military transmission systems from jamming. With the information in the message spread over a wide spectrum, the jammer would need to maintain a substantial energy density over the entire relevant bandwidth, requiring a far greater overall jamming power than to jam a conventional transmission.

In a highly-idealized system model, the three multiple access techniques—FDMA, TDMA, and CDMA—would theoretically yield identical capacities. Taking into account the practical factors of real systems, there may be substantial differences.

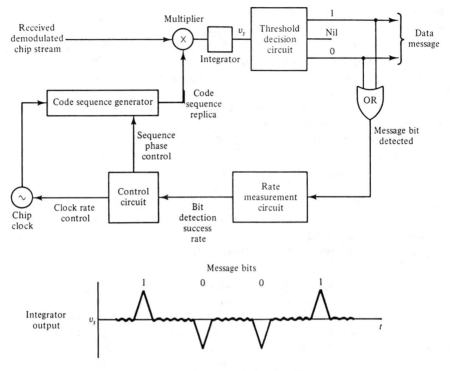

Figure 8-15 Cross-correlation decoding.

Where operation may be limited by interference considerations, as is often the case in today's satellite environment, CDMA operation may be especially attractive. Stations having small antennas, whose large beamwidths raise the prospect of interference with adjacent satellites, may become acceptable owing to the low power density of their spread-spectrum outputs and their decreased susceptibility to interfering inputs.

8.7 COMPARISON OF MULTIPLE ACCESS TECHNIQUES

The wide variety of multiple access techniques provides great flexibility in satellite networking. Table 8-1 summarizes the characteristics of the multiple access systems described in this chapter. These systems provide either analog or digital transmission in continuous or burst mode. Capacities range from 14 to 28 channels per megahertz of transponder bandwidth. Additional capacity can be accommodated through the use of signal processing. For example, digital speech interpolation, when applied with TDMA, can result in capacities approaching 56 channels per megahertz of transponder bandwidth.

Taken as a group, these multiple access systems provide many choices for the system designer, and each technique seems to apply best for a particular type of network. For example, SCPC techniques operate best in networks consisting of a large number of users, each with a relatively small traffic

TABLE 8-1 Characteristics of Three Multiple Access Types

Characteristic	FDMA		TDMA	CDMA
	SCPC	MCPC		
Transmission	Analog or digital	Analog or digital	Digital	Digital
Multiplexing	None	FDM or TDM	TDM	TDM
Modulation	FM or PSK (continuous or voice-activated)	FM or PSK	"High-speed" PSK (burst mode)	Chip encoded, AM or PSK
Carrier bandwidth	$0.7 \times$ bit rate	Depends on frequency plan	Full transponder (typical) or narrowband	Full transponder
Capacity (per megahertz of transponder bandwidth)	22 channels/MHz (voice only)	16 to 25/MHz channels (typical)	28 channels/MHz	
Primary applications	Many low-traffic stations	Heavy point-to-point links	Intermediate number of stations, moderate traffic	Interference-sensitive applications

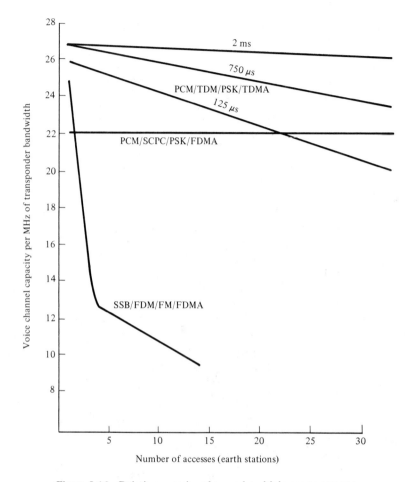

Figure 8-16 Relative capacity of several multiple-access systems.

density. SCPC systems provide multiple access at the individual channel level, thereby providing the small user with the advantage of multiple access, even though the user may not have the traffic density necessary to support more complex approaches. MCPC, either analog or digital, operates very efficiently in heavy point-to-point link applications with few (one or two) wide-bandwidth carriers occupying the transponder. This, of course, limits multiple access capability, but does provide a large number of channels per transponder. As we increase the number of carriers in the system, the multiple access penalties come into play, and the MCPC system capacity is correspondingly reduced. TDMA, on the other hand, provides a good compromise for those networks with an intermediate number of stations (perhaps 15 or 20), and moderate traffic at each station. It provides excellent interconnectivity and networking capacity for these systems. Applications of TDMA systems are growing far more rapidly than the other techniques used in modern satellite communications.

Figure 8-16 is a plot of the capacity per megahertz of transponder bandwidth versus the number of accesses or earth stations in a particular network. This figure illustrates the relative capacities of each multiple access system, depending on network size, expressed in terms of number of users. Notice that SCPC capacity is essentially insensitive to the number of users in the network. It is therefore well suited to a network consisting of a large number of small-capacity users. MCPC systems, on the other hand, do well as long as the number of accesses is very small. As the number of accesses increases (increasing the number of carriers in the FDMA transponder), the multiple access penalties begin to erode the capacity rapidly. TDMA provides excellent capacity as a function of the number of accesses as long as the frame length is long enough to provide high frame efficiency. Notice that with relatively short frame times, the capacity of the TDMA system degrades with the number of accesses because of the erosion of capacity due to the overhead associated with each new burst added to the system. However, for longer frame times, the TDMA capacity curve essentially flattens and becomes relatively insensitive to the number of users.

REFERENCES

CACCIAMANI, E.R., JR., "The SPADE System As Applied to Data Communication and Small Earth Station Operation," *COMSAT Technical Review*, Vol. 1, No. 1, Fall 1971, pp. 171–182.

CAMPANELLA, S.J., ET AL., "The INTELSAT TDMA Field Trial," *COMSAT Technical Review*, Vol 9, No. 2, Fall 1979, pp. 293–340.

CAMPANELLA, S.J., and K. HODSON, "Open-Loop Frame Acquisition and Synchronization for TDMA," *COMSAT Technical Review*, Vol. 9, No. 2, Fall 1979.

CAMPANELLA, S.J., and ROGER J. COLBY, "Network Control for Multibeam TDMA and SS/TDMA," *IEEE Transactions Communication*, Special Issue on Digital Satellite Communications, 1983.

"Construction Details for an INTELSAT Demand Assigned Multiple Access Terminal (SPADE)," INTELSAT, ICCSC/T-31, 20E w/6/69, Washington, D.C.

COOK, C., and MARSH, H., "An Introduction to Spread Spectrum," *IEEE Communications Magazine*, March, 1983.

DEAL, J., "Open Loop Acquisition and Synchronization," Joint Automatic Control Conference, San Francisco, CA, June 1974, *Proc.*, Vol. 2, pp. 1163–1169.

"Digital Interface Characteristics between Satellite and Terrestrial Networks," CCIR, Rep. No. 707, Vol. 4, International Telecommunications Union, Geneva, 1978a.

DILL, G.D., Y. TSUJI, and T. MURATANI, "Application of SS-TDMA in a Channelized Satellite," International Conference on Communications, Philadelphia, PA, 1976, Vol. 3, pp. 51-1–51-5.

EDELSON, B.I., and A.M. WERTH, "SPADE System Progress and Application," *COMSAT Technical Review*, Vol. 2, No. 1, Spring 1962, pp. 221–242.

"Energy Dispersal Techniques for Use with Digital Signals," CCIR, Annex III to Rep. No. 384-3, CCIR Vol. 4, Geneva, 1978b.

FEHER, K., *Digital Communications: Satellite and Earth Station Engineering*, Prentice-Hall, Inc., 1981.

GOLD, R., "Optimal binary sequences for spread spectrum multiplexing," *IEEE Trans. Inform. Theory*, vol. IT-13, 1967.

GOODE, B., "Demand Assignment of the SBS TDMA Satellite Communications System," EASCON '78, Washington, D.C.

"INTELSAT TDMA/DSI System Specification (TDMA/DSI Traffic Terminals)," INTELSAT, BG-42—65E B/6/80, Intelsat, Washington, D.C., June 26, 1980.

JEFFERIES, A., and K. HODSON, "New Synchronization Scheme for Communications Satellite Multiple Access TDM Systems," *Electronics Letters*, Vol. 9, No. 24, Nov. 29, 1973.

KWAN, R.K., "Modulation and Multiple Access Selection for Satellite Communications," IEEE, NTC-78, Vol. 3, Birmingham, AL, Dec. 3–6, 1978.

LUNSFORD, J., "Satellite Position Determination and Acquisition Window Accuracy in the INTELSAT TDMA System," COMSAT Lab. Tech. Memorandum CL-28—81.

MCCLURE, R.B., "The Effect of Earth Station and Satellite Parameters on the SPADE System," IEE Conf. Earth Station Technol., IEEE Conf. Publ. 72, London, Oct. 1970.

NUSPL, P.P., R.G. LYONS, and R. BEDFORD, "SLIM TDMA Project-Development of Versatile 3 Mb/s TDMA Systems," Proc. 5th Int. Conf. Digital Satellite Commun., Genoa, Italy, March 1981.

PERILLAN, L., and T.R. ROWBOTHAM, "INTELSAT VI SS-TDMA System Definition and Technology Assessment," Proc. 5th Int. Conf. Digital Satellite Commun., Genoa, Italy, 1981.

PICKHOLTZ, R., SCHILLING, D., and MILSTEIN, L., "Theory of Spread Spectrum Communications—A Tutorial," *IEEE Trans. Commun.*, vol. COM-30, no. 5, May 1982.

PONTANO, B., G. FORCINA, J. DICKS, and J. PHIEL, "Description of the INTELSAT TDMA/DSI System," Proc. 5th Int. Conf. Digital Satellite Commun., Genoa, Italy, 1981.

PUENTE, J.G., and A.M. WERTH, "Demand-Assigned Service for the INTELSAT Global Network," *IEEE Spectrum*, Vol. 8, No. 1, Jan. 1971, pp. 59–69.

SCHMIDT, W.G., "The Application of TDMA to the INTELSAT IV Satellite Series," *COMSAT Technical Review*, Vol. 3, No. 2, Fall 1973, pp. 257–276.

SEKIMOTO, T., and J.G. PUENTE, "A Satellite Time-Division Multiple-Access Experiment," *IEEE Transactions on Communications Technology*, COM-16, No. 4, Aug. 1968, pp. 581–588.

"SCPC System Specification," INTELSAT, BG/T-5–21E, w/1/74, Jan. 7, 1974, Washington, D.C., 1974.

SCHMIDT, W.G., "The Application of TDMA to the INTELSAT IV Satellite Series," *COMSAT Tech. Rev.*, Vol. 3, No. 2, Fall 1973, pp. 257–276.

9

Satellite Transponders

9.0 INTRODUCTION

A communications satellite may be considered as a distant repeater whose function is to receive uplink carriers, process them, and retransmit the information on the downlink. Modern communications satellites contain multichannel repeaters made up of many components, including filters, amplifiers, frequency translators, switches, multiplexers and hybrids. These repeaters function in much the same way as a line-of-sight microwave radio relay link repeater does in a terrestrial transmission.

9.1 FUNCTION OF THE TRANSPONDER

Figure 9-1 shows a generalized block diagram of a multichannel repeater as implemented in a typical modern communications satellite. The path of each channel from receiving antenna to transmit antenna is called a *transponder*.[1] It is through its transponders that a communication satellite earns its living. The basic functions of each transponder are isolation of neighboring RF channels, frequency translation, and amplification.

Table 9-1 summarizes the frequency bands commonly used for satellite communications. Generally, the higher the frequency, the higher the suscep-

---

[1] The term arises from earlier usage in aeronautics where it was applied to a device which received an interrogation signal from a ground station and returned a response signal.

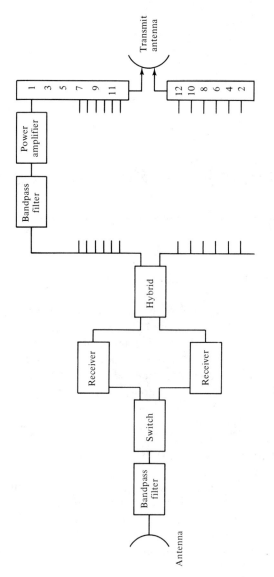

Figure 9-1 Multitransponder repeater.

280

Sec. 9.1 Function of the Transponder

TABLE 9-1 Communications Satellite Frequency Allocations (MHz)

Use	Downlink Frequency (MHz)	Uplink Frequency (MHz)
	Fixed Service	
Commercial (C-band)	3,700–4,200	5,925–6,425
Military (X-band)	7,250–7,750	7,900–8,400
Commercial (K-band)		
Domestic	11,700–12,200	14,000–14,500
International	10,950–11,200	27,500–31,000
	11,450–11,700	
	17,700–21,200	
	Mobile Service	
Maritime	1,535–1,542.5	1,635–1,644
Aeronautical	1,543.5–1,558.8	1,645–1,660
	Broadcast Service	
	2,500–2,535	2,655–2,690
	11,700–12,750	
	Telemetry, Tracking, and Command	
	137.0–138.0, 401.0–402.0, 1,525–1,540	

tibility to rain attenuation and the more expensive the equipment required. However, general congestion at lower frequencies continues to promote higher-frequency operation.

Uplink frequencies are separated from downlink frequencies to minimize interference between transmitted and received signals. The downlink commonly uses the lower frequency, which suffers lower attenuation and thus eases the requirement on satellite output power. The available bandwidth of each transponder may be used either by multiple carriers as in FDMA or by a single carrier as with TDMA.

The system depicted in Figure 9-1 is characterized as a *quasilinear repeater*. This means that the repeater provides almost linear response as long as it is not operated too close to its maximum power output. The signals transmitted from earth stations pass through this repeater with negligible distortion. It should be noted that another type of transponder using a *hard-limiting receiver* was used in early designs of commercial communications satellites. This approach, in which the output is virtually independent of the input, and the limiting occurs just marginally above the noise level, has also been used in military satellite transponders to reduce sensitivity to jamming. Yet another type of transponder, called a *regenerative repeater*, utilizes on-board signal processing of digital signals to achieve improved end-to-end performance.

Satellite transponders are subject to transmission impairments which are functions of available power and bandwidth as well as of system operating mode. Some transmission impairments that are of major consequence in satellite systems have not usually been considered important in terrestrial microwave systems because of the availability of essentially unlimited power on the earth. Because of the high cost of power and transponder mass in orbit, systems engineers must deal with the problem of balancing the cost of available power, bandwidth, and reliability against the impact of distortion. That distortion produces such impairments as intermodulation distortion, AM-to-PM conversion, impulse noise, and interference. In addition to these "nonlinear" impairments, operation through satellite transponders is a subject to the typical linear impairments such as thermal noise and imperfections in the amplitude- and phase-response versus frequency.

The next section of this chapter deals with implementation of both quasi-linear and regenerative transponders. Block diagrams are shown and discussed, and performance criteria are outlined. This is followed by a discussion of the characteristics of various devices used in transponders such as filters, high-power amplifiers, and oscillators. The next section describes transmission impairments and presents a development of quantitative performance measures. Section 9.4 deals with further quantification of the impairments for systems engineering calculations. Section 9.5, the final section of this chapter, deals with systems aspects of transponders, including a brief discussion of the methods used to accommodate various multiple access techniques.

9.2 TRANSPONDER IMPLEMENTATIONS

9.2.1 Quasilinear Repeaters

The term *quasilinear* describes the tendency of satellite transponder amplifiers, like all amplifiers, to exhibit nonlinear response close to their maximum power output and more linear response at lower power levels. Choosing an operating point for such an amplifier is a major concern of the

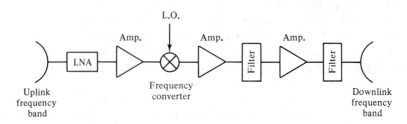

Figure 9-2 Quasilinear single conversion repeater.

communications systems engineer. Figure 9-2 shows a generic block diagram of a quasilinear repeater. This repeater receives, separates, and amplifies its assigned uplink carriers, translates the frequency to the downlink band, and amplifies the signal for retransmission on the downlink. This transponder design is sometimes referred to as a single-conversion type because it translates from the uplink to the downlink band in one step. The final high-power output stage of a transponder is often constructed using a traveling-wave tube amplifier. The characteristics of these amplifiers are important to the communications systems engineering problems and to analysis of the nonlinear impairments discussed in Sections 9.2.2 and 9.4.

Typically, common equipment is used in the earlier stages of the repeater for more than one radio-frequency channel, often for the entire bandwidth (in a given band) used by the whole ensemble of transponders. These earlier stages usually comprise a filter to eliminate energy wholly outside the operating band, a low-noise amplifier (LNA) to increase the signal power, and a broadband frequency converter to shift the entire operating band from uplink to downlink frequency.

The full operating bandwidth is then separated by filters into the individual transponder channel bands (often 36 MHz wide each). Each channel band signal is then amplified by an individual high-power amplifier (HPA), possibly preceded by a driver amplifier. The output of each HPA is passed through a bandpass filter which eliminates out-of-band products of the amplifier's nonlinearity. The outputs of several channels' HPAs are then combined in an output multiplexer (typically employing microwave circulators) and fed to a common antenna system for transmission.

In practice, a single antenna system may be used for both reception and transmission. In that case, a *duplexer* is used to separate the receiving and transmission paths. There is some leakage of the transmitted signal into the receiving path, but owing to the substantial frequency difference between downlink and uplink bands, this leakage is effectively blocked by the input filter.

Another transponder design is the dual-conversion type, which is useful in certain applications. Figure 9-3 depicts a system operating in two bands, A and B. Each band taken separately could be amplified and transmitted using simple frequency translation between the associated uplink and downlink bands. Assume, though, that there is a need for interconnectivity between the two frequency bands. This is met by first translating the band B uplink frequency to the downlink frequency of band A which is used as a universal "intermediate frequency." When both bands A and B have corresponding channelization and frequency plans, any particular channel can either be through-connected to its own downlink frequency or cross-connected to the corresponding downlink frequency in the other band. This kind of implementation may be accomplished by a set of C-switches as shown in Figure 9-4. A C-switch is a two-by-two switch providing alternatively through- or

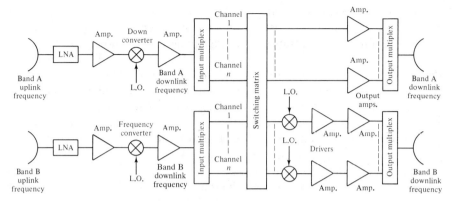

Figure 9-3 Dual conversion repeater.

cross-connections of its paired inputs and outputs. They are used extensively in virtually all transponder designs. The switch outputs which are destined for the band B downlink must be frequency converted from band A to B before final amplification, as shown in Figure 9-3. It should be noted that there are many other applications for the dual-conversion approach. For example, design trade-offs in the implementation of filters, oscillators, down-converters, and so on, may favor the double-conversion design.

Many of the transmission impairments, particularly those due to AM and FM conversion, intermodulation, and variations in group delay, can be considered at the system level as producing a form of noise, characterized by a carrier-to-impairment noise term $(C/N)_I$. This term can then be combined with the thermal noise terms due to the uplink and downlink as discussed in Chapter 6. The important transmission impairments are discussed in Sections 9.3 and 9.4.

9.2.2 Regenerative Repeater

In digital transmission applications a more complex type of satellite transponder may be employed to achieve improved performance. Figure 9-5 is a block diagram of a *regenerative repeater*. A regenerative repeater performs

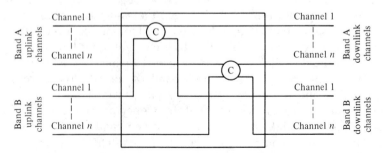

Figure 9-4 Switch matrix using C-switches.

the receiving and transmitting functions in the same manner as the quasilinear repeater. However, the regenerator contains in each transmission link a demodulator that demodulates the uplink signal to the digital baseband signal and a modulator that remodulates that signal on a downlink carrier. The demodulated digital signal is retimed and restored to standard form. This approach effectively isolates the uplink performance from the downlink performance, preventing the accumulation of noise and distortion over the two links. Contrast this with the use of a quasilinear repeater where the uplink impairments (typically thermal noise) directly affect the carrier-to-noise ratio in the received downlink.

In systems engineering calculations using a regenerative repeater the performance is measured in terms of the overall bit error rate in the end-to-end link. Let P_U and P_D be the probability of bit being in error on the uplink and downlink, respectively. Then the probability of a bit *not* being in error (i.e. correct reception) in the end-to-end link is given by

$$P_C = (1 - P_U)(1 - P_D) = 1 - (P_U + P_D) + P_U P_D \qquad (9\text{-}1)$$

Therefore, the probability of a bit error in the end-to-end link is

$$P_E = P_U + P_D - P_U P_D$$

As long as the error rate is fairly low, the overall error rate is essentially

$$P_E = P_U + P_D \qquad (9\text{-}2)$$

This equation illustrates the independence of the uplink and the downlink. Compare this to the traditional quasilinear repeater where uplink and downlink S/N ratios combine using the "resistors-in-parallel" formula.

9.2.3 Device Characteristics

Transponders are made up of a number of devices with special functions and characteristics. The following sections briefly describe some of the more important ones.

9.2.4 Filters

Filters are critically important in all communication circuits and their characteristics can have dramatic effects on the level of transmission impairments. Figure 9-6 shows the amplitude versus frequency charactcristic of an ideal filter with perfectly flat passband characteristics, infinite out-of-band rejection, and perfectly linear phase. Such a filter is not realizable but serves as the model of the synthesis techniques used to design approximations to these filters. For example, the Butterworth approximation filter illustrated in Figure 9-7 emulates the ideal filter in that it has a maximally flat passband, but it suffers from rather slow roll-off (6 dB/octave per pole) of the filter skirts. The Tschebychev approximation shown in Figure 9-8 solves the out-

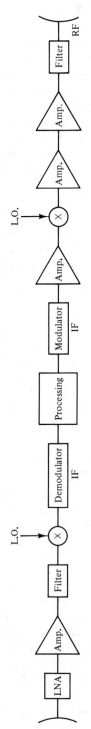

Figure 9-5 Remodulating or regenerative repeater.

Sec. 9.2 Transponder Implementations

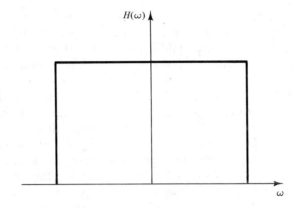

Figure 9-6 Frequency response of ideal filter.

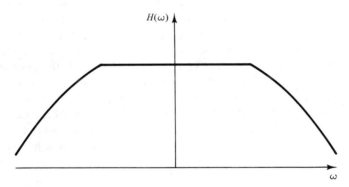

Figure 9-7 Frequency response of Butterworth filter.

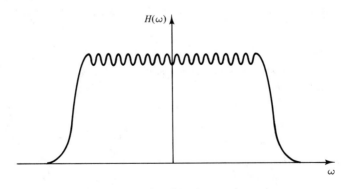

Figure 9-8 Tschebychev filter frequency response.

of-band rejection problem of the Butterworth design at the expense of passband ripple. The design trade-offs here are selectivity (sharpness of cutoff) versus passband ripple. The elliptic function filter of Figure 9-9 is a compromise that can deliver acceptably low passband ripple, good out-of-band rejection, but limited passband-to-stopband rejection ratio.

The phase characteristics of these filters are typically described in terms of group-delay distortion. Group delay is directly related to the rate of change of phase with frequency. Filters with rather sharp skirts often produce undesirable group-delay distortion at the band edges. This characteristic can cause intersymbol interference on digitally transmitted information (increasing the bit error rate) or phase distortion on an analog carrier. Typical filters used in satellite transponder designs are equipped with equalizing circuits sometimes designed to vary as the function of the phase roll characteristics of the communication chain, thereby minimizing group-delay distortion effects.

Chapter 10 contains a discussion of a figure of merit called *noise power ratio* (NPR) used to quantify system intermodulation performance. NPR is determined by loading a system with noise, uniformly covering the baseband bandwidth except for a small region (typically corresponding to one voice channel in an FDM baseband assembly). The noise simulates the loading of all remaining channels by actual voice signals. The NPR then is the ratio of the noise power appearing in the selected voice bandwidth (resulting from intermodulation) to the per-channel loading noise power. The NPR varies with the level of noise loading level and with the position of the measurement "slot" within the baseband spectrum. NPR results presented as a function of slot frequency for various noise loading levels provide system performance data in simple and repeatable form.

9.2.5 High-Power Amplifiers

All amplifiers, both low-level and high-power output types, exhibit an input/output characteristic like that shown in Figure 9-10. As the input drive level is increased, the amplifier reaches a saturated state which corresponds

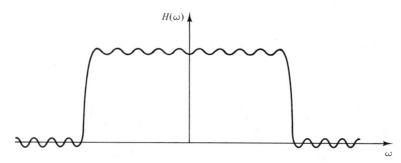

Figure 9-9 Frequency response of elliptic filter.

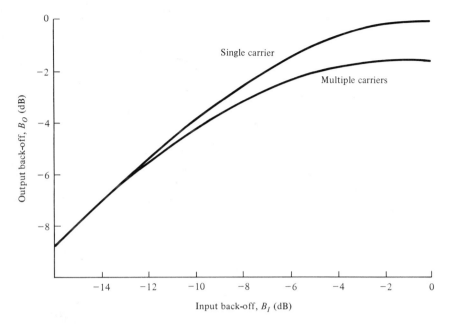

Figure 9-10 Typical HPA input–output characteristics.

to the maximum power output available. As long as the amplifier is processing only a single carrier (as in the case of TDMA), it can be operated close to saturation without causing serious impairments. However, in multicarrier operation (e.g., FDMA) impairments caused by the increasingly nonlinear response close to saturation result in undesired increases in intermodulation distortion. The impairments due to these nonlinearities are discussed in Sections 9.3 and 9.4.

In a typical multicarrier application, the RF link is designed to set the operating point of the HPA in a region linear enough to produce acceptably low nonlinear impairments. This is accomplished by reducing the input drive level (also known as increasing the input back-off) relative to saturation. This results in an output reduction (output back-off) from the saturated power output level to a point which reduces the nonlinear effects to an acceptable level. As discussed in Chapter 6, there is an optimal compromise between decreased intermodulation impairment level and the increase in noise impairment caused by decreased transmitter power. Figure 9-11 shows the amplifier transfer function over a wider operating range to illustrate the behavior of the intermodulation products as a function of backoff from saturation. Notice that the transfer characteristic near saturation depends on whether single-carrier or multiple-carrier operation is being employed. In the multiple-carrier case, a lower-power output results compared to the single-carrier case by an amount equal to the power loss due to intermodulation products that fall out of band and are thus not counted as "output."

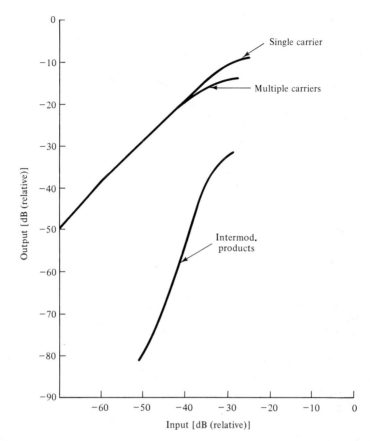

Figure 9-11 HPA input–output characteristic and intermodulation levels versus drive level.

HPAs for many years have used traveling-wave tubes (TWTs) as the output power amplifier. Because of their saturation characteristics, TWTs are the most significant contributors to nonlinear impairments. A TWT is also a life-limiting component of a spacecraft communications package due particularly to limited lifetime of the tube's cathode. Overall efficiencies of TWTs have reached levels as high as 45%. (Efficiency is defined as the ratio of the maximum RF output power to the input electric power required by the tube.) Gains as high as 70 dB have been achieved with maximum output power up to 30 W.

HPAs using solid-state devices have also been developed for satellite applications. Continuous improvements in life and performance of solid-state amplifiers are being developed. Although currently limited in total available output power at microwave frequencies, significant progress has been made in gallium arsenide and field-effect transistor amplifiers. Compared to TWTs, solid-state amplifiers tend to remain more linear and then to saturate abruptly. An optimal operating point can therefore be chosen relatively higher. Equally important, they seem to exhibit less AM-to-PM conversion than TWTs. Although they are not as efficient as TWTs, solid-state HPAs will tend to have longer life.

Low-noise amplifiers. The front end of a communications satellite repeater employs a specially designed low-noise amplifier. Such an amplifier must have an extraordinarily flat passband (less than 0.2 dB over 500 MHz) and an extremely low noise figure (less than 4 dB). This performance must be achieved while the amplifier is operated uncooled. The amplifier chain must provide a maximum gain of approximately 5 dB. Early transponder designs employed bipolar transistors and tunnel diode amplifiers as the basic components in the low-noise amplifier. Modern designs employ field-effect transistors using gallium arsenide technology (GaAsFET). The gain and noise-figure performance achievable with this technology has improved steadily and is comparable to TWT performance in lower-level amplifier applications.

Oscillators. Oscillators also play a very important role in transponder designs. In addition to the obvious characteristics of frequency stability, both short and long term, local oscillator harmonics and phase jitter are often a critical source of spurious outputs from satellite transponders. Careful design of oscillator circuits coupled with careful aging and selection criteria for crystals is used to reduce and minimize the long-term effects.

9.3 TRANSMISSION IMPAIRMENTS

In this section we present a discussion of the methods used to quantify the performance of satellite transponders in terms of some of the most important transmission impairments. Table 9-2 classifies transmission impairments into linear types and nonlinear types for both single- and multiple-transponder operation. Linear impairments relate primarily to the amplitude and phase response of the networks used in the transponder as well as the thermal noise usually dominating the front end of the satellite receiver. Nonlinear impairments consist mainly of intermodulation distortion as well as the related phenomenon called amplitude modulation-to-phase modulation (AM-PM) conversion. Also of concern are impulse noise and intelligible crosstalk. In the case of a multiple-transponder analysis we must be concerned with adjacent transponder effects and multipath. The following sections contain a discussion of these effects as well as methods used to quantify them.

TABLE 9-2 Classification of Transmission Impairments

	Single Transponder	Application	Multiple Transponders
Linear impairments	Thermal noise (uplink)		Dual-path group-delay delay distortion
	Group-delay distortion		
	Amplitude/frequency distortion		
Nonlinear impairments	Intermodulation (including AM-PM)		Adjacent transponder intermodulation
	Intelligible in-band crosstalk		Intelligible out-of-band crosstalk

9.3.1 Linear Impairments

As discussed in Chapter 6, thermal noise is typically the dominant linear impairment in a satellite link. Uplink performance is dominated by thermal noise which establishes the uplink carrier-to-noise ratio. If the thermal noise is high compared to the carrier level, the carrier-to-noise ratio of the uplink will be small and the overall carrier-to-noise ratio will then be dominated by the uplink noise.

As described in Chapter 6, this thermal noise can be treated as having two components. One is noise generated within the receiver itself by the random motion of electrons within the receiver components. This noise can be characterized by the receiver *noise temperature* or by the related receiver *noise figure*. The second component is that which emerges from the antenna, and largely reflects the temperature-dependent radiation of the "scene" which the antenna observes. It again is expressed as a noise temperature. The total noise can be expressed as a *system temperature*. If we were to replace the antenna by a resistor at that temperature, and replace the receiver with one making no internal contribution to the noise, the noise at the output of the receiver would be exactly that of our real receiving system.

Losses in the transmission line (typically waveguide and accessories) between the antenna and the receiver itself add to the system noise temperature and thus must be kept to a minimum. In contrast, noise contributed by elements further along the receiver chain contribute less to the overall noise figure since their noise contributions are, in effect, divided by the gain achieved by the chain to that point.

Typical satellite receivers achieve noise temperatures in the range 1000 to 3000 K, corresponding to noise figures of 5 to 10 dB. This should be compared to earth station front-end noise temperatures in the range 50 to 800 K resulting from more extensive use of passive and active cooling systems to minimize the noise temperature. Also, uncooled parametric amplifiers are better than FETs, but designers have been reluctant to use them in spacecraft because of their complexity and reliability characteristics. In an earth terminal, unlike a spacecraft, the weight of even cryogenic devices is not an important factor.

The amplitude and delay characteristics of filters used in satellite transponders can also produce distortion and must be considered carefully by the satellite systems designer. As discussed previously, different filter designs have differing amplitude and phase characteristics. Distortion may be caused by both amplitude response over the carrier bandwidth and by group-delay distortion (variations in delay through the filter as a function of frequency).

Group-delay variation occurs primarily at the band edge of the filter, where the amplitude characteristic is decreasing rapidly. The impact of delay distortion becomes more critical when in-band response is more important than out-of-band noise rejection. The tolerable distortion caused by amplitude and group-delay effects depends on the signal spectrum passing through this filter. For example, if the signal is a single wideband modulated carrier located at the center of the RF bandwidth, band edge effects (group-delay distortion) may not be significant. However, when the signal consists of multiple carriers spread over the entire bandwidth, band edge effects can become extremely important to the carriers located close to the band edges. Usually, satellite transponder filters are equipped with delay equalizers which are designed to minimize the distortion effects.

9.3.2 Nonlinear Impairments

Analyzing the response of nonlinear systems to arbitrary input signals has absorbed communications theoreticians for many years. The analysis usually starts with assuming a mathematical expression for the input signal (typically a Gaussian function response). The input autocorrelation function is then found by performing the inverse transform on the assumed power spectrum. This, combined with the nonlinear system response (often assumed to be an error function), is then used to compute the output autocorrelation and spectral functions. The output, although expressed in mathematical terms, is usually not calculable in closed form. Numerical methods using computer simulation in the time domain or fast Fourier transforms or other techniques can also be employed to obtain the desired results. These methods are appropriate in the more refined stages of system design when engineers are optimizing overall system performance. However, they are too tedious and expensive for use in the early stages of systems engineering and planning.

Other approaches utilize experimental measurements on actual systems and use the results directly in the calculations. This method is also often impractical because of the variation of experimental data from device to device as well as the usual unavailability of experimental data or facilities to obtain it. In this section we develop a practical approach to permit the systems engineer to estimate the total level of nonlinear distortion due particularly to intermodulation and AM-to-PM conversion. The mathematical development that follows is intended to produce results that can be used in conjunction with typical data provided by manufacturers of high-power amplifier devices. The accuracy of these methods is more than adequate for early systems planning and engineering.

Intermodulation distortion. All amplifiers, including solid-state amplifiers, exhibit nonlinear characteristics when operated close to their finite saturated power outputs. As illustrated in Figure 9-12 this nonlinear behavior may be of little consequence when operating with a single carrier. However, when two or more different carrier frequencies are present in any nonlinear device, harmonics of each are produced in addition to sum and difference frequencies, sometimes called *beats*. The presence of several different input frequencies in a nonlinear transmission system (whether they be multiple carriers or the spectral components of individually modulated carriers) produces a complex spectrum of beats called *intermodulation products*. In many practical applications the number of spurious frequencies generated is so high

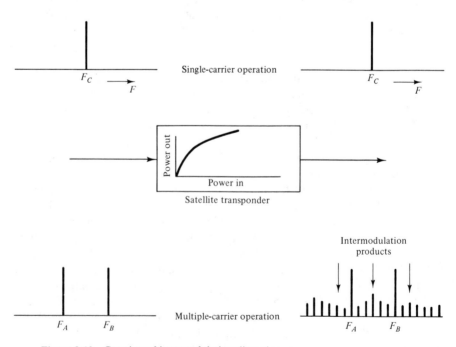

Figure 9-12 Creation of intermodulation distortion.

Sec. 9.3 Transmission Impairments

and their individual amplitudes so low that they can be treated as if they were incoherent and additive, like thermal noise components. In the following paragraphs we present arguments that will culminate in the development of a mathematical approximation which can be used to estimate the level of intermodulation distortion as a function of the number of carriers and the measured value of carrier to intermodulation using two carriers.

The simplest mathematical approach to modeling transponder nonlinearities involves assuming that the instantaneous transfer characteristic of the transponder can be written as a Taylor series:

$$e_0 = a_1 e + a_3 e^3 + \cdots + a_k e^k \tag{9-3}$$

where e is the instantaneous input voltage and e_0 is the resulting output voltage. The coefficients a_k alternate in sign. For an input of n equal carriers:

$$e = \sum_{i=1}^{n} A \cos \omega_i t \tag{9-4}$$

with a total input power P_i (normalized to 1Ω circuit impedance) of $\frac{1}{2}nA^2$. By substituting Eq. (9-4) into (9-3) and after much routine algebra and trigonometry, one can derive the following expressions.

For each of n equal carriers:

$$A_n = a_1 \sqrt{\frac{2P_i}{n}} \left[1 + 3 \frac{a_3}{a_1} \frac{P_i}{n} \left(n - \frac{1}{2} \right) \right.$$
$$\left. + 15 \frac{a_5}{a_1} \left(\frac{P_i}{n} \right)^2 \left(n^2 - \frac{3}{2} n + \frac{2}{3} \right) + \cdots \right] \tag{9-5}$$

and as $n \to \infty$,

$$A_\infty = a_1 \sqrt{\frac{2P_i}{n}} \left(1 + 3 \frac{a_3}{a_1} P_i + 15 \frac{a_5}{a_1} P_i^2 + 105 \frac{a_7}{a_1} P_i^3 + \cdots \right) \tag{9-6}$$

The convergence to this form as n increases is rapid, so Eq. (9-6) is quite useful.

The term $a_1 \sqrt{2P_i/n}$ is the linear component of the transfer characteristic, and the term in parentheses represents the nonlinear or "compression" factor $F_n(P_i)$.

In addition to the carrier outputs, there are many intermodulation products, the most important of which are:

1. Products of the form $(2f_1 - f_2)$ whose amplitudes are:

$$I_n = \frac{3}{4} a_3 \left(\frac{2P_i}{n} \right)^{3/2} \left\{ 1 + \frac{2 a_5 P_i}{3 a_3 n} [12.5 + 15(n-2)] + \cdots \right\} \tag{9-7}$$

which reduces to

$$I_n = \frac{3}{4} a_3 \left(\frac{2P_i}{n}\right)^{3/2} F_n(P_i) \qquad (9\text{-}8)$$

2. Products of the form $(f_1 + f_2 - f_3)$ whose amplitudes are

$$I'_n = \frac{3}{2} a_3 \left(\frac{2P_i}{n}\right)^{3/2} \left\{1 + 10\frac{a_5 P_i}{a_3 n}\left[\frac{3}{2} + (n-3)\right] + \cdots\right\} \qquad (9\text{-}9)$$

which reduces to

$$I'_n = \frac{3}{2} a_3 \left(\frac{2P_i}{n}\right)^{3/2} F'_n(P_i) \qquad (9\text{-}10)$$

The terms outside the parentheses are the contributions of the cubic term in the series (9-3) and the terms in parentheses represent a kind of intermodulation compression attributable to the higher-power terms in the transfer characteristic. Note that the a_3 term changes by 3 dB for each 1-dB change of input power. There are also terms of the form $(3f_1 - 2f_2)$ due to the a_5 term, but they have been neglected for our purposes. Note that the terms of the form $(f_1 + f_2 - f_3)$ are 6 dB higher in amplitude than the terms of the form $(2f_1 - f_2)$.

In addition to the amplitude of each product, we are also concerned with their number. Westcott (1967) gives expressions for calculating the number and distribution of each type of intermodulation product. Within the transponder bank, the products fall on the same frequencies as the carriers. If there are n carriers, the number of products of each type falling on the rth carrier is given by:

For $(2f_1 - f_2)$ type:

$$v_{Dn} = \frac{1}{2}\{(n-2) - \tfrac{1}{2}[1 - (-1)^n](-1)^r\} \qquad (9\text{-}11)$$

For $(f_1 + f_2 - f_3)$ type:

$$v'_{Dn} = \frac{r}{2}(n - r + 1) + \tfrac{1}{4}[(n-3)^2 - 5] - \tfrac{1}{8}[1 - (-1)^n](-1)^{n+r} \qquad (9\text{-}12)$$

The maximum number of products falls on the center carriers. Equations (9-11) and (9-12) simplify rather nicely for the center pair where $r = n/2$.

$$v_{Dn} = \frac{n-2}{2} \qquad v'_{Dn} = \frac{(n-2)(3n-4)}{8} \qquad (9\text{-}13)$$

Carrier to total intermodulation ratio. We assume that the intermodulation products are incoherent (not exactly true) and hence can be added on a power basis. Thus we can write in general:

$$\left(\frac{C}{T}\right)_n = \frac{A_n^2}{v_{Dn}I_n^2 + v'_{Dn}I_n'^2} \tag{9-14}$$

Substituting for A_n, we can write

$$\left(\frac{C}{I}\right)_n = \frac{4n^2(a_1/a_3)^2 F_n^2}{9 P_i^2(v_{Dn}F_n^2 + 4v'_{Dn}F_n'^2)} \tag{9-15}$$

using Eqs. (9-5), (9-8), and (9-10).

Only the second term in the denominator is really important, both because of the factor of 4 which is attributable to the higher amplitudes of the $(f_1 + f_2 - f_3)$ products and because v'_{Dn} is much greater than v_{Dn} for $n > 3$. If we neglect terms in the series expansion above third order, the compression factors F_n and F_n' for the intermodulation products can be taken equal to unity and the term $(v_{Dn} + 4v'_{Dn})$ simplifies to $3(n-1)(n-2)/2$. Equation (9-15) becomes

$$\left(\frac{C}{I}\right)_n = \frac{8n^2 (a_1/a_3)^2 F_n^2}{27 P_i^2(n-1)(n-2)} \tag{9-16}$$

The amplitude compression factor F_n also depends on P_i and a_1/a_3, so any method for inferring the ratio a_1/a_3 permits Eq. (9-16) to be evaluated. Section 9.5 describes several other common measures of nonlinearity from which we can draw a value of a_1/a_3 to a first approximation. The specific expression for $(C/I)_2$, a factor frequently measured by amplifier manufacturers, is readily found from Eqs. (9-5) and (9-8). It is, neglecting terms above third order:

$$\left(\frac{C}{I}\right)_2 = \frac{A_2^2}{I_2^2} = \frac{16}{9}\left(\frac{a_1}{a_3}\right)^2 \frac{F_2^2}{P_i^2} \tag{9-17}$$

Note that we use the notation C/I_n for the ratio of a single carrier to a single intermodulation product for n carriers, whereas $(C/I)_n$ has been used for the ratio of a single central carrier to the total of all the products falling on it.

Note also that F_2, the series in parentheses in Eq. (9-17) with $n = 2$, also contains a_1/a_3. Equation (9-17) is readily solved for a_1/a_3 in terms of a measured value of $(C/I)_2$. The result is

$$\frac{a_1}{a_3} = \frac{3}{4} P_i \left[\sqrt{\left(\frac{C}{I}\right)_2} - 3\right] \tag{9-18}$$

Substituting Eq. (9-18) into (9-16) yields, after some reduction,

$$\left(\frac{C}{I}\right)_n = \frac{n^2}{6(n-1)(n-2)}\left[\sqrt{\left(\frac{C}{I}\right)_2} + \left(\frac{n-2}{n}\right)\right]^2 \tag{9-19}$$

If $(C/I)_4$, the ratio of the power output in one carrier to the total intermodulation power on a central carrier is given, the foregoing procedure yields

$$\left(\frac{C}{I}\right)_n = \frac{n^2}{6(n-1)(n-2)} \left[\frac{3}{2}\sqrt{\left(\frac{C}{I}\right)_4} + \frac{n-4}{2n}\right]^2 \quad (9\text{-}20)$$

The asymptotic result for a large number of carriers is

$$\left(\frac{C}{I}\right)_\infty = \frac{1}{6}\left[\sqrt{\left(\frac{C}{I}\right)_2} + 1\right]^2 \quad (9\text{-}21)$$

Data sheets on amplifiers for transponders, both low and high level, usually quote values of the third-order intermodulation products $(2f_1 - f_2)$ or the equivalent $(C/I)_2$. These products normally fall outside the transponder band and are not significant in two carrier operation. Nonetheless, they are a measure of the amplifier nonlinearity and thus can be used with Eqs. (9-19) and (9-21) to infer usably accurate values of C/I for larger numbers of carriers.

The expressions tend to give conservative results for C/I and are most accurate a little away from saturation because they ignore the higher terms in the series necessary to describe the transfer characteristic in its entirety. They also become poorer approximations for very high values of C/I, where they yield values of $(C/I)_n$ that are too low. This is also not critical to practical system design.

AM-to-PM conversion. Another nonlinear impairment closely related to intermodulation distortion is caused by AM-to-PM conversion. Many amplifiers, particularly traveling-wave tube amplifiers, have a total phase shift that is a function of the input level. TWTAs are particularly susceptible to this effect since it is inherent in the mechanism of cavity-coupled amplifiers. In such a device any amplitude modulation of the input signal will cause frequency and phase modulation components to appear at the output. Amplitude variations may be present even when using a constant amplitude modulation technique because of the amplitude variations caused by the ripple in the passband of the bandpass filters used in the transponder. Also, in a multiple-carrier case, the input signal is composed of many different incoherent carrier frequencies and the amplitude of the composite envelope will fluctuate in accordance with the Rayleigh distribution. These amplitude variations are then converted into phase modulation which manifests itself as a spectrum of spurious frequencies, located in fact at the same frequencies as those created by intermodulation distortion. For practical purposes, we may then treat the AM-to-PM effects in much the same fashion as intermodulation distortion was treated in the preceding section.

Systems-level approximations to the distortion caused by AM-to-PM conversion can be made using data typically provided by the manufacturer. A data sheet on a TWTA usually contains a value for a parameter known as the AM-to-PM conversion coefficient, K. This parameter is specified in degrees of phase shift per decibel of amplitude change and is a function of the input level. Therefore, K is effectively the slope of the phase shift versus amplitude curve. The amplitudes of the distortion products resulting from AM-to-PM conversion are proportional to this parameter and distributed in frequency, number, and relative amplitude exactly as intermodulation distortion products.

We may proceed in a manner analogous to that used in the preceding section to develop Eq. (9-16). This results in the following expression for the carrier-to-interference ratio due to AM-to-PM conversion as a function of the number of carriers given by

$$\left(\frac{C}{I_{AM/PM}}\right)_n = \frac{n^2}{0.1516K^2} \frac{8}{3(n-2)(n-1)} \tag{9-22}$$

This provides a good approximation for a given value of K. Note that as n increases the expression quickly converges to

$$\frac{1}{0.1516K^2}$$

The factor 0.1516 converts degrees per decibel to radians per volt using the approximation that $\ln(1 + X) = X$ if X is small. Figure 9-13 shows the typical variation of K as a function of back-off. Note that K does not decrease rapidly with reduced power.

Using the resistors-in-parallel formula, Eq. (9-22) may be combined with Eq. (9-20) to produce a composite carrier-to-interference ratio repre-

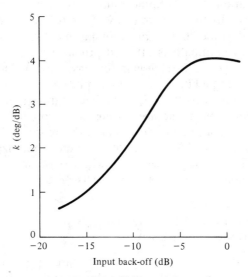

Figure 9-13 AM to PM conversion coefficient versus input back-off.

senting the effects of both intermodulation distortion and AM-to-PM conversion.

Other impairments. In addition to the impairments already discussed, the systems engineer must also deal with several other impairments in designing the system. For example, impulse noise can create service degradations which are difficult to quantify from a design standpoint. Impulse noise can be generated from a number of different sources. In FM transmission, when the noise vector momentarily exceeds the size of the carrier vector, impulsive noise can be generated, often referred to as *click noise*. Impulsive noise can also be generated from interference of adjacent carriers within the same transponder bandwidth. Also, the band-limitation function of bandpass filters separating carriers can help to create additional impulsive-noise components.

In considering multiple-transponder operation within a single satellite the systems engineer must also deal with the problems of multipath propagation plus the intermodulation and intelligible crosstalk distortions caused by interfering signals in adjacent transponders. For example, a signal near the edge of a transponder band can be a source of interference into the adjacent transponder, and by being amplified by it, can interfere with itself. Passing through the desired bandwidth, a carrier is attenuated by the bandpass characteristics of its own input channel and then amplified. A portion of the carrier energy may also pass through the adjacent transponder and be attenuated by the substantial rejection characteristic of that channel and then amplified by the same amount as in the desired transponder. After some additional attenuation at the output circuits, these signals are added in a phase relation determined by the relative delay of the two independent paths. This multipath effect degrades the carrier-to-impairment ratio. Although the C/I component due to multipath is affected primarily by the sharpness of the input filters, the use of adjustable gain settings, cross-polarization frequency reuse, and higher-frequency operation complicate the problem further. For example, in cross-polarized operation an adjacent transponder is isolated only by the polarization from an adjacent transponder operating at the same or an overlapping frequency band. At Ku band, severe signal depolarization can occur during heavy rain which can reduce the cross-polarization isolation to less than 15 dB. These effects, combined with unwise or unfortunate channel gain settings, complicate the multipath effects.

9.4 USING MANUFACTURERS' DATA AND EXPERIMENTAL RESULTS

Sometimes, instead of quoting $(C/I)_2$, manufacturers provide two other measures of nonlinearity particularly for low-level devices. They are the 1-dB compression point $P_{1.0dB}$ and the intercept point P_I. $P_{1.0dB}$ is the output power

Sec. 9.4 Using Manufacturers' Data and Experimental Results

for which the power transfer characteristic departs by 1.0 dB from linear. P_I is the output power, in watts, at the intersection of the extended linear portions of the transfer characteristic and the third-order intermodulation output curve for two carriers. These definitions are illustrated in Figure 9-14.

From Eqs. (9-5) and (9-7) with $n = 2$, we can write the amplitude expressions, neglecting terms above a_3, as

$$A_2 = a_1 \sqrt{P_i} \left(1 + \frac{9}{4}\frac{a_3}{a_1}\right) \qquad (9\text{-}24)$$

$$I_2 = \frac{3}{4} a_3 P_i^{3/2} \qquad (9\text{-}25)$$

Note that a_1^2 is the low-level (linear) power gain G_0 of the amplifier. These equations can be applied to determine the relations among P_I, $P_{1.0dB}$, and the series coefficients a_1 and a_3. If we assume that power measurements are made into 1-Ω loads and noting that the output carrier power is then $A_2^2/2$ and the intermodulation power $I_2^2/2$, we can derive routinely:

$$P_I = \frac{2}{3} G_0 \frac{a_1}{a_3} \quad \text{W} \qquad (9\text{-}26)$$

$$P_{1.0} = \frac{2}{3} K^2(K-1) G_0 \frac{a_1}{a_3} \quad \text{W} \qquad (9\text{-}27)$$

where $K = 0.89$ to correspond to the 1.0-dB compression point. G_0 is usually given so that these equations can be used to find values for a_3 from the nonlinear characteristics.

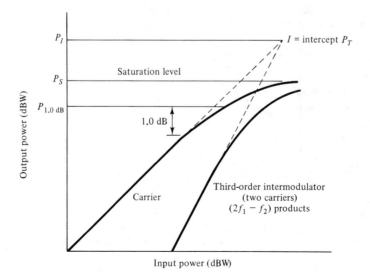

Figure 9-14 Definitions of intercept point and 1.0 dB compression point.

These power expressions assume the convenient arbitrary 1Ω load impedance. Actual measurements are usually taken into 50-Ω loads or higher. Thus, measured values of P, in watts, should be multiplied by the load impedance before entering the values into the equation. The term a_3 can be used to estimate $(C/I)_n$ for any number of carriers using Eq. (9-16). Assuming a slope of 3.0 dB/dB for the intermodulation curve, we can also show that

$$\left(\frac{C}{I}\right)_2 = 2(P_I - P_0) \qquad (9\text{-}28)$$

This equation is correct only on the linear part of the curves, but is a fair approximation even close to saturation.

In addition to transfer characteristics of the kind shown in Figure 9-10, it is often possible to obtain data on carrier-to-intermodulation ratios as a function of power output—either from experiments or computer simulations.

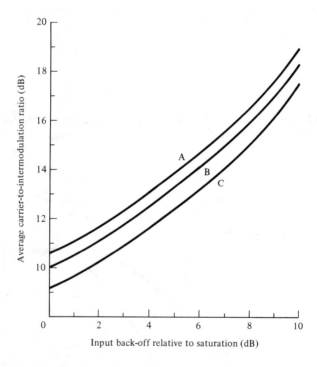

Figure 9-15 Intermodulation distortion versus back-off for 6 carriers (A), 12 carriers (B), and 500 carriers (C). (From Beretta et al., "Improvements in the Characterization of High Power Simplifiers in Multicarrier Operations," ESA *Scientific and Technical Review*, Vol. 2.)

Examples of such curves are shown in Figures 9-15 through 9-17. Figure 9-15 gives C/I_n for nonlinear intermodulation versus input back-off with n as a parameter. Figure 9-16 shows results for a very large number of carriers. It also separates nonlinear amplitude intermodulation from AM-to-PM conversion—the result of variation in phase shift with power level. The nonlinear intermodulation curves agree reasonably well and can be considered as typical for TWTAs.

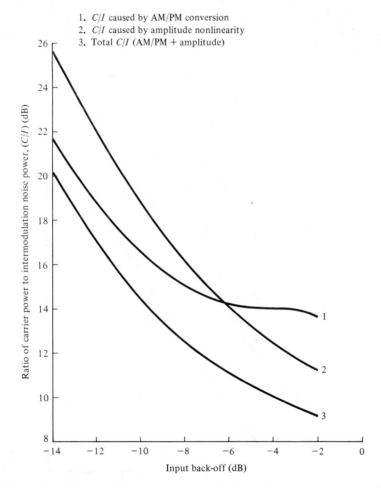

Figure 9-16 Transponder carrier-to-intermodulation ratio versus input back-off (large number of carriers): 1, C/I caused by AM-to-PM conversion; 2, C/I caused by amplitude nonlinearity; 3, total C/I (AM/PM + amplitude). (Reprinted with permission from Communications Satellite Corp., COMSAT Labs Technical Memorandum, CL-12-71 by R. McClure, 1971.)

Figure 9-17 is rather interesting and important. It shows the result when there are a large number of carriers which are not necessarily all active at the same time. This would be the case if a transponder were carrying many voice channels—SCPC—with each carrier voice activated. Because each talker is totally silent half the time statistically, and because of intersyllabic pauses, a typical channel is, conservatively, active only 40% of the time. As can be seen from the experimental curves, this leads to a 3.0- to 4.0-dB improvement in overall C/I_n. A similar advantage occurs in total available power per channel. In a sense, the available transponder transmitter power is "overbooked" to exploit the statistics of voice channels. The ability to do this is one of the

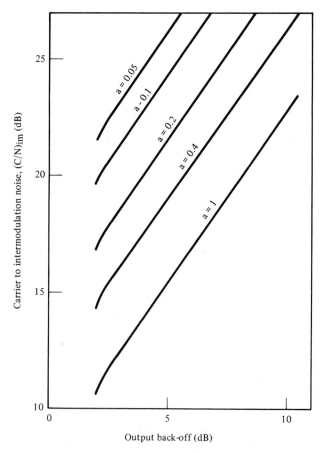

Figure 9-17 Effects of voice activation of intermodulation in SCPC voice circuits (a = activity).

notable advantages of SCPC multiple-access systems. Note that the advantage is much less with data channels. Their activity factors run much higher—perhaps between 0.7 and 0.9.

Beretta's intermodulation curves can be approximated linearly with

$$n = 500 \quad C/I_n = 8.60 + 0.82 BO_i$$

$$n = 12 \quad C/I_n = 9.48 + 0.82 BO_i \quad (9\text{-}29)$$

$$n = 6 \quad C/I_n = 10.00 + 0.82 BO_i$$

The curves of Figure 9-16 can be approximated by

$$\text{composite } C/I = 5.94 BO_i \; 0.40$$

including AM–PM, and (9-31)

$$\text{nonlinear } C/I_n = 7.15 BO_i \; 0.43$$

for intermodulation only.

9.5 OTHER ASPECTS OF TRANSPONDERS

9.5.1 Frequency Reuse and Transponder Gain Adjustment

Modern communications satellites often provide the capability to double the use of the available satellite bandwidth (typically 500 MHz) by providing dual-polarized transmit and receive antennas and two sets of transponders, one set for operation with each polarization. Additionally, the gain of each transponder chain can often be adjusted by ground command over a wide range of values as discussed in Section 6.4.3. This allows optimization of the earth station transmitter power and costs over a much wider range of earth station sizes and capacities than is possible with a fixed-gain transponder. The technique can double the capacity of the satellite system and is effective as long as the polarization isolation exceeds about 30 dB. Typical measured values are 33 to 35 dB.

9.5.2 Cross-Band Operation

In some satellite applications, a transponder must interconnect stations operating in two different frequency bands. An example of such operation is the maritime satellite service operating in L-band having to interconnect with the fixed satellite service operating in C-band. Transmissions from a ship at sea to shore are received at the satellite in the 1.6-GHz band uplink and transmitted to shore in the 4-GHz band. Transmission from shore to a ship at sea are received at the 6-GHz band and transmitted to the ship in 1.5-GHz band. In such a case, separate transponders are always provided for each direction of transmission.

REFERENCES

BELL TELEPHONE LABORATORIES, Members of the Technical Staff, *Transmission Systems for Communications*, 5th Ed., Bell Telephone Laboratories, Inc. 1982.

BERMAN, A., and C. MAHLE, "Nonlinear Phase Shift in Traveling-Wave Tubes as Applied to Multiple Access Communications Satellites," *I.E.E.E. Transactions on Communications Technology*, Vol. COM 18, No. 1, Feb., 1970, pp. 33–47.

BERRETTA, G., R. GOUGH, and J. T. B. MUSSON, "Improvements in the Characterization of High-power Amplifiers in Multicarrier Operation," *ESA Scientific and Technical Review*, Vol. 2, No. 2, 1976, p. 104.

BOND, F., and H. MEYER, *Intermodulation Effects in Limiter Amplifier Repeaters*, Aerospace Corporation Report, Contract No. FO4695-67-C-0158, Aerospace Corporation, 1967.

HEITER, G. L., "Characterization of Nonlinearities in Microwave Devices and Systems," *I.E.E.E. Transactions on Microwave Theory and Techniques*, Vol. MTT-21, December, 1973, pp. 797–805.

LORENS, C. S., *Intermodulation of Saturating Transfer Devices*, Aerospace Corporation Report TR0059(6510-06)-1, Aerospace Corporation.

MCCLURE, RICHARD B., *Link Power Budget Analysis for SPADE and Single-Channel PCM/PSK*, COMSAT Labs Technical Memorandum CL-12-71, INTELSAT R&D Task No. 211-4021, Communications Satellite Corporation, 1971.

MINKOFF, J. B., "Wideband Operation of Nonlinear Solid-State Power Amplifiers—Comparisons of Calculations and Measurements," *AT&T Bell Laboratories Technical Journal*, February, 1984, Vol. 63, No. 2, pp. 231–248.

PRITCHARD, W. L., unpublished notes.

SHIMBO, O., "Effects of Intermodulation, AM-PM Conversion, and Additive Noise in Multicarrier TWT Systems," *Proceedings of the I.E.E.E.*, Vol. 59, 1971, pp. 230–238.

Wass, J., "A Table of Intermodulation Products," *Journal of the I.E.E.*, 1948 No. 95, Pt. III, Pg. 31.

Westcott, J., "Investigation of Multiple f.m./f.d.m. Carriers Through a Satellite t.w.t. Near to Saturation," *Proceedings of the I.E.E.*, Vol. 114, No. 6, June, 1967.

10
Earth Stations

10.0 INTRODUCTION

We call the collection of equipment on the surface of the earth for communicating with the satellite an *earth station*, regardless of whether it is a fixed, ground mobile, maritime, or aeronautical terminal. We recognize that, with our broad concept of communications satellites, earth stations can be used in the general case to transmit to and receive from the satellite, but in special applications only to receive or only to transmit. Receive-only stations are of interest for broadcast transmissions from a satellite, and transmit-only stations for the still much less developed application of data gathering. Figure 10-1 is a quite general block diagram of an earth station capable of transmission, reception, and antenna tracking. We identify the following major subsystems:

> *Transmitter:* There may be one or many transmit chains, depending on the number of separate carrier frequencies and satellites with which the station must operate simultaneously.
>
> *Receiver:* Again, there may be one or many receiver/down-converter chains, depending on the number of separate frequencies and satellites to be received, and various operating considerations.
>
> *Antenna:* Usually one antenna serves for both transmission and reception, but not necessarily. Within the antenna subsystem are comprised the antenna proper, typically a reflector and feed; separate feed systems to permit automatic tracking; and a duplex and multiplex arrangement

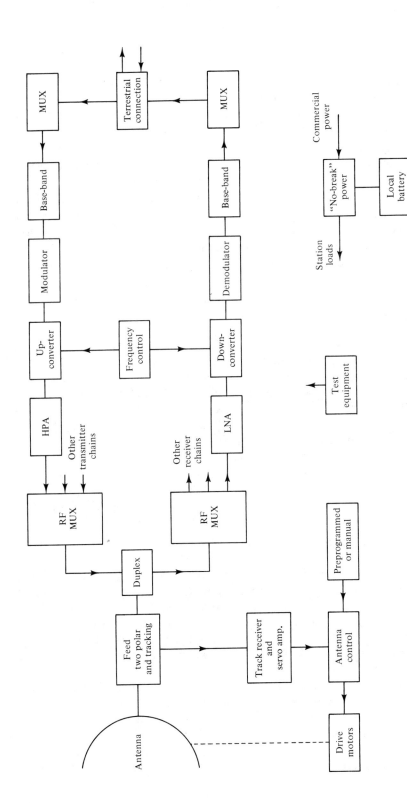

Figure 10-1 General earth station.

to permit the simultaneous connection of many transmit and receive chains to the same antenna.

Tracking system: This comprises whatever control circuit and drives are necessary to keep the antenna pointed at the satellite.

Terrestrial interface: This is the interconnection with whatever terrestrial system, if any, is involved. In the case of small receive-only or transmit-only stations, the user may be at the earth station itself.

Primary power: This system includes the primary power for running the earth station, whether it be commercial, locally generated, battery supplied, or some combination. It often includes provision for "no break" changeover from one source to another.

Test Equipment: This includes the equipment necessary for routine checking of the earth station and terrestrial interface, possible monitoring of satellite characteristics, and occasionally for the measurement of special characteristics such as G/T.

In the following sections we will deal, to varying extents, with each of these subsystems, viewing the earth station from the point of view of the complete system designer. We are interested in those aspects of the earth station that affect its communication link to the satellite and its ability to resist interference from other satellites and terrestrial systems. We are not concerned with the design of earth stations and certainly not with the detailed design of individual subsystems. Nonetheless, some knowledge of each subsystem is required in order intelligently to specify earth stations within a complete system and to ascertain what can be expected of them. Table 10-1 lists typical characteristics of some important categories of earth stations.

10.1 TRANSMITTERS

Transmitter subsystems vary from very simple single transmitters of just a few watts for data-gathering purposes to multichannel transmitters using 10-kW amplifiers, such as those found in Intelsat standard A stations. When multiple transmitter chains are required, common wideband traveling-wave tube amplifiers can be used, such as the arrangement shown in Figure 10-2, or each channel can use a separate high-power amplifier, typically a klystron, as shown in Figure 10-3.

Two-for-one redundancy switching is shown, by way of example, with the TWTAs. Numerous methods and levels of redundancy (e.g., three-for-two, four-for-three, etc.) exist. Similarly, multiplexer and filter arrangements are also multitudinous and only one scheme is shown. The common wideband amplifier is the more usual type, despite its suffering from the familiar problem of intermodulation when nonlinear amplifiers handle more than one carrier

TABLE 10-1 Typical Earth Station Characteristics

	\multicolumn{7}{c}{CATEGORY OF STATION}

	Intelsat Standard A	Intelsat Standard C	Domestic Systems	TVRO for Cable Distribution	Direct Broadcast Terminal	Maritime Mobile Terminal
Antennas	29–35 m, auto tracking	15–20 m, auto tracking crossed linear polarizers	5–15 m, step track	Multiple beam, 4–6 m, fixed, pointable	0.5–1.5 m, fixed, pointable	1–2 m (coupled to ship motion) auto track
G/T	40–44	35–50	C 30–34 K 40–44	C 18–24 K 10–15	K 6–14 S −6 to 0	−4.0
Frequencies	C-band Broadband	K-band Broadband	C-band K-band Broadband	C-band K-band Broadband	K-band S-band	L-band Broadband Tunable
Transmitters	10 kW, up to 24 carriers	2 kW, many carriers	1–8 kW, many carriers	—	—	10–200 W
Receivers	Cooled parametric-amps, up to 72 carriers	Cooled parametric-amp LNA, many carriers	Cooled parametric-amp LNA, many carriers	Transistor LNA	Transistor LNA	Transistor LNA
Multiple access	FDMA, TDMA	FDMA, TDMA, SCPC	FDMA, TDMA, SCPC	FM video, FDMA	FM video	SCPC, TDMA

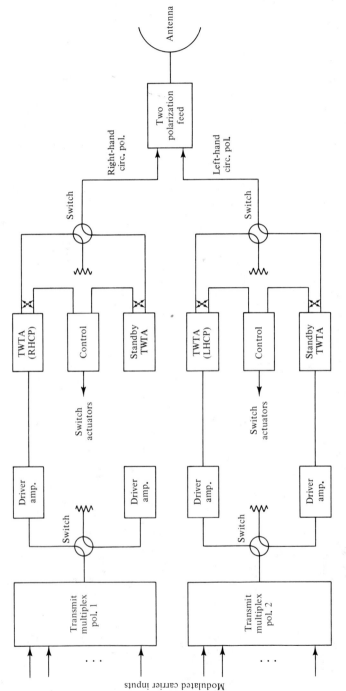

Figure 10-2 Common TWTA transmitter with redundancy (two polarizations).

Sec. 10.1 Transmitters

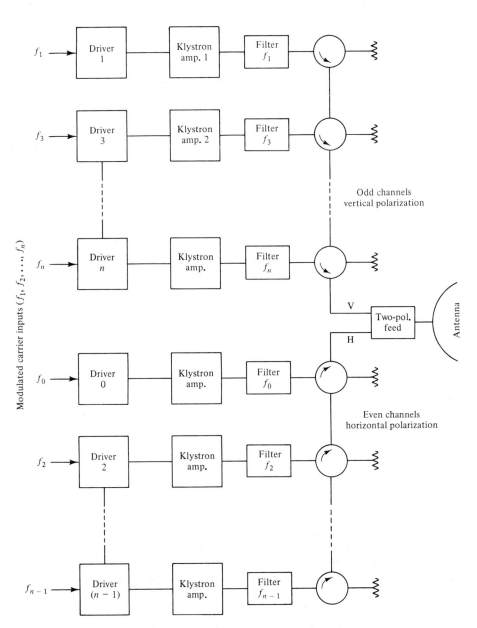

Figure 10-3 Multiple klystron transmitter.

simultaneously. Note that a transmitter carrier-to-intermodulation ratio that is not very high must be considered in calculating the overall $(C/N)_T$. It is added to the result using the formula derived in Chapter 6; that is, the intermodulation noise is assumed to be additive to the channel thermal noise. This problem is commonly dealt with by the use of back-off, but with considerable reduction in output power. Systems using feedback to reduce the nonlinearity effect are coming into use and allow greater power output. Another increasingly popular technique for improving carrier-to-intermodulation performance in high-power amplifiers uses predistortion. In this method, a low-level nonlinear amplifier of characteristics similar to those of the high-power amplifier is introduced ahead of the HPA, together with appropriate amplitude and phase equalizing, and a reference linear signal in order to generate an inverse, predistorted signal that, when applied to the high-power amplifier, will result in substantially decreased overall third-order intermodulation products. Miya (1982) gives results for three carriers which show that reductions of as much as 10 dB in third-order intermodulation products for three carriers can be achieved using this method. Because typical back-offs in high-performance earth stations can be of the order of 7 to 10 dB, this technique, by allowing substantial reductions in that back-off, can yield good improvements in efficiency.

The alternative of using separate amplifiers is less flexible in operation. Usually, the separate amplifiers are narrowband and require retuning to change frequencies. The problems of multiplexing many chains on an antenna without interaction among the amplifiers become still more complicated. The simple ways of effecting such combinations, such as the use of hybrids, also produce power losses, typically of the same order of magnitude as those involved in backing-off wideband amplifiers. Nonetheless, such systems are used from time to time since klystron amplifiers are currently simpler and cheaper than TWTAs. One can also argue that the reliability of such systems is higher because there are fewer single-point modes of failure. A few typical high-power amplifier specifications are shown in Table 10-2.

10.2 RECEIVERS

To receive a signal from a satellite, several distinct operations must be performed. The signal must first be amplified, then reduced to a frequency low enough for convenient further amplification and demodulation, then demodulated and delivered to whatever baseband processing equipment is needed. The signal may be used either at the earth terminal itself, say in the case of a home TV receive-only (TVRO) terminal, or converted into a form suitable for transmission elsewhere. When we speak of the receiver chain, we refer here specifically to the low-noise amplifiers, down-converters, and demodulators. Down conversion can be accomplished either in one step, going directly from the satellite downlink carrier frequency to the intermediate

TABLE 10-2 High-Power Amplifier Characteristics

Type	C-Band TWT (low power)	K-Band TWT (high power)	C-Band TWT (high power)
Frequency range (GHz)			
Lower limit	5.925	14.0	5.925
Upper limit	6.425	14.5	6.425
Bandwidth (MHz)	500	500	500
Power output (W)			
Typical	42	250	400
Minimum	35	225	350
Gain (dB)			
At rated output, minimum	42	60[a]	40[a]
Typical	45	—	70
Gain variation (dB) in 500 MHz	2	1.5	±1.0
Gain slope maximum (dB/MHz) in bandwidth	0.05	0.02[b]	0.02[c]
Gain stability (dB/day ±)	0.25	0.25	0.25
Noise figure (dB) maximum	35	37	37
VSWR (load)	1.5	2.0	2.0
Harmonic output at rated power (dBc)	8	−60	−60
Residual AM			
f_1 (kHz)	0.3	10	10
f_2 (kHz)	3.4	500	500
dBc below f_1 (kHz)	—	−40	−40
dBc between f_1 and f_2	—	−20	−20
dBc above f_2		$1 + \log f$	$1 + \log f$
dBc above f_2	60	−80	−80
At line-frequency harmonics	40	—	—
Spurious output (dBW/4 kHz)			−130
	—	−65	−65
Group delay over center			
Bandwidth of MHz	36	50	40
Linear (ns/MHz)	0.1	0.1	0.1
Parabolic (ns/MHz)	0.05	0.05	1.05
Ripple (ns peak to peak)	1.0	1.0	1.0
AM to PM conversion (single carrier)			
At output power (W)	42	250	400
0/dB	5	5	5

[a] With single-sideband driver.
[b] Over 50-MHz band.
[c] Over 40-MHz band.

demodulator frequency (characteristically 70 MHz), or it can be accomplished in several steps. Two-stage down conversion is often done when the same receiver is to be tuned to a multiplicity of channels. Figure 10-4 is a prototypical receiver chain for a quite general case. LNA redundancy is shown to illustrate the switching, but other redundancies, such as for the down-con-

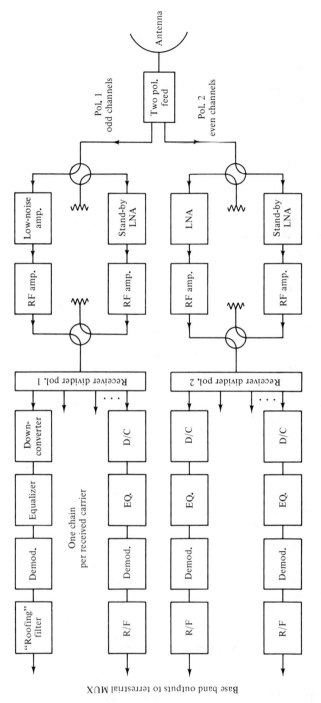

Figure 10-4 Receive subsystem for multicarrier earth station.

verter, are not indicated, although they are common. Again, as in the case of transmitters, the variety of possibilities for switching and multiplexing is considerable. Figure 10-5 is the interesting special case of a TVRO station. Such stations can be used to feed cable and local TV redistribution systems and in the extreme case as terminals for the direct reception of TV in the home. Double conversion is generally used.

The low-noise amplifier is one of the critical elements in determining the earth station performance as a system element. This performance is characterized by the familiar figure of merit, G/T, as shown in Section 6.2. It is determined by the antenna gain, discussed in Section 10.3, and the system temperature, the expression for which was developed in Chapter 6 and is repeated here for convenience.

$$T_s = T_a + (L - 1)T_0 + LT_R + \frac{L(F - 1)}{G_R} T_0 \qquad (6\text{-}52)$$

where all terms are expressed in the "dB" form.

The second term in this equation is the clue to the reason for many decisions in earth station design. Even if the antenna temperature is very low, as is normally the case in clear weather, and the excess temperature of the receiver T_R is also low, perhaps 50 K, the system temperature can be surprisingly high if there is even a small loss in the transmission line between the antenna and low-noise amplifier. The term $(L - 1)T_0$ for a loss of 0.5 dB is 35 K, as high as the receiver itself in low-noise cryogenic systems. This makes it a necessity in high-performance earth stations for this transmission line to be made as short as practical. It explains the almost universal use of Cassegrainian antennas in large, high-performance systems, especially at C-band and below, where antenna temperatures can be expected to run low even in the presence of rain. On the other hand, at higher frequencies, designed of necessity to have a high rain margin, the antenna temperature under those rainy circumstances will be several hundred degrees; a certain amount of waveguide loss is now tolerable since it produces less proportionate deterioration. Additionally, at these higher frequencies, the antennas are smaller since higher levels of transmitter power are normally used in the satellites. Physically smaller antennas are also desirable at higher frequencies to keep the beamwidths from becoming too small. These small antennas make it practical to use prime-focus-fed reflectors (see Section 10.3) and still have adequately short lengths of waveguide to receiver locations.

In addition to the excess temperature of the receiver, a number of other characteristics are of importance in determining the station performance, notably those that affect the degree of transmission impairments, such as group delay, gain stability, gain flatness, and AM-to-PM conversion factors. Their effects are handled in the assessment of overall performance very much as discussed in Chapter 9. A set of typical low-noise amplifier characteristics is shown in Table 10-3. By way of clarification, the term "intercept point" in

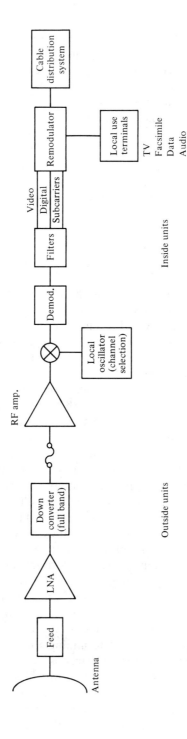

Figure 10-5 General TVRO station—direct reception or cable distribution.

Sec. 10.2 Receivers

TABLE 10-3 LOW NOISE AMPLIFIER CHARACTERISTICS

Type:	K-Band Parametric amp	K-Band Parametric amp	C-Band Parametric amp
Cooling method	Thermoelectric	Cryogenic	Thermoelectric
Frequency range (GHz)			
Lower limit	11.7	3.7	3.7
Upper limit	12.2	4.2	4.2
Bandwidth (MHz)	500	500	500
Noise temperature (K)			
Typical	100	13	40
Maximum	110	16	—
Gain (dB)			
Minimum	—	—	55
Typical	0.55	60	—
Gain flatness (dB)			
10 MHz	0.2	0.3	—
500 MHz	0.5	0.5	0.5
Gain slope (dB/MHz)	0.02	0.02	0.02
Output power			
Compression (dB)	0.5	0.5	0.5
Power (dBm)	−5	0	−5
VSWR			
Input	1.2	1.25	1.2
Output	1.2	1.25	1.2
Intermodulation (intercept point)			
dB below two carriers	58	51	58
At output level (dBm)	−64	−63	−64
Group delay (over 40 MHz)			
Linear (ns/MHz)	0.1	0.1	0.1
Parabolic (ns/MHz^2)	0.03	0.01	0.03
Ripple (ns peak to peak)	0.5	0.5	0.5
AM-to-PM conversion level (dBm)	−60	−70	−60
Conversion (0/dB)	0.5	0.5	0.5
Temperature range	0–50°C	0–50°C	0–50°C

connection with intermodulation is worth a note. Low-noise amplifiers, like all amplifiers, saturate and thus have the standard problems of intermodulation. It has become common to specify this intermodulation by the "intercept point," the point at which the extended linear portion of the curve third-order intermodulation products extended would intercept the extended linear portion of the amplifier's power transfer characteristic, as seen in Figure 10-6. It is not difficult to derive an expression for carrier-to-third order modulation products, the more useful measure in computation, from this diagram. The result is

$$(C/I)_3 = 2(P_x - P_0) \tag{10-1}$$

P_0 is the saturated output power and P_x is the intercept point, both usually

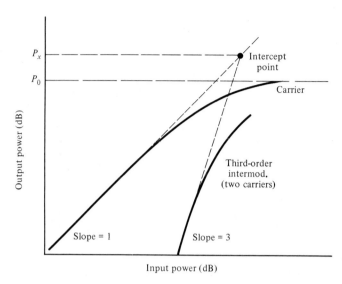

Figure 10-6 Intercept point as a measure of third-order intermodulation level.

taken in dBm. This result depends on the fair assumption that the total of the third-order modulation products varies with the cube of the input power.

10.3 ANTENNAS

10.3.1 General

The parabolic reflector antenna has become the symbol for a satellite communication earth terminal. The synecdoche "dish" for an entire earth terminal has become universal in popular usage. This symbolism is reasonable, not only because the antenna so distinguishes the terminal physically, but also because its characteristics are by far the most important of all in determining the overall earth station performance, both on the uplink and downlink. As we have seen in Chapter 6, the carrier-to-noise ratios achievable on these links, given fixed transmitter powers and geographical coverages, are directly determined by the physical size of the earth station antenna. By now, no one should need further convincing that, for a given system and desired performance, antennas at K-band must be larger than those at C-band, not smaller. This is because of the rain attenuation. The antenna sizes would otherwise be independent of frequency as long as the noise levels are not determined principally by interference. The reason that K-band antennas are generally found to be smaller is simply that more power is used in the satellite transmitters.

Sec. 10.3 Antennas

The antenna electrical performance is involved in the system planning in many ways, the most important are the following:

Characteristic	Affects:
Overall gain, G	System G/T_s
Antenna temperature, T_a	G/T_s
Sidelobe level (including spillover)	Interference (C/I), antenna temperature
Cross-polarized response	C/I and C/N for entire system
Beamwidth	Geographical coverage (satellite antennas), tracking requirement

The next section will consider some of the important ideas in antenna design. The relation of these characteristics to antenna geometry is the province of the antenna designer—a major engineering specialty in itself.

The electrical characteristics are by no means independent of each other, and their optimization for a particular system is a joint effort of the antenna designer and systems engineer. For system planning, a generalized antenna pattern is often useful. A good pair of equations for such use is:

On main lobe:

$$\frac{G}{G_m} = \left[\frac{\sin 1.39\ (\theta/\theta_0)}{1.39\ (\theta/\theta_0)}\right]^2 \quad (10\text{-}2)$$

Far from main lobe:

$$\frac{G}{G_m} = \frac{1}{1 + (\theta/\theta_0)^{2.5}} \quad (10\text{-}3)$$

Note that θ_0 is half the half-power beamwidth. The beamwidths of the antenna are also related to its gain in a geometric fashion, quite independently of the antenna geometry. Gain is defined as the ratio of radiation intensity in a given direction to that it would have were the total radiated power to be radiated isotropically. It comprises two elements: The *directivity*, the component of the ratio determined by the geometry of the antenna system, and The effect of *losses* due to such factors as dissipation and spillover. The "directivity" part is the more important and thus, as a surprisingly good working relation, one can use

$$G \simeq \frac{4\pi}{\theta_1 \theta_2} \simeq K \frac{41{,}253}{\theta_1^0 \theta_2^0} \quad (10\text{-}4)$$

where K is a factor to allow for energy not in the main beam (it is about 0.65). θ_1 and θ_2 are the antenna beamwidths in radians or degrees, as appropriate. The equation follows from the assumption that the radiated power is confined principally to the main lobe, instead of being radiated over 4π

steradians isotropically, and is modified approximately by a factor to allow for total energy in all the side lobes.

Although the parabolic reflector is by far the most important kind of antenna that we find, both in earth stations and on the satellite, nonetheless other types are important, particularly *horns*. They are used widely as primary feeds for reflectors, and occasionally as principal radiators themselves. Two other kinds are occasionally seen, especially in spacecraft. They are *lenses* (either the dielectric or waveguide types) and *phased arrays*. The latter is not a different type of radiator per se, but simply a controlled combination of any kind of individual element. For instance, horn feeds, dipoles, and even parabolic reflectors can be used in arrays with the composite pattern determined by conventional antenna array theory. The array is controlled by varying the phase and amplitude of the excitation to the individual elements. We do expect that this kind of antenna will become increasingly important in spacecraft design as the carrier frequencies get lower. This will be the case for satellite service to small mobile terminals because of the requirement for frequency reuse to provide many channels and the concomitant narrow beamwidths. Arrays will be the easiest way to achieve the large apertures. The theory of arrays of arbitrary elements is classical and well developed in many texts (Jasik, 1961; Silver, 1949; Kraus, 1950).

10.3.2 Horn Antennas

Horn antennas are commonly used as primary radiators in reflector systems and sometimes as complete radiators when wide beamwidths are required. Frequently, one finds horn antennas on board the satellite to provide earth coverage beams. That angle is about 18° from geostationary orbit and simply achieved with horns.

We find two kinds of horns in common use: the *pyramidal horn* as an extension of rectangular waveguide, and the *conical horn* as an extension of circular guide. Pyramidal horns are easily designed and the following equations are applicable for those horns long compared to a wavelength:

$$G = 10 \frac{AB}{\lambda^2}$$

$$= 51 \frac{\lambda}{B} \qquad (10\text{-}5)$$

$$= 70 \frac{\lambda}{A}$$

where A is the longer dimension of the horn aperture. If it is desired to have the shortest length possible, that length, L_1, is given by

$$L_1 = L\left(1 - \frac{a}{2A} - \frac{b}{2B}\right) \tag{10-6}$$

Figure 10-7 permits the design of an optimum-length horn. Conical horns, which are the natural extension of circular waveguides, are often used and typically exploit higher-mode propagation. If the TM_{11} and TE_{11} modes in circular waveguide are superimposed on each other with suitable control of the relative amplitude and phase, the composite radiation pattern can be

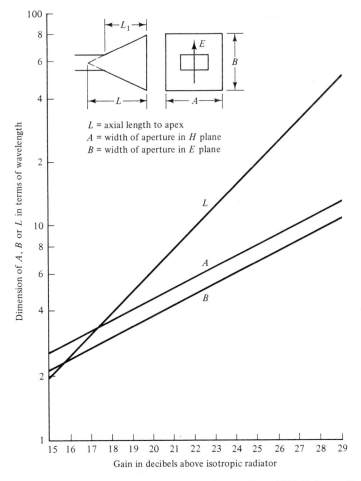

Figure 10-7 Design of electromagnetic-horn radiator. (From ITT, *Reference Data for Radio Engineers*, reprinted by permission of Howard W. Sams & Company, Inc., Publishers, Indianapolis, Ind., 1975.)

improved over that of the single-mode horn in either rectangular or circular guide. Another variation of the horn feed very much used in primary feeds for big earth stations is a *hybrid-mode* horn. Annular corrugations are placed on the inner wall of a circular waveguide in such a way that neither TE or TM modes can be propagated, but instead a hybrid mode is generated. These antennas can be used to improve cross-polarization and sidelobe performance, and also to achieve axially symmetric beamwidths. Miya (1982) gives considerable detail on the theory of this particular kind of horn. Conventional single- and multiple-mode horn feeds are discussed in both Silver (1949) and Jasik (1961).

10.3.3 Reflector Antennas

We divide the reflector antennas broadly into two categories: those using a single reflector and horn feed and those using multiple reflectors. In the first category, we have the familiar prime focus feed (Figure 10-8) and the offset-fed parabolic reflectors; in the second, we have a family of antennas developed by analogy to astronomical telescopes and thus called Newtonian, Cassegrainian, and Gregorian. The latter categories depend on whether the

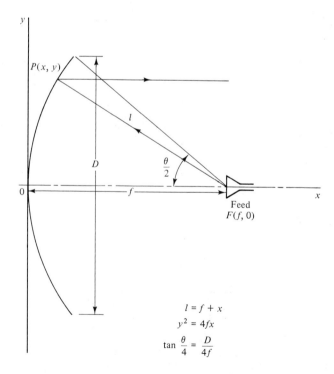

$$l = f + x$$
$$y^2 = 4fx$$
$$\tan \frac{\theta}{4} = \frac{D}{4f}$$

Figure 10-8 Basic geometry: prime-focus-fed parabolic reflector.

Sec. 10.3 Antennas

subreflector is plane, hyperbolic, or ellipsoidal. These antennas are shown in a convenient summary chart in Figure 10-9 and in more detail in Figures 10-10 through 10-12. Literally dozens of variations on these arrangements are possible. Several themes can be kept in mind to understand the variations. The first is that a paraboloidal reflector will take spherical waves emerging from a point source at the focus of the paraboloid and convert them into the desired plane wavefront. The distance from the focus on the paraboloid to the reflector, plus the distance from the point of reflection to any plane surface normal to the axis, must be a constant. This is in fact the fundamental optical requirement of all reflector antenna systems. If subreflectors are used, the multiple reflected distances must be added in satisfying the requirement. Another notion that helps in understanding is that whereas the paraboloidal reflector will convert a spherical wave into a plane wave, hyperboloidal and ellipsoidal reflectors will leave spherical waves emerging from a focus unchanged. They simply seem to have originated at a different focus. This explains, for instance, the modified Cassegrainian antenna (Figure 10-12) seen occasionally in which the horn feed located at the vertex of the main reflector is paraboloidal. As a result, it is necessary to use a paraboloidal subreflector to convert the plane waves from the prime source back into spherical waves, which in turn get converted back into plane waves by the main paraboloid.

An important effect of the secondary reflector on a Cassegrainian or Gregorian antenna is to increase the apparent focal length of the antenna. This increase is called *magnification* by analogy to the optical case, and proceeding from the geometric definitions—that the ellipse and hyperbola are, respectively, the loci of points for which the sum of or difference between the distances to two fixed points is a constant (the two fixed points being called the foci)—one can demonstrate that the equivalent focal length of the Cassegrainian reflector system is given by

$$f_e = mf = \frac{e+1}{e-1}f \tag{10-7}$$

The geometry is seen in Figure 10-10. The angle subtended at the focal point F_2 is very much less than it would be if the feed were located at the virtual focus, F_1. The feed located at F_1 can be designed as if the focal length, and thus the ratio f/D, were longer by the factor m. This makes the realization of high aperture efficiencies and low cross-polarization components much easier. Values of m typically range from 2 to 6.

If several horn feeds, with emerging beams at different angles, are to be used, it is possible to use a main reflector that is circular in one cross-section and parabolic in the other (Figure 10-13). This kind of toroidal antenna was first used in large early-warning radars to permit rapid beam scanning over perhaps a 70° sector. As such, it is also useful when one antenna is to be used with several satellites. The circular cross-section produces spherical aberration in one axis which is correctable in the horn feed design. It is

Type	Ray diagram	Optical elements	Pertinent design characteristics
Paraboloid		Reflective M_p = paraboloidal mirror	1. Free from spherical aberration 2. Suffers from off-axis coma 3. Available in small and large diameters and f/numbers 4. Low IR loss (reflective) 5. Detector must be located in front of optics
Cassegrain		Reflective M_p = paraboloidal mirror M_s = hyperboleidel mirror	1. Free from spherical aberration 2. Shorter than Gregorian 3. Permits location of detector behind optical system 4. Quite extensively used
Gregorian		Reflective M_p = paraboloidal mirror M_s = ellipsoidal mirror	1. Free from spherical aberration 2. Longer than Cassegrain 3. Permits location of detector behind optical system 4. Gregorian less common than Cassegrain
Newtonian		Reflective M_p = paraboloidal mirror M_s = reflecting prism or plane mirror	1. Suffers from off-axis coma 2. Central obstruction by prism or mirror

Type	Ray diagram	Optical elements	Pertinent design characteristics
Herschelian	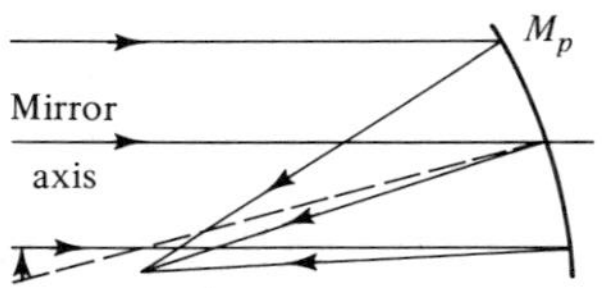	Reflective M_p = paraboloidal mirror inclined axis	1. Not widely used now 2. No central obstruction by auxiliary lens 3. Simple construction 4. Suffers from some coma
Fresnel lens	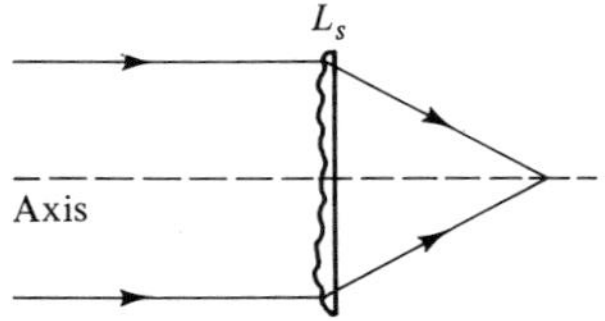	Refractive L_p = special Fresnel lens	1. Free of spherical aberration 2. Inherently lighter weight 3. Small axial space 4. Small thickness reduced infrared absorption 5. Difficult to produce with present infrared transmitting materials
Mangin mirror	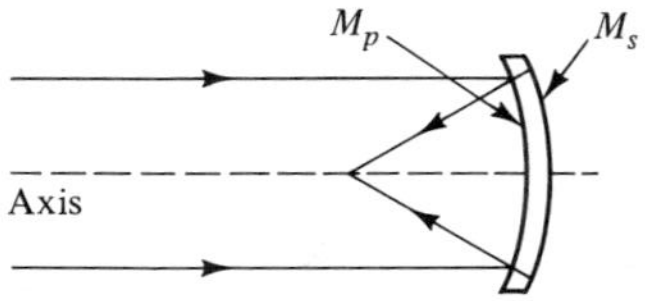	Refractive–reflective M_p = spherical refractor M_s = spherical reflector	1. Suitable for IR source systems 2. Free of spherical aberration 3. Most suitable for small apertures 4. Covers small angular field 5. Uses spherical surfaces

Figure 10-9 Quasi-optical apertures. (Reprinted with permission from G.F. Levy, "Infrared System Design," Electrical Design News, May 1958, Table I.)

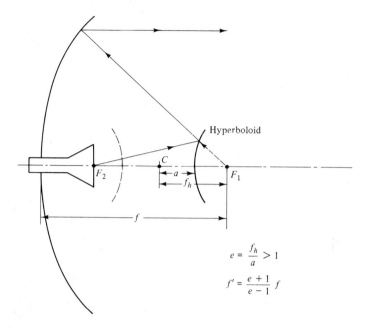

Figure 10-10 Basic Cassegrainian antenna.

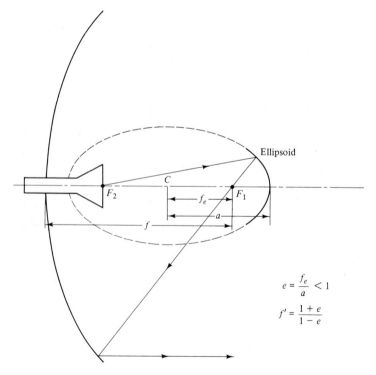

Figure 10-11 Basic Gregorian antenna.

Sec. 10.3 Antennas

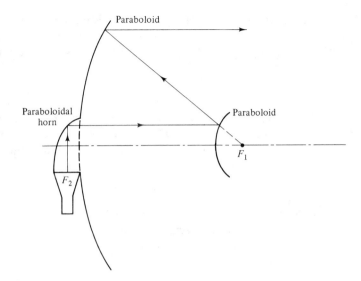

Figure 10-12 Near-field or modified Cassegrainian antenna.

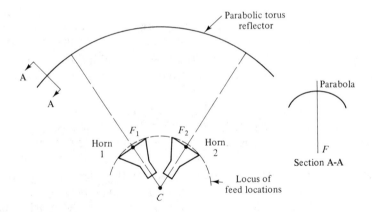

Figure 10-13 Multifeed toroidal antenna.

important to note that there is no saving in total aperture, but there can be a saving in cost and complexity as the number of beams (feeds) is equal to three or more.

10.3.4 Antenna Performance

The easiest way to compare antennas for system performance is by considering them as illuminated apertures. The secondary radiation pattern from an illuminated aperture can be shown to be a Fourier transform of the primary pattern and quite general relations can be found among such critical parameters as size, illumination taper, sidelobe level, directivity, and beamwidth. The universal antenna formula relating the effective area (or capture cross-section) of the antenna A_e and its gain and wavelength is the familiar

$$A_e = \frac{G\lambda^2}{4\pi} = \eta A \qquad (10\text{-}8)$$

The effective or capture area is related to the physical area A by the overall efficiency η. This overall efficiency η, which must be used in calculating received carrier level, is itself the product of several constituent efficiencies. Thus

$$\eta = \eta_a \eta_b \eta_s \eta_p \eta_e \eta_L \qquad (10\text{-}9)$$

where η_a = *aperture efficiency*, the result of nonuniform illumination, phase errors, and so on; it increases as the sidelobe level increases

η_b = *blockage efficiency*, resulting from blockage of main reflector by the subreflector or feeds

η_s = *spillover efficiency*, the loss of energy because the subreflectors and main reflector do not intercept all the energy directed toward them

η_p = *cross-polarization efficiency*, the loss of energy due to energy coupled into the polarization orthogonal to that desired

η_e = *surface efficiency*, the loss in gain resulting from surface irregularities, the statistical departure from a theoretically correct surface

η_L = *ohmic and mismatch efficiency*, the loss from energy reflected at the input terminals (VSWR > 1.0) and that dissipated in ohmic loss in the conducting surfaces, dielectric lenses, and so on

Aperture efficiency is an important but subtle concept. It describes the degree to which an illuminated aperture achieves the *directivity*, and thus the *gain*, which its area would imply (Kraus, 1950; Silver, 1949). The aperture efficiency, η_a, is equal to unity for an aperture which is illuminated uniformly in amplitude and phase, in which case the directivity is the maximum for the given area. In this case, however, the gain of the first sidelobe is only 13 dB below that of the main lobe.

By "tapering" the intensity of illumination toward the edge of the aperture, the relative sidelobe gain decreases, but so also does η_a. Thus, the on-axis gain decreases, and, as a corollary, the width of the main beam increases (See Eq. (10-4)). At the same time, the taper of illumination likely reduces the spillover (a cause of loss of overall efficiency). The taper, however, must be produced by narrowing the beam of the feed system. This in turn implies an increase in the size of the feedhorn (or a subreflector) and thus may increase blockage, giving decrease in overall efficiency. The reduced

Sec. 10.3 Antennas

spillover will not only reduce the value of η_s but also may improve the antenna temperature, T_a (see Chapter 6), since less energy is interchanged with the ground. Thus the compromise to achieve optimum G/T is indeed complex. To assist the system designer in knowing the possibilities, we have Table 10-4, which gives beamwidths, aperture efficiencies, and sidelobe levels for several common primary patterns—achievable with horn feeds. The chart assumes constant phase across the aperture. Since this is not easy to achieve perfectly, the actual aperture efficiencies will be slightly lower.

In practice, the compromises are made by controlling the edge illumination or taper. For a "cosine"-type horn feed, practical designs can be arrived at in a simpler way. We start with the desired efficiency and a certain edge taper, say from Figure 10-14.

Note that this "taper" in reflector illumination has two components: one due to the horn feed pattern, as shown in Table 10-4, and one due to the inherent reflector geometry that would be present even with a uniform primary pattern. The second term is sometimes called *space attenuation* and is simply the difference in inverse square-law loss between the edge and center of the aperture. From the geometry of the parabola, it can be shown that this loss is given by

$$\text{space attenuation} = \left(\frac{R}{f}\right)^2 = \sec^4 \frac{\theta}{4} \qquad (10\text{-}10)$$

where θ is the full angle subtended by the reflector at the horn.

A good approximation to a "cosine" horn pattern (in decibels) is simply $10(\theta/\beta_{10})^2$, where β_{10} is the horn beamwidth at the tenth power point. For preliminary planning, it can be used as a universal feed pattern. Thus the net

TABLE 10-4 Aperture Characteristics versus Illumination Pattern

Illumination across Aperture		Aperture Efficiency, η_a	Half-Power Beam Width[a]	First Null[b]	First Sidelobe dB[c]
Uniform	$n = 0$	1.00	50	57	−13.2
	$n = 1$	0.810	69	86	−23
$\cos^n \frac{\pi x}{2a}$	$n = 2$	0.667	83	115	−32
	$n = 3$	0.575	95	143	−40
	$n = 4$	—	111	11	−48
$1 - \varepsilon x^2$	$\varepsilon = 1$	0.833	66	82	−20.6
	$\varepsilon = 0.5$	0.970	56	65	−17

[a] as a multiple of λ/D, where D is the physical diameter of the aperture.
[b] from axis as a multiple of λ/D.
[c] with respect to on-axis gain.

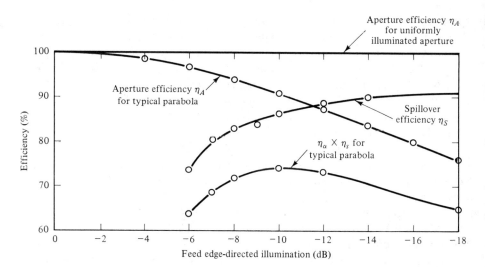

Figure 10-14 Typical spillover and aperture efficiencies.

edge taper T is

$$T = 10 \log \sec^4 \frac{\theta}{4} + 10\left(\frac{\theta}{\beta_{10}}\right)^2 \qquad (10\text{-}11)$$

and the horn feed beamwidth β_{10} can be chosen so as to achieve the desired taper. θ is determined from the f/D ratio and the reflector system geometry. The main reflector beamwidth is calculated using the beamwidth factor that goes with the chosen aperture efficiency in Table 10-4.

In addition to the aperture efficiency just discussed, we have a variety of other factors that contribute to the overall capture efficiency. Some of them, notably spillover, tend to vary contrary to the aperture characteristic as seen in Figure 10-14. The optimum system performance may be achieved using a slightly higher taper than shown, since the spillover has the doubly bad effect of both reducing that constituent of the gain and deteriorating the antenna temperature.

Aperture blockage is a significant problem, especially in Cassegrainian and Gregorian antennas. As a good approximation, the related efficiency η_b is given by $[1 - \eta_a(A_B/A)]^2$, where A_B is the blocked area and A is the total aperture area.

Cross-polarization efficiency η_p is another important one in satellite antennas. For a symmetrically illuminated antenna, it depends on the curvature of the reflector and thus on the f/D ratio. Cassegrainian antennas with magnifications of 2 or greater are extremely good in this respect, but focal point feeds can be somewhat poorer. Off-center feeds are still worse. The loss will depend on the fourth power of the subtended angle at the primary feed.

Sec. 10.3 Antennas

There is always a fundamental loss in efficiency because of random surface irregularities. Ruze (1952) in a classic paper developed the following equation for the effect of surface variation:

$$\eta_e = \frac{G}{G_0} = e^{-k(4\pi\delta/\lambda)^2} \qquad (10\text{-}12)$$

$$k = \frac{1}{1 + (D/4f)^2} \simeq 1 \qquad (10.13)$$

These equations hold for Gaussian distribution of phase errors due to surface imperfections. Correlation intervals should be small compared to aperture and comparable to or greater than a wavelength. δ is the mass surface deviation and G_0 is the gain of a perfect surface reflector. It is a good, practical equation. For multiple reflectors, δ^2 should be the sum of squares of the values for the various surfaces. The effect of Ruze's equation is to put a practical upper limit on the gain of a constant-size antenna as the wavelength decreases. This limit seems to be somewhere around 70 dB today, and any system plan calling for an antenna gain in that region should be reviewed carefully with antenna experts.

An often overlooked loss in gain is that due to resistive losses and antenna mismatch. This loss is not part of the radiation characteristic of the antenna, but is nonetheless effective in reducing the system performance. Not only do resistive losses reduce the gain, but they also increase the noise temperature.

The overall gain, taking into account all these effects, is the one to be used in system planning. The composite of all the efficiency factors for large and expensive earth stations today seem to run between 0.6 and 0.7; for smaller, cheaper stations between 0.55 and 0.6. Home terminals could well be assumed conservatively at 0.5 for system planning.

10.3.5 Cross-Polarized Systems

Often, the satellite channel capacity is limited by the available bandwidth and, if sufficient power is available, it is desirable that the assigned frequency band be used as many times as possible. This can be achieved in the satellite by spot beam antennas, as discussed in Chapter 9, but in addition, and now almost universally, two polarizations are used in satellite communication systems to achieve at least a two-for-one frequency reuse. Polarizations used can be either crossed linear, that is, vertical vs. horizontal, or counterrotating (left- vs. right-hand circular polarization). Keep in mind that circular polarization may be considered simply the combination of two linearly polarized waves in both axis and time quadrature. Typical dual-polarization channelization plans are shown in Figure 10-15. As a first approximation, both systems are equally effective at isolating two beams at the same frequency from each

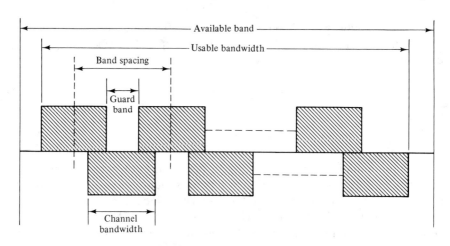

Figure 10-15 Interlaced, cross-polarized channelization.

other, and the choice between the systems becomes a matter of practical technique. Circular polarization systems have the advantage of not requiring polarization orientation, which can be important in simple systems, but they suffer more from depolarization during heavy rain. Conversely, crossed linear systems require polarization orientation and sometimes readjustment as satellite pointing is changed, but tend to perform slightly better in the presence of rain. The crossed-polarization isolation of an antenna is thus an important parameter. In this respect, the Cassegrainian family of antennas is extremely good because of both the long effective focal length and the axial symmetry. Offset antennas, sometimes used for small terminals and multiple beams, suffer from a lack of axial symmetry and therefore have poor crossed-polarization characteristics. On the other hand, such antennas are free of the aperture blockage problems that plague the Cassegrainian family. As a result, their sidelobe levels can be made extremely low, not much above the theoretical level attributable to the illumination taper and aperture size.

To produce circularly polarized waves, it is necessary to shift one linear component (either vertical or horizontal) by 90° in time relative to the other and to maintain this phase shift over a wide band of frequencies. This is not easy, and the antenna curvature itself disturbs the phase relationship. If the required axis and phase quadratures are not maintained, or if the two components are not of equal magnitude, an *elliptical* rather than circular polarization is created. This can be shown to be resolvable into two circularly polarized signals of opposite "hand," with field strength magnitudes E_L and E_R, respectively. Assume that for this "channel" left-hand polarization was intended. The E_R component will now represent a type of crosstalk into the receiver channel which is intended for our other signal (with right-hand polarization intended). Thus, the systems' discrimination between polarizations is reduced by "ellipticity" to the *circular polarization ratio*.

$$p = \frac{E_R}{E_L} \qquad (10\text{-}14)$$

which in dB form is called the *cross-polarization* discrimination

$$\text{XPD} = 20 \log p \qquad (10\text{-}15)$$

The ratio between the amplitudes of the electric fields along the major and minor axes of the polarization ellipse, the *axial ratio*, is

$$r = \frac{E_L + E_R}{E_L - E_R} \qquad (10\text{-}16)$$

The two ratios, r and p are related by

$$r = \frac{p+1}{p-1} \qquad p = \frac{r+1}{r-1} \qquad (10\text{-}17)$$

Imperfect cross-polarization discrimination, whether for linear or circular polarization systems, effectively yields another carrier-to-interference ratio that must be considered in assessing overall performance.

10.3.6 Antenna Mounts

The antenna must be pointed at the satellite. Rarely, this pointing is fixed permanently; sometimes it is occasionally adjusted, and in some installations it is continually driven by a tracking system. Such tracking systems will be discussed in Section 10.4, but in the meantime we note that every earth station antenna must be capable of some adjustment in pointing, even if only for initial setup. Such adjustments come in three categories geometrically. The simplest, and indeed the most flexible, is the *elevation over azimuth system*, usually called "Az-El" (Figure 10-16a). In this system the azimuth is determined by rotation about a vertical axis normal to the local horizontal plane, and the elevation is adjusted about a horizontal elevation axis which is in turn normal to the vertical azimuth axis. This system is simple, effective, and capable of use almost anywhere. There is some difficulty with automatic tracking systems if the earth station is to be located near the equator. Tracking through the zenith is awkward with Az-El systems, and if that is required, it is common to use a system in which one axis is parallel to the ground and another axis is normal to it, as shown in Figure 10-16b. This kind of mount, commonly called *XY*, can also be used to point anywhere, but is awkward for tracking close to the horizon. An interesting third possibility, borrowed from astronomy, is the equatorial mount (Figure 10-16c). One axis, the *hour-angle axis*, is parallel to the axis of the earth, and the other axis, normal to it, produces the desired *declination*. This mount is used universally

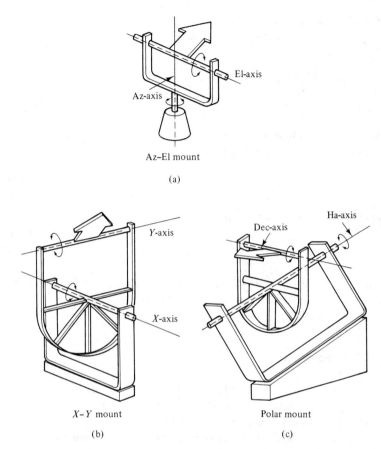

Figure 10-16 Antenna mounts: (a) azimuth over elevation; (b) Y over X; (c) equatorial or polar. (Reprinted with permission from K. Miya, *Satellite Communications Engineering*, 2nd ed., Lattice Company, Tokyo, p. 232.)

for astronomical telescopes since it permits the automatic tracking of a celestial object despite the rotation of the earth. It is easily applicable to satellite systems. An interesting complication appears if the mount is to be sufficiently flexible, as most mounts are, to point anywhere on the visible geostationary arc. If the earth station were located on the equator, it would be necessary only to make an adjustment in the hour-angle axis. The declination would be everywhere 0°. Even for locations at latitudes as high as 40° or 50°, the change in declination as the hour angle is varied to point along the geostationary arc is rather small. If the beamwidth of the antennas is wide compared to the associated change in declination, as is often the case with small terminals, it is possible to point to satellites at different longitudes by varying only one axis. The error is small enough to be acceptable in many systems. We expect that this approach will be used increasingly in small- and medium-sized antenna terminals at latitudes below 45° or 50°. The equations in Chapter 3 can be used to calculate the geometry for any of these pointing systems, and in particular, the error to be dealt with in using single-axis pointing. The easy way is to use the astronomer's approach to calculating the parallax error in declination because his telescope is not located at the center of the earth but at a particular latitude. The equation for that error is given at the end of Chapter 3 in the discussion on the nonspherical earth and can be used to calculate the declination parallax error in a single-axis drive.

10.4 TRACKING SYSTEMS

Not only is it necessary to be able initially to point the antenna in any desired direction, but it is frequently necessary to change this pointing, not only to switch from one satellite to another, but also to follow the residual orbital motions of the satellite and to allow for wind deflection of the antenna. Clearly, the necessity for such tracking increases as the beamwidth of the antenna gets narrower. We identify a hierarchy of pointing and tracking necessities as follows:

1. No tracking is necesaary and only initial fixed-pointing adjustment is required.
2. Repointing of the antenna is needed to switch from one satellite to another and possibly to correct for satellite motion. This repointing can be needed rarely or frequently.
3. Tracking is required, but it is satisfactory to drive the antenna in two axes and to preprogram this drive in accordance with the calculated satellite drift.
4. Automatic tracking is necessary but can be achieved by a simple *step-tracking* system.
5. Fully automatic continuous tracking is necessary.

Some comment on each type is useful.

Fixed-pointing only. Fixed-pointing systems are usually restricted to small antennas. The geometry of the mounts is as discussed in the preceding section and screw drives are available for initial adjustments.

Occasional repointing. The adjustments are flexible enough so that they can be changed manually without difficulty. Simple motor drives may be added to do it remotely. One-axis mounts are common.

Preprogrammed. Once motor drives are available for one- or two-axis control, a variety of methods, both automatic and preprogrammed, can be used. It must be remembered that the satellite is not drifting at random, but is in an elliptical orbit, ever so slightly eccentric, inclined, and nonsynchronous. As such, once the parameters of this orbit are measured, the methods of Chapter 3 can be used to calculate its position at any time. The antenna can be preprogrammed to track it "open loop." Often the principal apparent satellite motion is that due to imperfect inclination control. This motion, for small inclinations and otherwise perfect orbits, is a figure eight with a period of one sidereal day. Its vertical height is twice the orbital inclination and its width is only a fraction of that value (i 2/4, for i in radians). The methods of Section 3.5 are easily used to calculate the figure eight in detail, if necessary. If the orbit has zero inclination but has a small eccentricity e, the amplitude of the maximum longitudinal departure is $2e$ radians. One can calculate the satellite orbital position as a function of time, considering both effects, and then correct to azimuth and elevation. This tracking method has been used frequently in large stations.

Step tracking. Step tracking uses a primitive servomechanism in which the antenna is moved a discrete amount in a step, and if the signal level increases, it is moved again in this direction. As soon as the signal level does not increase, it returns to the previous position. The "fineness" of this method obviously depends on the size of step. Nonetheless, it is satisfactory for all but rather demanding applications.

Fully automatic. Fully automatic tracking can be provided using methods originally developed for the pointing of radar antennas. The most common is the *monopulse* or *simultaneous lobing system*, in which four beams are generated in an auxiliary feed, and combinations of the signals from these four beams provide left–right and up–down error signals. These error signals are detected, amplified, and used to generate control signals for driving the antenna. A block diagram of a general automatic tracking system is seen in Figure 10-17. Such systems are complicated and expensive, and are required only for narrow beamwidths, typically less than one-fourth of a degree. Note that it requires four extra antenna feeds in addition to the electronics and precise two-axis drives. Needless to say, such precision usually precludes the use of single-axis mounts. It is possible to derive the error signals either with multihorn systems or by the use of higher modes in the main antenna feed. The multiple horn feeds use four horns grouped together, or sometimes four horns grouped around a single larger horn, whereas the higher-mode error-determining signals use circular waveguide modes such as TM_{01} or TE_{01}, which have no field component on the axis. The secondary radiation patterns produced by such modes thus have a null on the main axis of the antenna rather than a peak, and departures from this null can be used to generate an error

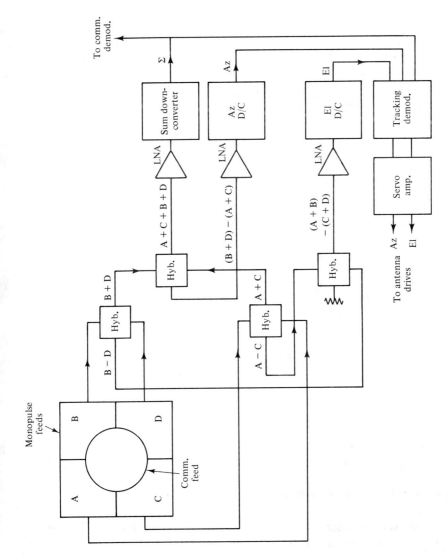

Figure 10-17 Tracking system.

signal. Many variations using different modes and combinations are possible and the choice is the antenna designer's and depends on the polarization method. Horn feed design for tracking and cross-polarized systems is complicated. One possible block diagram is shown in Figure 10-18.

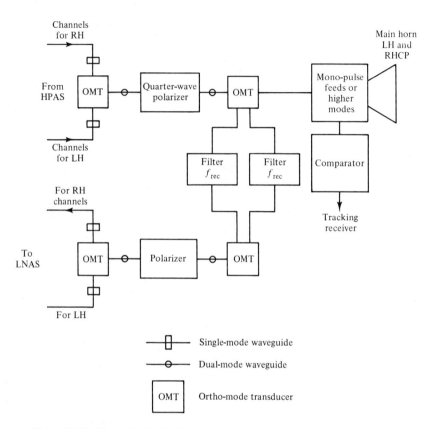

Figure 10-18 Two-polarization horn feed system.

10.5 TERRESTRIAL INTERFACE

The terrestrial interface comprises a wide variety of equipment. At one extreme, when the terminal is a mobile or receive-only station, there may be no terrestrial interface equipment at all. The operating devices, such as TV receivers, telephones, data sets, and so on, are used right at the earth station. At the other extreme, we find the interface equipment necesary in a large commercial satellite system for fixed service. In such cases, hundreds of telephone channels, together with data and video, are brought to the station by microwave and cable systems using either frequency- or time-division terrestrial multiplex methods. The signals must be changed from those formats into formats suitable for satellite transmission. In an easy case, frequency-division multiplex groups and supergroups, as brought in from terrestrial transmission facilities, can be transmitted directly or with simple translation in basebound frequency from the satellite after modulation and up-conversion, but in many cases it is necessary to reformat extensively for terrestrial circuits. Individual telephone channels, for instance, may all be transmitted on the same carrier, which is received by many earth stations in the network. The return channels for particular conversation circuits will be coming in on various carrier frequencies, depending on their source, and they must be tagged and put together with the corresponding outgoing circuit to make up a terrestrial circuit. This can be a complex process. The presence of video and data complicates matters further.

If the satellite transmission is single channel per carrier, it is necessary to bring each terrestrial carrier down to baseband before remodulation. The interfaces between terrestrial time-division and satellite frequency-division systems, and vice versa, are complicated and can be accomplished in a variety of ways. Television video signals must often be separated from order wire channels, program sound channels, cueing channels, and so on, and then matched up again at the proper point.

The underlying theory for understanding what must be done and how it must be done is presented in Chapters 7 and 8. Usually, in the systems engineering and programming planning phase it is only necessary to be alert to the problems and possibilities; the detailed design can be saved until later in the program.

10.6 PRIMARY POWER

Primary power systems vary from plain battery- or solar-cell-operated remote transmitters for data gathering to huge combined commercial power and diesel generator systems for large stations. Almost all transmit and receive earth stations require some kind of "no-break" power system, that is, emergency power to continue the communications during commercial power outages. Such power outages are frequent, even in highly organized industrial areas, if for no reason other than thunderstorms. The no-break transition derives its name from the necessity to make the change over from one power system to another without any interruption in service. Almost all systems today use batteries to effect this transition. Some systems have been devised in which motor generators store enough energy in flywheels to permit a smooth mechanical transition.

10.7 TEST EQUIPMENT

10.7.1 Noise Power Ratio (NPR)

Earth stations are typically provided with complex test equipment, ranging from that necessary for routine measurements of voltage, power, temperature, and so on, to sophisticated and specialized measurements unique to satellite communication. We address only a few of the latter class in this section. One of these is *noise power ratio* (NPR)—the traditional measure of intermodulation noise for FDM systems in the communications field. The principle of NPR measurement involves loading the entire baseband spectrum, save for the one voice-frequency channel "slot," with noise, simulating in total the loading of the system by actual voice traffic in all but that channel. Noise appearing in the unloaded slot is a manifestation of intermodulation. The ratio of that noise power to the per-channel "loading" noise power is the NPR. NPR is measured by a setup as shown in Figure 10-19. The "system" can be between any two points of interest. The noise generator *band* is limited by filters to the baseband and the noise generator *level* is set to simulate full load according to the CCIR formulas

$$P = -15 + 10 \log N \text{ dBmO} \quad N > 240$$
$$P = -1 + 4 \log N \text{ dBmO} \quad N \leq 240$$

These CCIR expressions give equivalent Gaussian noise to simulate N speech channels during busy hours (see Chapter 7). The filter has a stopband corresponding to the selected voice channel. The receive passband corresponds to that same channel. The difference between the noise receiver readings with the filter in and out is the NPR.

Measurement is typically made at low, center, and highest telephone channels. If they are carried out at different levels (and deviations for FM), they have a shape indicative of the system nonlinearities and frequency response. NPR is usually converted to an equivalent per-channel signal-to-noise ratio.

$$\text{BWR} = 10 \log \frac{\text{baseband total bandwidth}}{\text{single channel bandwidth}} \quad (10\text{-}18)$$

$$\text{NLR} = 10 \log \frac{\text{baseband noise test power}}{\text{test-tone power per channel}} \quad (10\text{-}19)$$

$$= \text{dBmO of loading calculation}$$

The equivalent baseband signal-to-noise ratio due to intermodulation is then

$$S/N = \text{NPR} + \text{BWR} - \text{NLR} \quad (10\text{-}20)$$

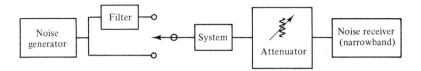

Figure 10-19 Noise power ratio—test setup. Noise generator band is limited by filters to baseband; noise generator level is set to simulate full load.

If

$$960 \text{ channels}$$
$$\text{NPR} = 55 \text{ dB in 3-kHz slot}$$
$$B = 4028 - 60$$

then

$$\text{BWR} = \frac{4028 - 60}{3} = 31.2 \text{ dB}$$

$$\text{NLR} = 10 \log 960 - 15 = 14.8 \text{ dBmO}$$

$$(S/N)_{\text{equiv}} = 71.4 \text{ dB}$$

10.7.2 The Measurement of G/T

System temperature T_S can be determined by conventional laboratory noise generator measurement of receiver noise figure and radiometric measurements of antenna temperature. The basic system parameter G/T_S also requires a knowledge of antenna gain, and as the antennas get larger, this characteristic is not so easy to get. The gain of smaller antennas, say less than 7 or 8 m, can be found from pattern measurements on a range or by comparison to a gain standard, but these methods are cumbersome and may be quite impractical for larger antennas.

Large earth stations, with antenna sizes up from 10 m, can sometimes use a carefully calibrated satellite signal to measure G/T_S. In effect, G/T_S is calculated from the link equation, knowing the other variables. This method is often used with intermediate-sized antennas (i.e., from 5 to 15 m).

An ingenious method has been developed for the measurement of G/T_S for large antennas using the known radio noise characteristics of stellar sources, usually called *radio stars*. These characteristics, particularly S, the flux density of the source in W/m²·Hz, have been accurately measured by radio astronomers and are shown in Table 10-5.

The accurate implementation of the method requires care and many detailed corrections (Wait et al., 1974; Price, 1982) too lengthy for this section. Nonetheless, understanding the idea is important.

TABLE 10-5 Radio Stars

Radio Source	Location[a] / Perturbation — Right Ascension	Location[a] / Perturbation — Declination	Shape	Size	Spectral Index[b]	Polarization Position Angle 4.170 MHz	Polarization Position Angle 6.390 MHz	Flux Density ($W/m^2 \cdot Hz \times 10^{-26}$) 4.000 MHz	Flux Density ($W/m^2 \cdot Hz \times 10^{-26}$) 6.390 MHz
CasA	$\frac{23h\ 21m\ 11.4s}{2.71s}$	$\frac{58°31.9'}{0.33'}$	Annular	Diameter 4 ft	−0.792	—	—	1.067[c]	774
TauA	$\frac{05h\ 31m\ 30s}{3.61s}$	$\frac{21°59.3'}{0.04'}$	Elliptical (Gaussian)	Major axis 4.3 ft Minor axis 2.7 ft	−0.287	$\frac{5.7\%}{143°}$	$\frac{7.0\%}{147°}$	679	604
CygA	$\frac{19h\ 57m\ 44.5s}{2.08}$	$\frac{40°35.8}{0.16'}$	Dual point source	Separation 2 ft	−1.198	$\frac{3.0\%}{160°}$	$\frac{5.7\%}{148°}$	483	297

[a] Perturbation per annum. Location (1950 + x) = location (1950) + perturbation × x.
[b] 1–16 GHz.
[c] Value for January 1965.

Source: Reprinted with permission (Miya, 1982).

Sec. 10.7 Test Equipment

The basic measurement is that of *Y factor*. By definition, *Y* factor is the ratio of the output noise measured when the receiver is connected to a hot noise source (T_h), to the output noise measured when connected to a cold source (T_c). This is a familiar measurement in receiver work and is discussed at length in Mumford and Scheibe (1968). Using the equations of Chapter 6, it is straightforward to show that the receiver excess noise T_e is related to the *Y* factor by

$$T_e = \frac{T_b - YT_c}{Y - 1} \qquad (10\text{-}21)$$

If the cold source is the normal sky and the hot source the radio star, the operating system temperature T_S, the sum of T_c and T_e given above, is easily shown to be

$$T_S = \frac{T_h - T_c}{Y - 1} = \frac{\Delta T_a}{Y - 1} \qquad (10\text{-}22)$$

where ΔT_a is the increase in *antenna* temperature when changing from a radio source to the cold sky. This apparent increase in antenna temperature is related to the noise density increase by Boltzmann's constant k. If S is the randomly polarized flux density for the given star in W/m²·Hz, only one polarization is received by an antenna of gain G, and a is a factor to allow for atmospheric loss ($a > 1$), then, from the universal antenna formula,

$$\Delta T_a = \frac{S}{2ak} \frac{G\lambda^2}{4\pi} \qquad (10\text{-}23)$$

and

$$\frac{G}{T_s} = \frac{G(Y - 1)}{\Delta T_a} = \frac{8\pi k}{S\lambda^2 a}(Y - 1) \qquad (10\text{-}24)$$

If a is the atmospheric absorption at the zenith, then at an elevation angle θ,

$$\frac{G}{T_s} = \frac{8\pi k}{S\lambda^2 a}(Y - 1) \sin\theta \qquad (10\text{-}25)$$

If the stellar source is not randomly polarized, another correction factor is needed. Cassiopeia A, the most commonly used source, does not need this correction.

Some further correction may be necessary if the beamwidth of the antenna under test is narrow compared to the stellar radio source. An extended source of varying brightness can be considered as equivalent to a Rayleigh–Jeans blackbody radiator and the brightness integrated over the extent of the source. Correction factors can be arrived at (see Figure 10-20). Note that they are significant for narrowbeam antennas.

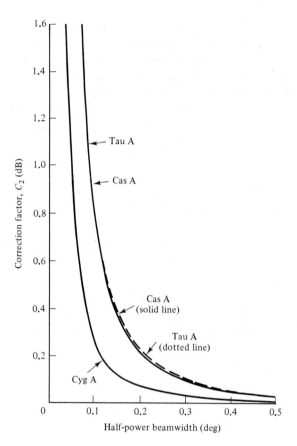

Figure 10-20 Correction factor for G/T measurement using extended sources. (Reprinted with permission from K. Miya, *Satellite Communications Engineering*, 2nd ed., Lattice Company, Tokyo, 1982.)

If the method is to be used at higher frequencies than given in Table 10-5, corrected values of S must be used. Wait et al. (1974) present data to justify assuming that S varies as $f^{-0.875}$.

REFERENCES

COMSAT/INTELSAT "Edited Lectures," United States Seminars on Communication Satellite Earth Station Technology, Comsat Corporation, Washington, D.C., 1966.

ITT, *Reference Data for Radio Engineers*, 6th ed., Howard W. Sams & Company, Inc., Publishers, Indianapolis, 1975.

JASIK, H., ed., *Antenna Engineering Handbook*, McGraw-Hill Book Company, New York, 1961.

Kraus, J. D., *Antennas*, McGraw-Hill Book Company, New York, 1950.

Miya, K., ed., *Satellite Communications Engineering*, 2nd ed., Lattice Company, Tokyo, 1982.

Mumford, W. W., and E. H. Scheibe, *Noise Performance Factors in Communication Systems*, Horizon House–Microwave Inc., Dedham, Mass., 1968.

Price, R., "RF Tests on Etam Standard C Antenna," *Comsat Tech. Rev.*, Vol. 12, No. 1, Spring 1982.

Ruze, J., "Effect of Aperture Distribution Errors on the Radiation Pattern," *ASTIA Report AD202826*, Apr. 28, 1952.

Saad, T., ed., *Microwave Engineer's Handbook*, Artech House, Inc., Dedham, Mass., 1971.

Scientific-Atlanta, "Satellite Communications Symposium '81" (Seminar Notes), 1981.

Silver, S., *Microwave Antenna Theory and Design*, McGraw-Hill Book Company, New York, 1949.

Wait, D. F., et. al., "A Study of the Measurement of G/T Using Cassiopeia," *Technical Report ACC-ACO-2-74*, National Bureau of Standards, Boulder, Colo., 1974.

11

Interference

11.0 INTRODUCTION

Interference may be defined as the effect of an unwanted signal on the reception of a wanted signal. Interference is both inevitable and ubiquitous, given the wide variety of uses of the radio spectrum. We are concerned only with *detectable* interference; *undetectable* interference, like the sound of a tree falling in a distant forest, may be ignored.

The detectability of interference depends on the characteristics of the wanted signal, the unwanted signal, and the communications system. For analog systems, these characteristics include the received power levels, the spectral power distributions of the wanted and unwanted signals, and the spectral sensitivity of the system. For digital systems, the corresponding characteristics are the amplitudes and pulse shapes of the wanted and unwanted signals and the positions of the two signals relative to the system "eye" diagram. Whether the system is analog or digital, the subjective response to the interference must also be considered.

There are also special characteristics of communications systems which affect their sensitivity to wanted and unwanted signals; an example is the "capture" phenomenon in FM systems, in which the stronger signal seizes the system and dominates the weaker one.

We may then break the interference problem into two parts: the *relative strength* of the wanted and unwanted signals (usually expressed in terms of the *carrier-to-interference ratio* C/I) and the *tolerance* of the system to interference (including system sensitivity and subjective response). The latter part

Sec. 11.1 Calculation of C/I for a Single Interfering Satellite

of the problem is in general quite difficult; we shall concentrate primarily on calculation of C/I and on the application of system tolerance criteria (such as the *required C/I*, or *protection ratio*), but refer to the literature when necessary for the development of those criteria.

Interference is not the only degradation in a communications system; thermal noise and other impairments are present as well. Although both interference and thermal noise degrade the performance of a system, we often treat them separately, since in general their origins, their characteristics, and the tolerance of the system to their effects are all different. Then, with a mixture of candor and irony, we often bring them together again to simplify the analysis. We take that approach in this book, referring back to the treatment of thermal noise in Chapter 6.

Let us begin by calculating interference in the absence of thermal noise, and defer the matter of combining these degradations. Although it may seem logical to calculate the interference power I directly, it is usually more practical to calculate C/I instead.

The carrier-to-interference ratio may be calculated at any point in the system, but a convenient and useful point is the receiver input. We seek, then, the ratio of the wanted signal power C (given by the usual RF link equation and the geometry relating the desired satellite and the receiver) to the unwanted signal power I (given by a similar RF link equation and the geometry relating the interfering satellite and the receiver). This formulation makes it clear that our unwanted signal, or interference, is simply someone else's wanted signal.

In general, a carrier level may be calculated from

$$C = P_t + G_t - L + G_r \quad \text{dB} \quad (11\text{-}1)$$

or

$$C = E - L + G_r \quad \text{dB} \quad (11\text{-}2)$$

where P_t = transmitted power, dB
G_t = transmit antenna gain, dB
L = free-space loss, dB
G_r = receive antenna gain, dB
E = e.i.r.p., dB

11.1 CALCULATION OF C/I FOR A SINGLE INTERFERING SATELLITE

Let us consider a single receive earth station E that receives signals from a wanted satellite S; this earth station also receives signals (interference) from a single unwanted satellite S' (see Figure 11-1). This is often called *single-*

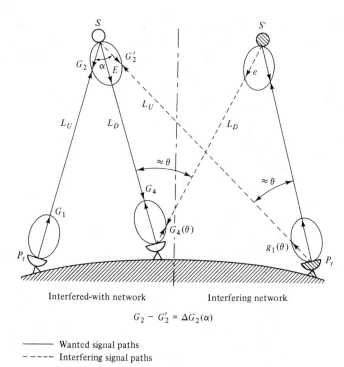

Figure 11-1 Satellite interference geometry: solid lines, wanted signal paths; dashed lines, interfering signal paths. (Reprinted from Vol. IV, XVth, *ITU Plenary Assembly*, Geneva, 1982, p. 327.)

entry interference. Subscripts are used to denote the various parts of the signal path: 1 for earth station transmit; 2 for satellite receive; 3 for satellite transmit; and 4 for earth station receive. Upper case letters are used to denote powers and antenna gains associated with the wanted earth station and satellite, and lower case letters are used for the interfering earth station and satellite. (*Note:* This notation, although cumbersome, is worth using because it is found in various CCIR reports (e.g. Ref. 1, Vol. IV, Report 455-2, Geneva 1978) which must often be consulted for more detail when working on specific problems. The notation is expanded in the following examples, and certain changes have been made for clarity.)

The wanted carrier C is given by

$$C = E - L_{dw} + G_4(0) \quad \text{dB} \quad (11\text{-}3)$$

where
C = wanted carrier power, dBW
E = e.i.r.p. (wanted), dBW
L_{dw} = space loss (downlink) in direction of wanted satellite, dB
$G_4(0)$ = earth station gain in direction of wanted satellite, dB

Sec. 11.1 Calculation of C/I for a Single Interfering Satellite

Similarly, the interfering carrier power I may be obtained from

$$I = e - L_{di} + G_4(\theta) \quad \text{dB} \quad (11\text{-}4)$$

where e = e.i.r.p. (interfering), dBW
L_{di} = space loss (downlink) in direction of interfering satellite, dB
$G_4(\theta)$ = earth station gain in direction of interfering satellite, dB

If polarization discrimination is used to reduce the interfering signal by an amount Y_d decibels, the interfering carrier power becomes

$$I = e - L_{di} + G_4(\theta) - Y_d \quad \text{dB} \quad (11\text{-}5)$$

Then the carrier-to-interference ratio is simply

$$C/I = E - e - (L_{dw} - L_{di}) + G_4(0) - G_4(\theta) + Y_d \quad \text{dB} \quad (11\text{-}6)$$

or

$$C/I = \Delta E - \Delta L_d + \Delta G_4 + Y_d \quad \text{dB} \quad (11\text{-}7)$$

Similarly, for the uplink, recalling that subscript 1 refers to the earth station transmit signal path and subscript 2 refers to the satellite receive signal path,

$$C/I = [P_t + G_1(0)] - [p_t + g_1(\theta)] - (L_{uw} - L_{ui})$$
$$+ (G_{2w} - G_{2i}) + Y_u - M_u \quad \text{dB} \quad (11\text{-}8)$$

or

$$C/I = \Delta(P + G_1) - \Delta L_u + \Delta G_2 + Y_u - M_u \quad \text{dB} \quad (11\text{-}9)$$

where
P_t = earth station transmit power (wanted), dB
p_t = earth station transmit power (interfering), dB
$G_1(0)$ = wanted earth station transmit antenna gain in direction of wanted satellite, dB
$g_1(\theta)$ = interfering earth station transmit antenna gain in direction of wanted satellite, dB
G_{2w} = wanted satellite receive antenna gain in direction of wanted earth station, dB
G_{2i} = wanted satellite receive antenna gain in direction of interfering earth station, dB
M_u = uplink margin, dB

11.2 CALCULATION OF C/I FOR MULTIPLE INTERFERING SATELLITES

If we wish to calculate the effect of multiple interfering satellites (*multiple entry* or *total entry interference*), we must first consider how we might combine these effects into an overall measure of interference. This problem is in general intractable, and (as promised earlier) we make the usual simplifying assumption that the interferers may be added on a power basis. This assumption requires that:

1. The interferers are uncorrelated.
2. The interferers are sufficiently numerous that each contributes only a small portion of the total power.
3. The total interfering power is much lower than the carrier power.

With these assumptions, we may write

$$I_t = I_1 + I_2 + \cdots + I_n \tag{11-10}$$

and

$$(C/I)_t^{-1} = (C/I_t)^{-1} = (C/I_1)^{-1} + (C/I_2)^{-1} + \cdots + (C/I_n)^{-1} \tag{11-11}$$

where $(C/I)_t$ is the total entry or multiple-entry carrier-to-interference ratio. Also, if the thermal noise N meets the foregoing criteria,

$$N_T = N + I_t \tag{11-12}$$

and

$$(C/N)_T^{-1} = (C/N)^{-1} + (C/I)_t^{-1} \tag{11-13}$$

where $(C/N)_T$ is the composite carrier-to-noise ratio including the effects of both interference and thermal noise.

Now let us consider a set of $2K$ interfering satellites which cause total entry interference power I_t to be delivered to the input of the wanted receiver. Then by reference to Eq. (11-5), and converting from decibels, the jth interfering satellite delivers interference power

$$I_j = \frac{e_j G_4(\theta_j)}{L_j Y_j} \tag{11-14}$$

The total entry interference power is then

$$I_t = \sum_{j=-K}^{K} I_j = \sum_{j=-K}^{K} \frac{e_j G_4(\theta_j)}{L_j Y_j} \quad j \neq 0 \tag{11-15}$$

Using the expression above for I_t, we may calculate the total entry carrier-to-interference ratio $(C/I)_t$ from

$$(C/I)_t^{-1} = \frac{I_t}{C} = \frac{1}{C}\sum_{j=-K}^{K} I_j = \frac{1}{C}\sum_{j=-K}^{K} \frac{e_j G_4(\theta_j)}{L_j Y_j} \qquad j \neq 0 \qquad (11\text{-}16)$$

The composite carrier-to-noise ratio $(C/N)_T$ is then obtained from

$$(C/N)_T^{-1} = (C/N)^{-1} + \frac{1}{C}\sum_{j=-K}^{K} \frac{e_j G_4(\theta_j)}{L_j Y_j} \qquad j \neq 0 \qquad (11\text{-}17)$$

Having defined a composite carrier-to-noise ratio taking into account both thermal noise and total entry interference, we may now calculate their combined effect on the demodulated baseband signal. If we define a receiver transfer function R_t which relates the wanted carrier power and thermal noise alone to the resulting demodulated signal-to-noise ratio S/N such that

$$S/N = R_t(C/N) \qquad (11\text{-}18)$$

then

$$S/N = R_t (C/N)_T \qquad (11\text{-}19)$$

using the composite carrier-to-noise ratio defined above. In the case of FM, for example, R_t is the familiar $3m^2(1 + m)$ times the appropriate preemphasis and weighting factors. The use of the same R_t factor for both thermal noise and interference is equivalent to the assumption that a given noise power and interference power affect signal-to-noise ratio equally. This assumption is often useful but generally invalid, and in practical cases the tolerance of a system to noise and interference combined is evaluated by means of carefully controlled subjective testing. We may then find the composite signal-to-noise ratio from

$$(S/N)^{-1} = \frac{1}{R_t}(C/N)_T^{-1}$$

$$= \frac{1}{R_t}(C/N)^{-1} + \frac{1}{R_t C}\sum_{j=-K}^{K} \frac{e_j f_j G_4(\theta_j)}{L_j Y_j} \qquad j \neq 0 \qquad (11\text{-}20)$$

where we introduce the factor f_j to account for differences in the tolerance of the system to noise and different types of interference.

11.3 TWO TYPES OF SATELLITES

An interesting special case of the analysis above consists of alternating sets of satellites of two different types, a and b, causing two types of interference to the wanted system (see Figure 11-2). For example, we may consider a wanted satellite system receiving interference from K similar satellites plus

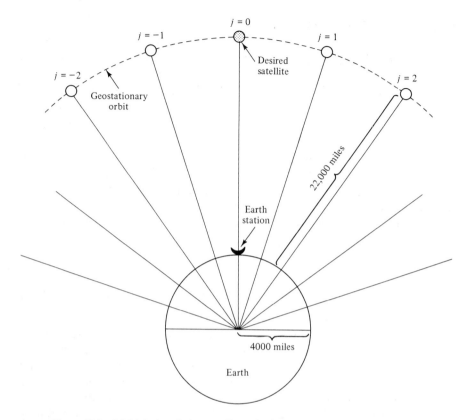

Figure 11-2 Multiple interfering satellites of two types.

K others providing a different service. In this example, the wanted satellite is placed at location $j = 0$ and the interfering satellites are distributed uniformly along the geostationary orbit in both directions. The number of interfering satellites must of course be finite because the capacity of the orbital arc is finite; also, blockage by the earth limits the number of satellites visible to the wanted earth station.

In the example shown, interfering satellites of type a occupy positions $j =$ even, and those of type b occupy positions $j =$ odd (we may use only positive values of j if we introduce factors of 2 in front of the summations). Let us assume also that the free-space loss L_j is equal for all interfering satellites ($L_j = L$ for all j), and that no polarization discrimination is available ($Y_j = 0$ for all j). We assume further that the tolerance of the wanted system to interference will take on only two values ($f_j = f_a$ for $j =$ even and $f_j = f_b$ for $j =$ odd). Finally, we assume that the e.i.r.p. e_j of each of the interfering satellites is equal to the e.i.r.p. E of the wanted satellite. Then the equation

Sec. 11.3 Two Types of Satellites

for the composite signal-to-noise ratio reduces to

$$(S/N)^{-1} = \frac{1}{R_t}(C/N)^{-1} + \frac{2f_a}{R_tC}\sum_{j=\text{even}}\frac{EG_4(\theta_j)}{L}$$

$$+ \frac{2f_b}{R_tC}\sum_{j=\text{odd}}\frac{EG_4(\theta_j)}{L} \quad (11\text{-}21)$$

Now recognizing that the desired carrier C is equal to

$$C = \frac{EG_4(0)}{L} \quad (11\text{-}22)$$

the expression for $(S/N)^{-1}$ further reduces to

$$(S/N)^{-1} = \frac{1}{R_t}(C/N)^{-1} + \frac{2f_a}{R_t}\sum_{j=\text{even}}\frac{G_4(\theta_j)}{G_4(0)} + \frac{2f_b}{R_t}\sum_{j=\text{odd}}\frac{G_4(\theta_j)}{G_4(0)} \quad (11\text{-}23)$$

We may now introduce the usual CCIR antenna sidelobe formulas:

$$G_4(\theta) = 32 - 25 \log \theta \quad \text{dB} \quad \frac{D}{\lambda} > 100$$

or

$$G_4(\theta) = \frac{10^{3.2}}{\theta^{2.5}} \quad (11\text{-}24)$$

Similarly:

$$G_4(\theta) = 52 - 10 \log \frac{D}{\lambda} - 25 \log \theta \quad \text{dB} \quad \frac{D}{\lambda} < 100$$

or

$$G_4(\theta) = \frac{(\lambda/D)10^{5.2}}{\theta^{2.5}} \quad (11\text{-}25)$$

Now let us recall that in this example the satellites have been assumed to be uniformly distributed over the orbital arc; thus the geocentric angle θ_j is just $j\theta_s$, where θ_s is the angular separation between adjacent satellites. Although these geocentric angles are not identical to the topocentric angles which should actually be used to compute the off-axis antenna gains, the approximation is good for all but the most exacting calculations. With the substitutions above, the equation for composite signal-to-noise ratio becomes

$$(S/N)^{-1} = \frac{1}{R_t}(C/N)^{-1} + \frac{2A}{R_tG_4(0)\theta_s^{2.5}}\left(\sum_{j=\text{even}}f_a j^{-2.5} + \sum_{j=\text{odd}}f_b j^{-2.5}\right) \quad (11\text{-}26)$$

The series above converges rapidly, and by numerical methods it can be shown that

$$(S/N)^{-1} = \frac{1}{R_t}(C/N)^{-1} + \frac{2A}{R_t G_4(0)\theta_s^{2.5}}(1.103 f_b + 0.231 f_a) \quad (11\text{-}27)$$

For convenience, we may also define an equivalent tolerance factor f such that

$$1.334 f = 1.103 f_b + 0.231 f_a \quad (11\text{-}28)$$

or

$$1.334 f = f_a\left(0.231 + 1.103 \frac{f_b}{f_a}\right) \quad (11\text{-}29)$$

11.4 HOMOGENEOUS SATELLITES

Another useful example is the further simplification of homogeneous satellites (i.e., $a = b$). In this case $f = f_a = f_b$, and

$$(S/N)^{-1} = \frac{1}{R_t}(C/N)^{-1} + \frac{2.67 A f}{R_t G_4(0)\theta_s^{2.5}} \quad (11\text{-}30)$$

Note that for this homogeneous system, we can write

$$(S/N)^{-1} = \frac{1}{R_t}(C/N)^{-1} + \frac{f}{R_t}(C/I)_t^{-1} \quad (11\text{-}31)$$

Eq. (11-30) can be solved for θ_s, yielding

$$\theta_s = \left[\frac{\dfrac{2.67 A f (S/N)}{R_t G_4(0)}}{1 - \dfrac{1}{R_t}\dfrac{S/N}{C/N}}\right]^{0.4} \quad (11\text{-}32)$$

It is often convenient to define the effects of interference and thermal noise in relative terms, by defining parameters such as

$$p = \frac{C/N}{(C/I)_t} = \frac{I_t}{N} \quad (11\text{-}33)$$

and

$$r = \frac{(C/N)_T}{(C/I)_t} = \frac{I_t}{N + I_t} \quad (11\text{-}34)$$

Sec. 11.4 Homogeneous Satellites

Then

$$p = \frac{r}{1-r} \tag{11-35}$$

and

$$r = \frac{p}{p+1} \tag{11-36}$$

With these substitutions, and recognizing that

$$C/N = (S/N)\frac{1+pf}{R_t} \tag{11-37}$$

Eq. (11-32) may be written

$$\theta_s = \left[\frac{2.67A}{R_t G_4(0)}\left(\frac{1+pf}{p}\right)\frac{S}{N}\right]^{0.4} \tag{11-38}$$

or

$$\theta_s = \left[\frac{2.67A}{R_t G_4(0)}\left(\frac{1-r}{r}+f\right)\frac{S}{N}\right]^{0.4} \tag{11-39}$$

Similarly, using Eq. (11-37), θ_s may be expressed in terms of C/N by

$$\theta_s = \left[\frac{2.67A}{G_4(0)p}\frac{C}{N}\right]^{0.4} \tag{11-40}$$

or

$$\theta_s = \left[\frac{2.67A}{G_4(0)}\left(\frac{1-r}{r}\right)\frac{C}{N}\right]^{0.4} \tag{11-41}$$

Finally, using Eqs. (11-19) and (11-38), we may write

$$\theta_s = \left[\frac{2.67A}{G_4(0)}\left(\frac{1+pf}{p}\right)(C/N)_T\right]^{0.4} \tag{11-42}$$

The equations above for the homogeneous satellite case may be used to develop an expression for the utilization of the geostationary orbit. The CCIR has proposed a measure of orbit utilization M, defined as

$$M = \frac{\text{total number of channels}}{\text{total usable orbital arc} \times \text{total bandwidth}} = \frac{\dot{n}}{\theta_s} \tag{11-43}$$

where

$$\dot{n} = \frac{\text{number of channels per satellite}}{\text{total bandwidth}}$$

and

$$\theta_s = \frac{\text{total usable orbital arc}}{\text{number of satellites}}$$

The number of channels n per satellite, in the power-limited case, is given by

$$n = \frac{S/N}{(S/N)_D} \qquad (11\text{-}44)$$

where S/N = total available signal-to-noise ratio
$(S/N)_D$ = required signal-to-noise ratio per channel

The term *channel* may apply, for example, to a carrier modulated by a single telephone channel, to a multiplexed baseband, or to a television signal. Then, for a given total bandwidth B, the orbit utilization is given by

$$M = \frac{(S/N)/(S/N)_D}{B\theta_s} = \frac{R_t(C/N)_T}{(S/N)_D B\theta_s} \qquad (11\text{-}45)$$

It is useful to seek the conditions under which the orbit utilization is a maximum. We may first combine Eqs. (11-19), (11-33) through (11-36), (11-41), and (11-45) to obtain

$$M = \frac{R_t(C/N)^{0.6}}{B(S/N)_D\,[2.67A/G_4(0)]}\, r^{0.4}(1-r)^{0.6} \qquad (11\text{-}46)$$

and then set $dM/dr = 0$. A complex but routine series of steps yields $r = 0.4$ for maximum orbit utilization, for this special case of homogeneous satellites. Recalling Eq. (11-34), we conclude that the orbit utilization is maximum when the interference power is 40% of the composite noise power. Under these conditions, Eq. (11-41) becomes

$$\theta_{so} = \left[\frac{2.67A}{G_4(0)} \left(\frac{3}{2}\right) \frac{C}{N}\right]^{0.4} \qquad (11\text{-}47)$$

and

$$M_o = 0.6 \left(\frac{2}{3}\right)^{0.4} \left[\frac{R_t(C/N)^{0.6}}{B(S/N)_D\,[2.67A/G_4(0)]}\right] \qquad (11\text{-}48)$$

11.5 PROTECTION RATIO

We have distinguished above between the objective analysis of the composite carrier-to-noise ratio, including the constituent carrier to noise and total entry carrier-to-interference ratios, and the subjective effects of these results on signal-to-noise ratio. In particular, we introduced the factor f_j in Eq. (11-20) to account for differences in the tolerance of communications systems to noise and different types of interference.

Sec. 11.5 Protection Ratio

Although the development of subjective tolerance criteria is usually not the task of the systems engineer, the application of these criteria is an essential and critical element in system design. The usual form in which these criteria are provided is a *protection ratio* curve or template, relating the subjectively perceived signal quality to the objectively determined carrier-to-interference ratio (usually in the presence of a specified level of thermal noise). The protection ratio is the carrier-to-interference ratio required to yield a chosen level of signal quality.

It is difficult to develop reliable methods to predict the subjective effects of various combinations of noise and interference; as a result, the subjective criteria usually apply only to the specific cases tested. A specific test is usually performed using a single interferer. This means that the systems engineer cannot usually estimate the subjective effects of a number of single interferers by means of the protection ratio criteria, and then combine these effects to predict the subjective effect of total entry interference. Systems engineers often apply these protection ratio curves to total entry interference, but this practice is valid only under the assumption of power addition of multiple interferers. This assumption has been shown by subjective tests to be valid in the case of TV for multiple cochannel interferers (Whyte, Cauley, and Groumpas, 1983) but not valid for adjacent channel interferers (Whyte, 1983).

An example of a protection ratio curve is given in Figure 11-3, which shows a typical CCIR defined subjective *impairment grade* as a function of the carrier-to-interference ratio. This example curve applies to the broad-

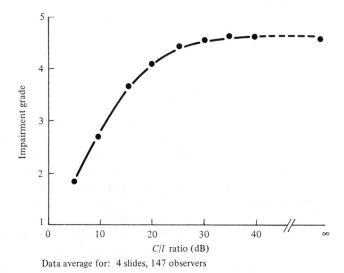

Data average for: 4 slides, 147 observers

Figure 11-3 Impairment grade versus carrier-to-interference ratio; data average for four slides, 147 observers. (Reprinted from Vols. X and XI, Part 2, *ITU Plenary Assembly*, Geneva, 1982, p. 148.)

casting satellite service, with frequency-modulated television as the wanted and unwanted signal. The five-point impairment-grade scale is shown in Table 11-1 as follows:

TABLE 11-1 Impairment Grade Scale

Impairment Grade Q	Subjective Description
5	Imperceptible
4	Perceptible but not annoying
3	Somewhat annoying
2	Severely annoying
1	Unusable

For many practical television designs, an impairment grade between 4 and 5 is chosen, and the median value of 4.5 is often suggested. This value corresponds approximately to an unofficial impairment grade of "just perceptible" interference, and also corresponds approximately to a total entry protection ratio of 30 dB, as shown in Figure 11-3.

The protection ratio may usually be reduced if the interfering carrier is displaced in frequency relative to the wanted carrier. This effect is illustrated by means of the normalized protection ratio template shown in Figure 11-4, again for the broadcasting satellite service.

The template is normalized in two ways. First, the carrier offset x is normalized to the Carson's rule bandwidth, given by

$$B = D_{vpp} + 2f_b \qquad (11\text{-}49)$$

where D_{vpp} = peak-to-peak video deviation
f_b = highest baseband frequency

Second, the *cochannel* protection ratio PR_0, shown on the template for $x = 0$, is normalized to the total entry protection ratio given above in Figure 11-3 for the chosen impairment grade. The *adjacent channel* protection ratio PR is shown at all other carrier offsets $x \neq 0$. Note that the template is symmetrical about $x = 0$; this is a simplification adopted for the particular case shown, but in general there are differences between adjacent channel protection ratios for interfering carriers located above and below the wanted carrier. [This template is shown because it illustrates some useful ways to express protection ratio data, and because of historical interest. Although it was adopted by the Conference Preparatory Meeting (CPM) for the RARC '83 conference, the RARC '83 conference itself abandoned this template in favor of a fixed template].

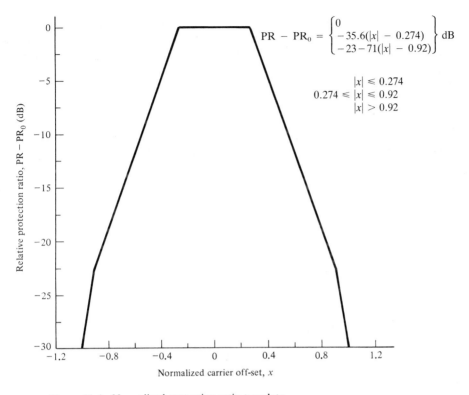

Figure 11-4 Normalized protection ratio template.

The cochannel protection ratio has been expressed analytically in the form

$$PR_0 = C - 20 \log \frac{D_{vpp}}{12} - Q + 1.1Q^2 \qquad (11\text{-}50)$$

where C takes on values from 12.5 to 18.5 dB, depending on the particular television standard used. C is equal to 13.5 dB for 525-line standard M/NTSC, and the value 12.5 dB has been adopted for general use in planning the broadcasting satellite service. (Recommendations and Reports of the CCIR, Vols. X and XI, 1982).

Note particularly the dependence of PR_0 on D_{vpp}, which yields a decrease in cochannel protection ratio with increasing deviation. Moreover, since the adjacent channel protection ratio is directly related through the normalized template to cochannel protection ratio as well as to bandwidth, the adjacent channel protection ratio increases, on balance, by only a very small amount. Thus a higher deviation makes a system less susceptible to cochannel interference, and only marginally more susceptible to adjacent channel interference.

An example may help to illustrate the magnitudes of the various parameters. Assuming that $D_{vpp} = 12$ MHz and $Q = 4.5$, Eq. (11-50) yields $PR_0 = 30.3$ dB, approximately equal to the value found from Figure 11-3. If $f_b = 5.5$ MHz, then $B = 23$ MHz from Eq. (11-49). Now if the channel offset (channel spacing) is chosen as 13 MHz, $x = 13/23 = 0.57$ and $PR = 30.3 - 10.4 = 19.9$ dB from either Figure 11-4 or the associated equation.

It is interesting to compare the effects of noise and interference on the subjective quality of a practical system. Let us begin by rewriting Eq. (11-31) as

$$S/N = R_t(C/N) \left[\frac{1}{1 + f[(C/N)/(C/I)_t]} \right] \quad (11\text{-}51)$$

Now if $(C/I)_t = \infty$ (no interference),

$$(S/N)_\infty = R_t(C/N) \quad (11\text{-}52)$$

Then

$$S/N = (S/N)_\infty \left[\frac{1}{1 + f[(C/N)/(C/I)_t]} \right] \quad (11\text{-}53)$$

or, converting to decibels,

$$S/N = (S/N)_\infty - 10 \log \left[1 + f \frac{C/N}{(C/I)_t} \right] \quad \text{dB} \quad (11\text{-}54)$$

Now let us assume a system having an impairment grade Q of 4.0 for $C/I = \infty$ as perceived by the median viewer; thus $(S/N)_\infty = 46$ dB (see Figure 11-5). If we then let $C/I = 20$ dB and $C/N = 14.8$ dB, the impairment grade Q is reduced to 3.6 (see Figure 11-6), corresponding to an equivalent interference-free S/N of approximately 43 dB (see Figure 11-5). Thus

$$(S/N)_\infty - (S/N) = 46 \text{ dB} - 43 \text{ dB} = 3 \text{ dB} \quad (11\text{-}55)$$

or

$$10 \log \left[1 + f \frac{C/N}{(C/I)_t} \right] = 3 \text{ dB} \quad (11\text{-}56)$$

Thus for this example $f = 3.31$, and Eq. (11-54) becomes

$$S/N = (S/N)_\infty - 10 \log \left[1 + 3.31 \frac{C/N}{(C/I)_t} \right] \quad \text{dB} \quad (11\text{-}57)$$

Sec. 11.5 Protection Ratio

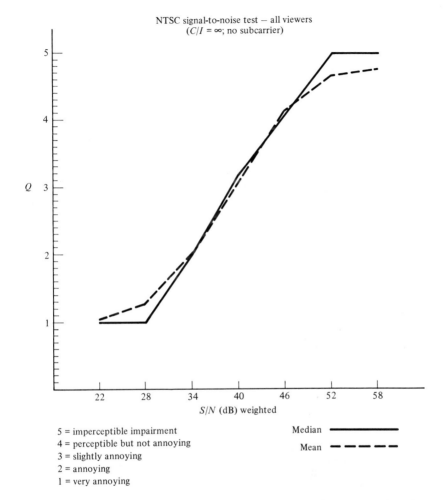

5 = imperceptible impairment
4 = perceptible but not annoying
3 = slightly annoying
2 = annoying
1 = very annoying

Median ——————
Mean   — — — —

Figure 11-5 NTSC signal-to-noise test-all viewers. (Reprinted by permission of CBS Technology Center.)

Now recalling Equations 11-33 and 11-35, we have

$$I_t = \frac{r}{1-r} N \tag{11-58}$$

For the homogeneous case, we have already shown that the orbit utilization is a maximum for r = 0.4; thus Eq. 11-58 yields

$$I_t = 0.67N \tag{11-59}$$

Now solving Eq. 11-56 for I_t, for the above example we get instead

$$I_t = 0.30N \tag{11-60}$$

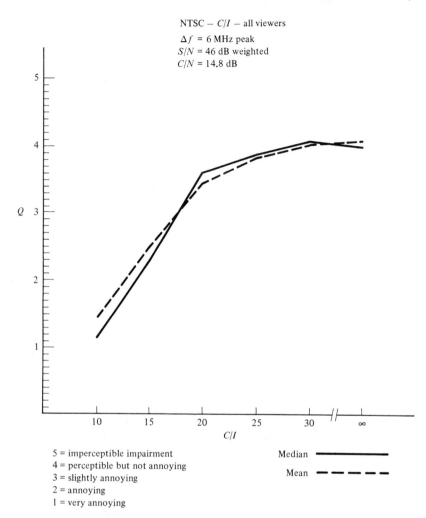

Figure 11-6 NTSC-C/I-all viewers. (Reprinted by permission of CBS Technology Center.)

Thus for C/I_t levels in the range of 20 dB and C/N levels in the range of 14.8 dB the broadcasting satellite service would be noise limited, representing a conscious tradeoff in favor of lower interference and hence higher potential S/N. Such a design would permit individual users to improve their quality of service by purchasing lower noise figure receivers, thus improving their overall performance as measured by the composite carrier-to-noise ratio $(C/N)_T$, up to the point where the interference contribution begins to dominate. Note that users may usually improve their performance even more by purchasing larger receive antennas, since the improvement in C/I_t due to receive antenna sidelobe performance depends on 25 log θ with θ proportional to D [see

Eqs. 11-24 and 11-25], while the improvement in C/N due to on axis gain depends on 20 log D. The effect of interference on $(C/N)_T$ thus decreases more rapidly than the effect of noise as the antenna size increases.

REFERENCES

INTERNATIONAL TELECOMMUNICATIONS UNION, *Recommendations and Reports of the CCIR*. Volume I, Spectrum Utilization and Monitoring, Geneva 1978, No. ISBN 92-00661-2; Volume IV, Fixed Service Using Communications Satellites, Geneva 1974, No. ISBN 92-61-00051-7; Volume IV, Fixed Service Using Communications Satellites, Geneva 1978, No. ISBN 92-61-00691-4; Volume IX, Fixed Service Using Radio-Relay Systems, Geneva 1978, No. ISBN 92-61-00741-4; Volumes X and XI, Part 2, Broadcasting-Satellite Service (Sound and Television), Geneva 1982, No. ISBN 92-61-01491-7.

WHYTE, W. A., M. A. CAULEY, and P. P. GROUMPOS, *The Subjective Effect of Multiple Cochannel Frequency Modulated Television Interference*, submitted for presentation at the 1983 GLOBCOM conference in San Diego CA, November 29 to December 1, 1983.

WHYTE, W. A., *Analysis of Adjacent Channel Interference for FM Television Systems*, NASA Lewis Research Center, June 6, 1983.

SCHWARTZ, MISCHA, WILLIAM BENNETT, and SEYMOUR STEIN, *Communications Systems and Techniques*, McGraw-Hill Book Co., New York, 1966.

12

Special Problems in Satellite Communications

12.0 BACKGROUND

The development of modern satellite communications technology using satellites in geostationary orbit has spawned many new services and capabilities not practical previously when using terrestrial systems. The unique geometric advantage of a satellite in stationary orbit allows multiple access by many earth stations on the earth's surface through a single satellite repeater. This high-altitude repeater instantly creates long-distance, wideband network facilities at low cost compared to other media. Unfortunately, the high altitude of the geostationary orbit also creates a rather long transmission-time delay which can accentuate undesirable subjective effects of echo on voice circuits, cause reduced throughput efficiency on data circuits, and create synchronization problems for digital transmission. At geostationary altitude, it takes about 270 ms for signals to travel from the transmitting earth station through the satellite to the receiving earth station. This delay is approximately 10 times that of the longest delays of modern domestic terrestrial circuits. Unfortunately, the echo control devices, modems, and protocols developed for terrestrial transmission are not suitable for this long delay. Not only is the absolute delay long, there is also a small variation in delay due to diurnal variations in the satellite orbit. This time variation creates a doppler-type effect that can cause problems for digital transmission between synchronous networks. In this chapter we examine the effects of the long time delay and time variation on various satellite communication services.

12.1 ECHO CONTROL

Telephone sets and the "loop" circuits which connect them to their serving central offices operate on a *two-wire* basis. That is, a single bidirectional medium (ordinarily, a cable pair) carries the voice signals in both directions. Until recently, the same was true of the trunk circuits interconnecting central offices in the same serving area. In the early days of telephony, even intercity circuits were of this type.

Multiplex systems, used today to derive most intercity circuits and an increasing portion of intracity, interoffice circuits, inherently utilize separate transmission paths for the two directions. The link is called a *four-wire* circuit, a term which was introduced when the two directions of transmission were first separated by using two separate conductor pairs.

The four-wire and two-wire circuits are interfaced by means of a four-port transformer circuit called a *hybrid coil*, shown in Fig. 12-1. One port is connected to the two-wire telephone circuit, or *line*. To the opposite port we connect a *balancing network* which simulates, over the range of voice frequencies, the impedance expected, looking into the line. The other two ports are connected to the sending and receiving ports of the four-wire circuit.

Assuming that the impedance of the "balancing network" is precisely

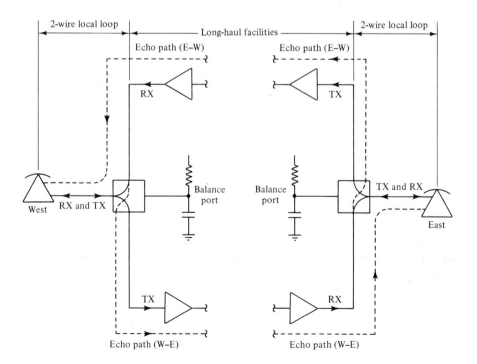

Figure 12-1 Full-duplex long-distance voiceband circuit.

that of the two-wire circuit, all the energy coming from the receiving port of the four-wire circuit divides between the line and the network. None passes "across" the hybrid to the sending port. Thus a circulating path, which would lead to echo and, if duplicated at the other end, to self-oscillation, is avoided. Signals arriving from the line divide evenly between the two four-wire ports. The energy which enters the receiving port is just lost.

The division of power when a signal passes through the hybrid represents a 3-dB loss, and dissipative loss in the transformers used in the hybrid adds about another 1 dB. Since the four-wire transmission system contains active devices, this loss is readily negated.

The two-wire line may of course pass through switching systems and then to a wide range of telephone stations and associated loop circuits. Hence its impedance may be quite variable, and cannot always be matched by the fixed balancing network. Therefore, in practice, the balancing network is a design compromise usually consisting only of a resistance and capacitance in series. As a result of imperfect balance between line and network impedances in any particular case, some of the signal arriving from the four-wire receiving port will cross the hybrid to the transmitting port, constituting an echo.

A measure of the balance of the hybrid in a particular connection is given by the *return loss*. This is the apparent loss encountered by a signal introduced into one four-wire port of a hybrid until it emerges from the other four-wire port, corrected for *twice* the inherent loss of a pass through the hybrid (typically 8 dB). This definition in effect treats the signal as though it passed through the hybrid to the line, was reflected, diminished by the return loss, and passed back through the hybrid to the conjugate port. *Echo return loss* is the value averaged over the voice band. It has a mean value of about 15 dB with a standard deviation of 3 dB, over all connections.

The level of echo (relative to the talker's speech) which can be tolerated by a talker depends on the time delay with which the echo returns. The more an echo is delayed, the more it is perceived, and so the lower in level it must be to be tolerable.

Based on this relationship, the basic method of echo control in the traditional telephone network depends on controlling the end-to-end loss of a connection, since that loss, doubled, attenuates any echo. This represents a necessary compromise with the desire for low transmission loss. The scheme by which this is accomplished is called the *via net loss* (VNL) plan. In effect, each link of a connection contributes a component of loss proportional to its round-trip transmission delay. The required increment of loss has been established as 0.1 dB per ms of round-trip delay. The scheme also results in a fixed overall loss component of 5.0 dB (measured between the local offices, and thus not inclusive of loop loss). An additional loss of 0.4 dB is contributed by each link to provide margin against downward loss drift.

For round-trip delays greater than about 45 ms, the required loss increment (resulting in a total loss of at least 10.7 dB between end offices) would not be acceptable from a pure transmission standpoint. Thus, for such circuits, devices known as *echo suppressors* are applied. Most long-haul terrestrial telephone circuits today use transmission media whose high propogation velocities keep round-trip delay below the 45 ms threshold. (In fact, when a connection uses both digital transmission and switching, a new loss plan is employed which results in a fixed total connection loss between end offices of 6.0 dB.) However, the emergence of satellite communication circuits, with round-trip delays easily reaching 540 ms, renewed the interest in these devices.

Figure 12-2 shows a simplified block diagram of an echo suppressor at the east side of the circuit. Note that an echo control device is required on each end of the link located in the four-wire circuit usually very close to the hybrid. Whenever speech from the distant telephone set is present on the receive side of the four-wire, a comparator compares the speech level to a preset threshold. If the threshold is exceeded, the echo suppressor inserts a high loss in the transmit path, thus blocking the receive-side speech from returning via the echo path through the hybrid to the distant telephone set. In blocking the transmission path in this manner, transmit signals from the near end (the east side) telephone set are also blocked from transmission to the distant end. Therefore, an echo suppressor must have a second operating mode that allows the transmission path to be reenabled whenever the near-end talker breaks into the conversation while the far-end talker is speaking. The near-end talker must exceed the relative level of the far-end talker to reenable the transmission path during this double talk. When the transmission path is reenabled, the echo path is also reenabled, so that during a double-talk condition not only are the speakers hearing each other but they are also hearing their respective echoes. Some echo suppressors insert modest loss in both directions during double-talk. When both parties are talking simulta-

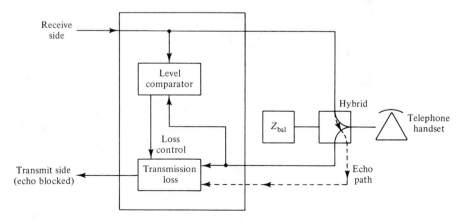

Figure 12-2 Simplified echo suppressor.

neously, the suppressor tends to insert and remove the loss rapidly, causing some speech to be blocked or chopped and in some cases, creating confusion.

On domestic terrestrial circuits the round-trip delay is rarely over 60 ms and echo suppressors adequately control the echo problem. However, the undesirable, subjective effects of echo are enhanced in the long delay environment of satellite communications. In fact, subjective studies at Bell Laboratories (Hatch and Ruppel, 1974) have shown conclusively that echo suppressors do not provide satisfactory performance in the long delay environment of satellite circuits. Therefore, in recent years a much more sophisticated echo control device called an *echo canceler* has emerged as a commercial reality.

Figure 12-3 shows a simplified block diagram of an echo canceler, which ideally is placed in the four-wire circuit near its junction with the two-wire circuit. An echo canceler controls echo by developing a replica of the echo signal and subtracting it from the rearward signal (comprising echo plus rearward speech), leaving only the desired rearward speech signal. In effect, the

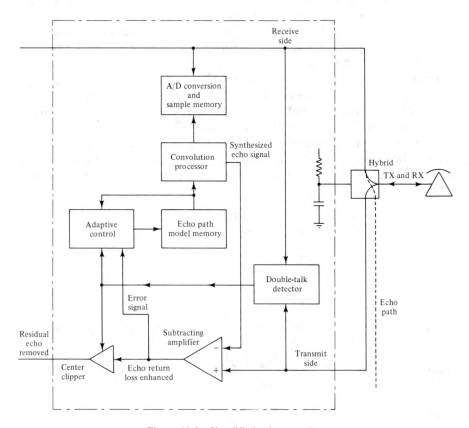

Figure 12-3 Simplified echo canceler.

echo canceler contains an "artificial line" which simulates the echo path in terms of delay and amplitude and phase response. The arriving inbound signal is passed through this "artificial line" and the resulting signal is the one subtracted.

The artificial line is in fact an adaptive linear filter, implemented with digital techniques. A convolution processor monitors the discrepancy between the actual echo and the echo replica and uses that information to refine the parameters of the filter's impulse response. This begins at the start of conversation with an arbitrary impulse response, and typically converges within about 1/2 s.

Typically, the cancellation process enhances the inherent echo return loss by about 18 dB so long as that inherent return loss is greater than 6 dB.

To improve the subjective quality, particularly during single talk, a center clipper device is normally added to the echo canceler to reduce effectively the small amount of residual echo to zero. This nonlinear clipping device produces some distortion, but it is acceptable as long as the echo return loss is in excess of 6 dB. During double-talk mode, the center clipper is removed to prevent distortion of the near-end talker's speech. Detection of double talk also stops the adaptive control processor to prevent contamination by the near-end talker's speech.

In satellite applications, the echo canceler must often be located at the interface between the satellite circuit and the general telephone network. This is not usually the end of the four-wire portion of the connection. Thus the delay in the echo path, as seen by the echo canceler, may be substantial. In turn, this implies that the "artificial line," or adaptive filter, in the echo canceler must have a significant delay capability. The use of VLSI technology makes the achievement of such delays with digital technique, feasible. The "end delay" accommodation of an echo canceler is an important specification to be considered in the selection of a unit for any given application.

Echo cancelers have clearly been accepted as the preferred method of echo control on satellite circuits. In fact, a great deal of development using a digital implementation of the echo canceler has been performed over the past decade and, at present, echo cancelers implemented in VLSI technology are available at prices comparable to those of echo suppressors. It is, therefore, very likely that echo cancelers will now be used not only on satellite circuits but also on terrestrial circuits requiring active echo control because of their high performance and affordable price.

For some time many people doubted the ability of satellite circuits to provide adequate voice communications because of the echo problem. However, because of the development of the echo canceler and its evolution into a commercial reality, the problem of echo control on voice circuits is now essentially solved. As we will see in the next section, data communications via satellite may also be impaired because of the long time delay, and good echo control also helps improve performance on data communications circuits.

12.2 SATELLITE DATA COMMUNICATIONS

The economics of long-distance telecommunications via satellite has created many new applications, services, and opportunities for data communication users. Unfortunately, the computer communications protocols, data modems, and echo control devices developed originally for terrestrial networks are not suitable for use in the long-time-delay environment of satellite communications. The impact of the delay on data circuits can be measured in terms of reduction in data throughput efficiency, plus potential modem and protocol malfunction due to mishandling of delayed echoes.

As illustrated in Figure 12-4, many new applications for data communications are available via satellite. Typical applications include resource sharing and load leveling, backbone networking, distributed processing, data base broadcast, and system backup and recovery. To accommodate these new services, much development effort has been expended in recent years to ensure that the long time delay and echo problems do not seriously impair the development of these applications.

12.2.1 Protocols

A set of rules that govern the exchange of data between information systems is called a *protocol*. There are a large number of protocols in use, many of which have been tailored to particular computer communications installations. Data are exchanged between business machines in block formats. That is, the data are grouped into blocks ranging in size from 1,000 to 100,000 bits. There are three basic classes of protocols in use today. The first class of protocol is the block-by-block transmission type illustrated in Figure 12-5. This class includes IBM's binary synchronous communications (BSC) protocol, which is by far the most prolific system in use today. In block-by-block protocols, data are transmitted in contiguous blocks with each block comprising a fixed number of bits. The block-by-block protocol employs a transmit, stop-and-wait error control technique with an automatic request for retransmission. As each block is transmitted to the distant end, it is checked for errors. If it is error-free, the reverse channel is used to acknowledge the receipt of an error-free block by transmitting an "ACK." If an error is found in the block, the reverse channel transmits a "NAK" signal, indicating detection of an error and requesting retransmission of that same block. As illustrated in Figure 12-5, waiting for the reverse channel acknowledgments (ACK/NAK signals) creates a large amount of idle channel time, which reduces throughput efficiency dramatically as the time delay increases. Throughput efficiency is defined as

$$\text{efficiency} = \frac{\text{number of bits received without error}}{\text{total number of bits transmitted}}$$

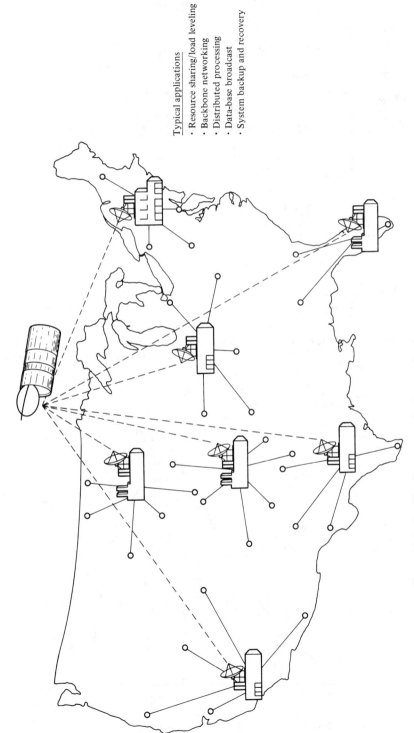

Figure 12-4 Satellite data communications applications.

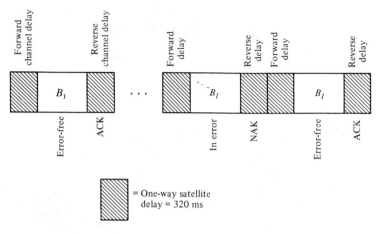

Figure 12-5 Block-by-block transmission protocol.

As shown in Cohen and Germano, 1970 (see references) the transmission efficiency can be improved by optimizing the block size as long as the error rate is not too high. Figure 12-6 shows a plot of transmission efficiency versus block size for the block-by-block transmission protocol for various error rates. Note that as long as the error rate is not too high ($< 10^{-6}$) we can achieve

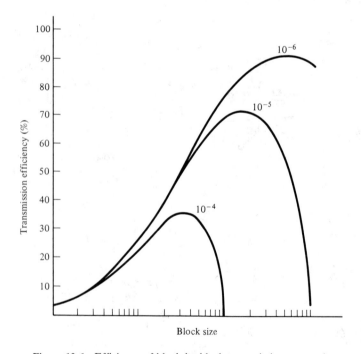

Figure 12-6 Efficiency of block-by-block transmission protocol.

Sec. 12.2 Satellite Data Communications

fairly high efficiencies by optimizing block size. However, if the block size is too small, even at low error rates, the waiting time is too large a percentage of the total transmission time. If the block is too large, the loss of a single block becomes a significant part of the total transmission time. Unfortunately, as the data rate increases, the optimum block size becomes impractically large, so that achieving high efficiency by simply selecting block size will work only for lower-speed applications.

A second class of protocols is illustrated in Figure 12-7. In this approach, blocks are transmitted continuously without waiting for the reverse channel to provide the acknowledgment signals. Note that as long as error-free blocks are received, blocks are transmitted contiguously and a high efficiency is achieved. Whenever a block is found in error (B_I in Figure 12-7), a NAK is received sometime after the end of that block. In this case the protocol simply completes the transmission of the block in process, stops transmitting, returns to the beginning of the block in error, and retransmits that block plus all succeeding blocks. This approach, called *continuous block transmission with restart after error detection*, can be implemented within the high-level data link control (HDLC) family of protocols. A typical example of such a protocol is the advanced data communications control protocol (ADCCP). As illustrated in Figure 12-7, the amount of idle channel time is substantially reduced compared to the block-by-block approach. In fact, the plot of transmission efficiency versus block size shown in Figure 12-8 shows that high efficiencies can be achieved even at relatively short block lengths. However, if the block length is too short (<100,000 bits), the amount of overhead information contained in the block becomes a significant portion of the total block, thus reducing efficiency significantly.

A third class of protocol uses continuous block transmission with selective block repeat. The HDLC family protocols can also be modified for this

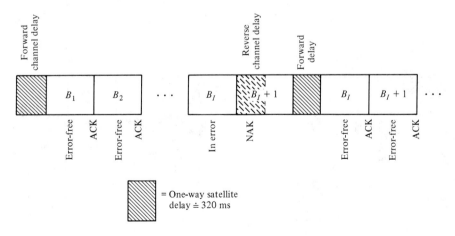

Figure 12-7 Continuous block transmission with restart after error detection.

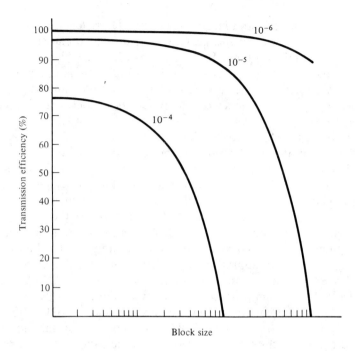

Figure 12-8 Efficiency of continuous block transmission protocol with restart after error detection.

technique. As illustrated in Figure 12-9, this method also transmits blocks contiguously without waiting for the acknowledgment signal. As long as blocks are error-free, block transmission continues with virtually no idle time. Whenever an error occurs, a NAK signal is received on the reverse channel. However, the protocol continues to transmit blocks one after another until the

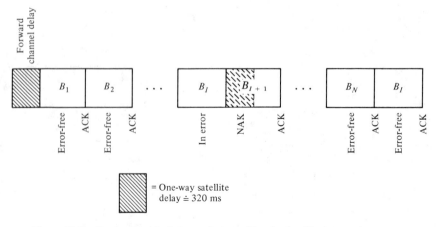

Figure 12-9 Continuous block transmission with selective block repeat.

end of that particular sequence of blocks. At the end of the block sequence, only the block(s) in error are repeated and a new sequence of block transmission begins. This protocol requires a block sequence numbering system which is usually included in the HDLC family protocols. This selective repeat protocol is indeed the most efficient, as illustrated in Figure 12-10. At virtually any error rate less than 10^{-5}, very high efficiencies are achieved with relatively short block lengths. Again, if the block size is too short, the block overhead becomes a significant portion of the block and reduces efficiency. Typical block sizes for this class of protocol are in the range of 10,000 bits.

12.2.2 Data Communications Efficiency

The achievable transmission efficiency of a data communications circuit is probably its most important attribute. There are, however, other factors which must be considered when evaluating its overall quality. In Owings' article (see references), results on efficiencies are provided for all three classes of data communications protocols for various data rates and error rates. These results are based on the following assumptions:

Physical file size = 10^{10} bits: Physical file sizes on disk or tape are now in the range 10^8 or 10^9 bits. In the near future 10^{10} bits will be commonplace.

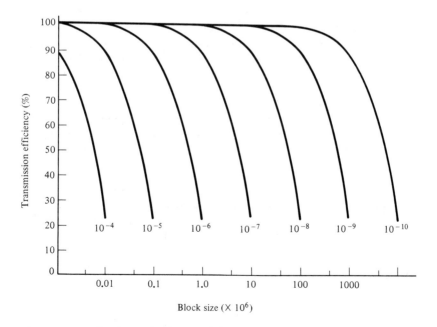

Figure 12-10 Continuous block transmission with selective block repeat.

Data rate range = 2.4 kb/s to 1.544 Mb/s: Although data rates are clustered toward the low end, there is a trend toward the higher rates.

One-way satellite transmission time delay = 320 ms.

Threshold bit error rate = 10^{-5}: Typical satellite circuits deliver error rates less than 10^{-6} more than 95% of the time. Typical rates observed are 10^{-7} or 10^{-8}. However, under worst-case conditions, error rates can drop as low as 10^{-5}.

Block overhead less than 1%: In each of the protocols in use today, there is a minimum amount of overhead data for housekeeping functions, such as block number identification, parity checks, and other error control bits. To achieve the 1% objective, block sizes greater than about 8,000 bits are required.

Based on these assumptions, we can set forth the following set of objectives and evaluate the relative performance of the three transmission protocols.

1. Probability that the entire physical file (10^{10} bits) is received with an undetected error must be less than 0.01.
2. Transmission efficiency must be at least 90%.
3. The efficiency must be independent of data rate.
4. Minimum sensitivity to error rate.

Tables 12-1 and 12-2 compare the three protocols relative to these performance objectives. Results are provided for three different data rates and two error rates. For voiceband rates (e.g., 4.8 kb/s) fairly high efficiencies are achievable even with the block-by-block protocol by optimizing the block size. However, as the data rate increases, the block size must also increase

TABLE 12-1 Protocol Efficiency Comparison

Protocol / Data rate / Error rate	Block-by-Block		Continuous with restart after error		Continuous with selective repeat	
	10^{-5}	10^{-6}	10^{-5}	10^{-6}	10^{-5}	10^{-6}
4.8 kb/s	72%	90%	97%	99%	90%	99%
56 kb/s	40%	70%	72%	97%	90%	99%
1,544 kb/s	5%	25%	10%	50%	90%	99%

Sec. 12.2 Satellite Data Communications

TABLE 12-2 Protocol Performance Comparison

Criteria \ Protocol	Block-by-Block	Restart after block error	Continuous with selective repeat
$P = 0.01$ for 10^{10} bits without error	No	Yes	Yes
Efficiency > 90% at BER = 10^{-5}	No	No	Yes
Efficiency independent of rate	No	No	Yes
Efficiency insensitive to error rate	No	No	Yes

dramatically and the efficiency degrades accordingly. The continuous transmission with restart-after-error protocol does a much better job than block-by-block transmission but still does not achieve the 90% efficiency objective, particularly at the higher data rates. Only the continuous block transmission system with selective repeat achieves at least 90% efficiency at an error rate of 10^{-5} over the full data-rate range.

12.2.3 Implementation

In the long term, the implementation and development of these protocols will be accomplished by software modification of existing protocols. However, in the interim there may also be a need to provide an external hardware solution in some cases, particularly in those older installations in which it is difficult to make major software changes in the protocol to accommodate satellite service. In these cases, a satellite delay compensator, illustrated in Figure 12-11, may solve the problem. This device is inserted between the terrestrial interface to the information system and the satellite data channel. In effect, it is a store-and forward data processor which interacts with the near-end information system, using the existing protocol (such as BSC) but with its counterpart at the distant end of the satellite circuit using a selective repeat or restart after error protocol. The device receives data from the information system and organizes it in blocks of the correct size. A multiple-block buffer stores transmitted blocks until the reverse channel indicates that a block was received in error. Whenever a request for retransmission occurs, the controller interrupts transmission from the information system and returns to the multiple-block buffer to retransmit either all of the previous blocks from the time the error occurred or a selected block, de-

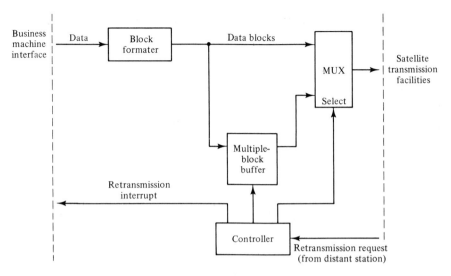

Figure 12-11 Simplified satellite delay compensator.

pending on which protocol is being implemented. This approach has the advantage of not disrupting the software and still solving the efficiency problem.

12.2.4 Forward Error Control

The use of forward error control (FEC) is also a potential solution to the problem of improving efficiency on satellite data communication circuits. By adding sufficient error correction coding bits at the transmit side, we can improve the overall error rate by correcting many errors at the receiver. Reduction of error rate automatically improves efficiency for any data communications protocol. However, this improvement in error rate with FEC must be balanced against the cost of the overhead bits used to perform the error correction function. In general, only highly efficient (high rate) codes can justifiably be used for this purpose. For example, a so-called 7/8 code uses one parity check bit for each seven information bits, thus resulting in a maximum achievable efficiency of 87.5%. Clearly, lower-rate codes, such as 3/4 or 1/2 codes, will not satisfy our efficiency criterion, despite the fact that the error rate can be reduced substantially. The decision to use FEC depends on the channel error rate, the protocol being used, and the desired efficiency objectives. Figure 12-12a shows the effect of 7/8 code on transmission efficiency, using a restart-after-error detection protocol. Notice that the FEC code provides an enhancement of efficiency as long as the error rate is less than 10^{-7} compared to the use of the protocol alone. Figure 12-12b shows the effect of the same 7/8 FEC code using the continuous protocol with selective repeat. In this case, use of coding provides no substantive advantage

Sec. 12.2 Satellite Data Communications

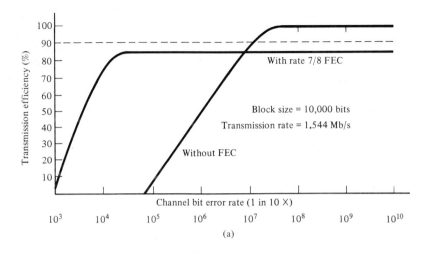

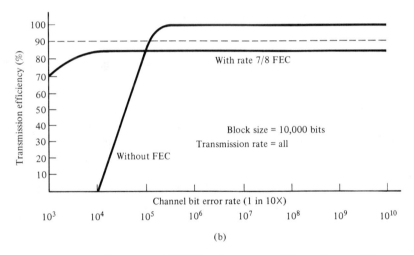

Figure 12-12 Effect of rate 7/8 FEC code on transmission efficiency: (a) with restart-after-error detection protocol; (b) with continuous protocol with selective repeat.

unless the error rate is less than 10^{-5}. This is true for all transmission rates and shows that the use of FEC may not always improve efficiency, depending on the desired rate.

12.2.5 Impact of Echo Control Devices and Data Modems

Uncontrolled echo on voice circuits can also be disruptive to data communications on voiceband circuits. A malfunctioning echo suppressor, for example, can cause clipping and chopping of the speech and echo signals.

This, in turn, may cause a data modem to misinterpret delayed echoes as data, and produce malfunctions such as shutdowns or impaired data. Modems designed for the typical delays of the terrestrial plant can also cause difficulty because of the long time delay. Such devices can trigger early timeouts because of expectations of return signals which do not arrive within the prescribed time period. In general, the introduction of echo cancelers on satellite circuits solves many of the problems produced by echo on data communications circuits. Improved modem design will also help.

12.3 ORBITAL VARIATIONS AND DIGITAL NETWORK SYNCHRONIZATION

A satellite in geostationary orbit does not remain exactly stationary. Departure from a stationary position is influenced by a number of factors, including the earth's nonsymmetrical gravitational field, and perturbations caused by the sun and moon. Inaccuracies in correcting satellite position and adjusting orbital parameters also contribute to the nonstationarity of the orbit. Although the total change in position translates into a change in transmission time of less than 0.5%, the variation is significant enough to require some compensation on satellite links exchanging digital data with synchronous terrestrial networks.

A synchronous satellite orbit departs from perfect stationarity because of its inclination angle relative to the plane of the equator and its orbital eccentricity (see Chapter 2). Figure 12-13 illustrates the variations in the satellite latitude relative to the equatorial plane. These variations are sinusoidal, with a period equal to the sidereal day. The north/south oscillation of the satellite relative to the equatorial plane has a peak-to-peak displacement of $2R$, where R is the nominal altitude of the satellite measured from the

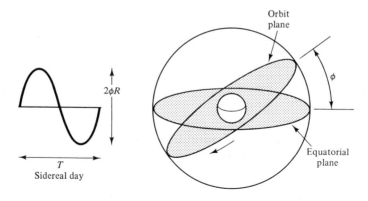

Figure 12-13 Orbital delineation (north-south oscillation).

Sec. 12.3 Orbital Variations and Digital Network Synchronization

center of the earth (42,164 km). Typically, the station-keeping limits of satellites in geostationary orbit today are about $\pm 0.1°$ in longitude and latitude. However, late in satellite lifetime, the station keeping may be relaxed to $\pm 0.5°$ in latitude. This worst-case condition results in a maximum peak-to-peak, north/south position variation of about 728 km.

A second component of orbital position variation is due to the eccentricity of orbit, as illustrated in Figure 12-14. The eccentricity causes an altitude oscillation and an east/west oscillation, both of which vary sinusoidally over a period of a sidereal day. Typical values of orbital eccentricity can reach 5×10^{-4}, resulting in an altitude variation of about 42 km and an east/west oscillation of about 84 km. In addition, there is also a slow drift in the east/west direction taking place over a period of many sidereal days which can be as much as 140 km.

The amount of range variation observed by any earth station of course depends on the exact position of the station on the earth's surface. For a station-keeping limit of $\pm 0.5°$ and an orbital eccentricity of 5×10^{-4}, the worst-case variation in one-way propagation time between the satellite and the earth is approximately 550 µs. Since this propagation-time variation is observed on both the uplinks and the downlinks, we must be able to compensate for a total peak-to-peak propagation time variation of approximately 1.1 ms. Although this is less than 0.5% variation compared to the nominal propagation time of 270 ms, it can be a significant factor in accommodating digital data transmission over satellite links.

These orbital variations have no impact on analog transmission, but digital data transmission may be dramatically affected by these time variations, which cause the data rate at the receiving station to vary over the period of the sidereal day about the nominal data rate. For example, at a data rate of

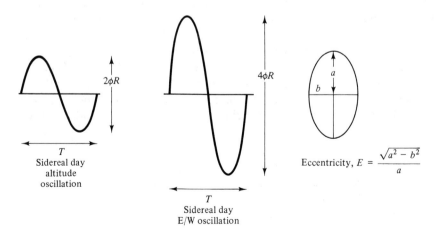

Figure 12-14 Effects of orbital eccentricity.

56 kb/s, a 1.1-ms transmission-path-length variation results in a peak-to-peak variation in data rate of 61.6 bits. Since terrestrial networks, using synchronous transmission, cannot accommodate data-rate variations of this magnitude, an elastic buffer must be inserted between the satellite facilities and the terrestrial network, as illustrated in Figure 12-15. An elastic buffer is essentially a first-in-first-out (FIFO) random-access memory which can be thought of as a bucket with a hole in its bottom. The rate is which water (bits) is poured into the bucket changes over the period of the sidereal day. However, the rate at which water (bits) leaves the bucket remains constant, independent of the fullness of the bucket. As the pouring rate increases, the bucket tends to fill, but the water output remains constant. When the pouring rate decreases, the bucket tends to empty, but again the output from the bottom remains constant. The elastic buffer operates analogously and the primary design consideration is to choose a buffer that is large enough to absorb the peak-to-peak variations in data rate. Table 12-3 shows buffer sizes for data rates from 2.4 kb/s to 6.3 Mb/s for a station-keeping limit of 0.5% and orbital eccentricity of 5×10^{-4}. Memory devices are generally built in sizes that are powers of 2. Table 12-3 also shows the minimum practical memory size for each rate. Since the use of these buffer memories inherently adds additional delay to the satellite link, it is important to choose the smallest memory consistent with the maximum expected range variation.

In addition to accounting for the range variations due to satellite position movement, the elastic buffer shown in the digital interface of Figure 12-15 can also be used to perform a second function. The necessity for it arises when the satellite is used to communicate between two separate digital networks, each operating with its own clock. Three methods have been proposed to accommodate this problem. The first is an asynchronous solution using bit stuffing and justification. The second is a synchronous method involving locking all clocks together. The third is the plesiochronous solution, in which the two clocks have almost the same frequency but are not actually synchronized. As shown in Figure 12-15, if the S clock and the T clock are independent clocks at the same nominal rate, the interface is said to be *asynchronous*. If the S clock and T clock are locked together or are, in fact, the same clock, the interface is called *synchronous*. If the S and T clocks are almost the same but not actually synchronized, the interface is referred to as *plesiochronous*. The plesiochronous solution to synchronization of two independent networks

Figure 12-15 Interface between satellite system and terrestrial facilities.

TABLE 12-3 Required Elastic Buffer Size to Accommodate
1.1-ms Peak-to-Peak Range Variation

Data Rate	Minimum Buffer Size (bits)	Minimum Practical Buffer
9600 b/s	10.56	16
56 kb/s	61.6	64
1.544 Mb/s (T1)	1698.4	2048
6.312 Mb/s (T2)	6943.2	8192

is the solution recommended by the CCITT. For example, assume that two independent networks are to be connected via the satellite and each network has a clock derived from an independent cesium beam oscillator with an accuracy of 1×10^{-11}, as recommended by the CCITT. Assuming nominal clock rates of 2..048 Mb/s (the first level in the European digital hierarchy) and the clock of one network is high by 1×10^{-11} in frequency and the other is low by 1×10^{-11}, a time-displacement error of one bit will accumulate in just under seven hours. As recommended in CCITT Recommendation G. 703, PCM multiplex system slips will be made in integral frame increments to avoid loss of frame in the multiplex equipment. With a frame size of 256 bits, slips will occur about every 72 days. Since the bit slips can occur in either direction, the function to be performed is essentially a first-in-first-out memory function. This can be accomplished by the elastic buffer of Figure 12-15, in addition to its orbital satellite range absorption function, as long as the buffer is of sufficient size.

REFERENCES

CAMPANELLA, S. J., H. C. SUYDERHOUD and M. ONFRY, "Analysis of an Adaptive Impulse Response Echo Canceler," *COMSAT Tech. Rev.*, Vol. 2, No. 1, Spring 1972.

CCITT Recommendation G.703, "General Aspects of Interfaces," November, 1980.

COHEN, L.A. and G. V. GERMANO, "Gauging the Effect of Propagation Delay and Error Rate on Data Transmission Systems," *ITU Telecommunication Journal*, Vol. 27, 1970, pp. 569–574.

DUTTWEILER, D. L., "Bell's Echo Killer Chip," *IEEE Spectrum*, Oct. 1980.

HARRINGTON, E. A., "Issues in Terrestrial/Satellite Network Synchronization," *IEEE Trans. Commun.*, Vol. Com-27, No. 11, Nov. 1979.

HATCH, R. W., and A. E. RUPPEL, "New Rules for Echo Suppressors in the DDD Network," *Bel Lab. Rec.*, Vol. 52, 1974.

NUSPL, P. P., and R. MAMEY, "Results of the CENSAR Synchronization and Orbit Perturbation Measurement Experiments," 4th International Conference on Digital Satellite Communications, Montreal, Canada, Oct. 1978.

OWINGS, JAMES L., "Satellite Transmission Protocol for High Speed Data," SBS, Applications and System Development.

ROSSITER, P., R. CHANG, and T. KANION, "Echo Control Considerations in an Integrated Satellite Terrestrial Network," 4th Int. Conf. Digital Satellite Commun., Montreal, Oct. 23–25, 1978.

SABLATASH, M. and J. R. STOREY, "Determination of Throughputs, Efficiencies and Optimal Block Lengths for an Error-Correction Scheme for the Canadian Broadcast Telidon System," *Can. Electr. Eng. J.*, Vol. 5, No. 4, Oct. 1980.

SCIULLI, J. A., "Data Communications Via Satellite—Problems and Prospects," Proc. of Interface '82, McGraw Hill, April 1982, pp. 103–105.

SETZER, R., "Echo Control for RCA Americom Satellite Channels," RCA Review 25—1, June–July 1979.

YEH, L. P., "Geostationary Satellite Orbital Geometry," *IEEE Trans. Commun.*, Col. COM-20, No. 4, April 1972.

Table of Useful Constants

Physical constants

Velocity of light	$c = 2.997\,924\,58 \times 10^8$ m/s (exact)
Boltzmann's constant	$k = 1.380\,622 \times 10^{-23}$ J/K
Gas constant	$R = 8.314\,34$ J/(K·mol)
Avogadro's number	$N_0 = 6.022\,169 \times 10^{23}$ mol^{-1}
Planck's constant	$h = 6.626\,196 \times 10^{-34}$ J·s
Stefan-Boltzmann constant	$\sigma = 5.669\,61 \times 10^{-8}$ W/(m^2·K^4)
Electron charge	$e = 1.602\,192 \times 10^{-19}$ C
Electron rest mass	$m = 9.109\,558 \times 10^{-31}$ kg
Permeability of free space	$\mu_0 = 4\pi \times 10^{-7}$ H/m (exact)
Permittivity of free space	$\varepsilon_0 = 1/\mu_0 c^2 = 8.854\,188 \times 10^{-12}$ F/m
Electrostatic constant	$k = 1/4\pi\varepsilon_0 = 8.987\,552 \times 10^9$ N·m^2/C^2
Gravitational constant	$G = 6.672 \times 10^{-11}$ N·m^2/kg^2
"Standard" acceleration of gravity	$g = 9.806\,65$ m/s^2 (exact)

Astronomical constants

Physical data for earth (IAU, 1976)	
Equatorial radius	$R_E = 6378.140$ m
Flattening factor	$f = 1/298.257$
Dynamical form-factor	$J_2 = 1.082\,63 \times 10^{-3}$
Gravitational constant	$\mu_\oplus = GM_\oplus = 3.986\,005 \times 10^{14}$ m^3/s^2
Mass Data	
Earth	$M_\oplus = 5.974 \times 10^{24}$ kg
Moon	$M_\mathrm{C} = 7.348 \times 10^{22}$ kg
Sun	$M_\odot = 1.989 \times 10^{30}$ kg
Orbital data for earth	
Semimajor axis	1.4953×10^{11} m
Eccentricity	0.016 74
Tropical year (equinox to equinox)	365.242 191 d
Sidereal year (fixed star to fixed star)	365.256 363 d
Orbital data for moon	
Semimajor axis	3.8440×10^6 m
Eccentricity	0.054 90
Inclination to ecliptic	5°.1454
Synodic month (new moon to new moon)	29.530 589 d
Tropical month (equinox to equinox)	27.321 582 d
Sidereal month (fixed star to fixed star)	27.321 662 d
Mean sidereal day	$23^h\,56^m\,04\overset{s}{.}09054 = 86\,164.0954$ ^{s}
Obliquity of the ecliptic (year 2000)	23°. 26′ 21″.488
General precession (year 2000)	5029″.0966/century

GLOSSARY OF ABBREVIATIONS

ACS	Attitude control system
ADC	Analog-to-digital coding
ADCCP	Advanced data communications control
ADM	Adaptive delta modulation
ADPCM	Adaptive differential pulse-code modulation
AFC	Automatic frequency control
AKM	Apogee kick motor
AM	Amplitude modulation
AT&T	American Telephone & Telegraph Co
BER	Bit error rate
BPSK	Binary phase-shift keying
BS	Broadcast satellite
BSC	Binary synchronous communication
CCIR	Comité Consultatif International de Radio (International Radio Consultative Committee)
CCITT	Comité Consultatif International de Téléphone et de Télégraph (International Telephone & Telegraph Consultative Committee)
CDC	Control and delay channel
CDMA	Code-division multiple access
C/N	Carrier-to-noise ratio
CS	Communications satellite
DBS	Direct broadcast satellite
DSCS	Defense service communications satellite
DTS	Digital termination service
ECS	European communications satellite
eirp	Equivalent isotropic radiated power
ESA	European Space Agency
FCC	Federal Communications Commission
FDM	Frequency-division multiplex
FDMA	Frequency-division multiple access
FEC	Forward error control
FET	Field-effect transistor
FIFO	First-in first-out
FM	Frequency modulation
GEO	Geostationary operating orbit
GMT	Greenwich mean time
GMST	Greenwich mean sidereal time
GST	Greenwich sidereal time
G/T	Gain-to-temperature ratio
GTO	Geostationary transfer orbit
HDLC	High-level data link control
HPA	High-power amplifier
IBM	International Business Machines

Glossary

IEEE	Institute of Electrical and Electronics Engineers
ISRO	Indian Space Research Organization
ITU	International Telecommunication Union
IUS	Interim upper stage
LEO	Low earth orbit
MCPC	Multiple channel per carrier
MSK	Minimum-shift keying
NASA	National Aeronautics & Space Administration (U.S.)
NASDA	National Aeronautics & Space Development Agency (Japan)
NBP	No baseband processing
NIC	Nearly instantaneous companding
NPR	Noise power ratio
NTIA	National Telecommunications & Information Administration
OOK	On/off keying
OTV	Orbital transfer vehicle
PAM	Payload-assist module
PAM	Pulse-amplitude modulation
PCM	Pulse-code modulation
PSK	Phase-shift keying
PTT	Postes, Téléphone & Télégraph
QPSK	Quaternary phase-shift keying
RARC	Regional Administrative Radio Conference
RCS	Reaction control system
RF	Radio-frequency
RTG	Radioisotope thermo-electric generator
SBS	Satellite Business Systems
SCPC	Single channel per carrier
SLV	Satellite launch vehicle
SSB	Single-sideband
STC	Satellite Television Company
TDM	Time-division multiplex
TDMA	Time-division multiple access
TDRS	Tracking and data relay satellite
TT&C	Telemetry, tracking and command
TVRO	TV receive-only
TWT	Traveling-wave tube
TWTA	Traveling-wave-tube amplifier
UT	Universal time
VNL	Via net loss
VHF	Very high frequency

Index

A

Abbreviations, 161, 246, 262
Advanced RCA Satcom, 118
Aerodynamic forces, 90
Aerospatiale (SNIAS), 11
A law, 210
American Satellite Company, 7
American Telephone & Telegraph Company (AT&T), 6
Amplifiers, 153, 282–4, 288, 294, 298
 back-off, 153
 field-effect transistor (FET), 18, 291
 gallium arsenide, 291
 high-power (HPA), 157, 283, 288–91, 314–15
 klystron, 313–14
 low-noise, 18, 283, 291, 317–19
 solid-state, 157, 290–1
 traveling-wave-tube (TWTA), 18, 283, 290–1, 298–9, 312–14
Amplitude modulation (*see* Modulation)
AM-to-PM conversion, 18, 157, 298–300
Angello, P.S., 217, 222, 238
Anik A, 8
Anik B, 8
Anik C, 8
Anik D, 8
Anomalies, 32–34

Antennas, 166–7, 292, 320–37, 343 (*See also* Receivers; Spacecraft)
 apertures, 330–1
 arrays, 322
 beams, 269–70
 Cassegrainian, 317, 324–6, 328–9, 332
 C-band, 320
 cross-polarization, 334–5
 earth station, 19, 317, 320–37
 Fresnel lens, 327
 Gregorian, 324–6, 328, 332
 Herschelian, 326
 horn, 322–4
 illumination, 331–2
 K-band, 320
 lens, 322
 Mangin mirror, 327
 mounts, 335–7
 Newtonian, 324, 326
 parabolic, 320, 324–6
 phased array, 322
 reflector, 324–9
 telemetry, tracking, and command, 134
 temperatures, 166, 317, 345
 toroidal, 325, 329
 tracking systems, 335, 337–40
Apogee kick motors (AKM), 98, 128, 137
Arab League, 11
Arabsat, 11

Index

Ariane, 96, 98, 99, 103, 106
ASC, 7
Astronomical Almanac, 65, 67, 68, 70–75, 81
Atlas SLV-3D/D1-A Centaur, 99, 102, 106
Attitude control, 111–20, 129, 137
 dual spin, 115
 momentum bias, 115
 sensors, 114
 spin-stabilized, 114–15
 three-axis, 115
Aussat, 11
Austin, M.C., 238
Australia, 11
Azimuth, 73, 77

B

Back-off, 153–9
Baseband (*See* Links)
Bate, R., 56
Batteries, 136
Bazovsky, I., 139, 143
Beam (*See* Antennas)
Bedford, R., 278
Bell Telephone Laboratories, 238, 306, 370
Bennett, W.R., 182, 186, 195, 217, 238
Bennett, William, 365
Berretta, G., 302, 306
Berman, A.I., 56, 85, 108, 306
Binary synchronous communications (BSC), 372, 379
Bit error rate (BER), 211, 214, 227–31, 285, 378 (*See also* Error)
Boeing, 105
Bond, F., 306
Bousquet, M., 144, 181
Brady, P.T., 253
Brazilsat, 11
British Aerospace, 13, 14
Broadcast signals, 186
Brouwer, D., 56
Brown, K., 92, 108
BS-2, 13
BS-2A, 52
Buffers, elastic, 384–5
Burst mode, 22
Bylanski, P., 182, 205, 216, 217, 238

C

Cacciamani, E.R., Jr., 277
Calendar date, 73
Campanella, S.J., 277, 385

Canadian satellites, 8
Carlson, J., 182, 195, 201, 205, 208, 217, 220, 238
Carrier:
 bandwidth, 275
 -to-impairment ratio, 300
 -to-interference ratio, 17, 243, 348–56, 358–9
 -to-intermodulation noise, 157, 314
 -to-intermodulation ratio, 302–5
 -to-noise ratio (C/N), 9, 16–17, 21, 149, 156–9, 191, 199, 212, 222–3, 237–8, 250–2, 257, 259–60, 266–7, 284, 358, 364–5
Cassiopeia, A, 345
Cauley, M.A., 359, 365
CBS, 12
Celestial mechanics, 29–32
Centaur, 102, 105 (*See also* Atlas, Titan)
Chang, R., 386
Channel, 1011, 358
 bands, 283
 capacity, 259–60, 266–7, 275–6
 coding, 21
 control and delay (CDC), 262–3
 cross-polarization, 334
 groups, 194–5
 protection ratio, 358–65
Chips, 271, 275
Circuits, 1011
 linear, 157
Clark, A.P., 186, 238
Clarke, Arthur C., 2
Clemence, G.M., 56
Climate, 174–7
Coaxial cable transmission, 25
Codecs, 205–6
Coding, 380
 analog-to-digital (ADC), 251–67
 rate, 257
 sequence, 272–3
 speech, 204–12
Cohen, L.A., 374, 385
Colby, Roger J., 277
Communications Act of 1934, 2, 27
Communications Satellite Act, 2
Communications Satellite Corporation (Comsat), 2, 346
Compandors, 201
Compression point, 300–1
Comstar, 6
Conservation of energy, 31
Conservation of momentum, 85

Cook, C., 270, 277
Cosmos 1546, 52
Courier, 1
CS-2, 10
Cuccia, C.L., 205, 220, 238

D

Data:
  circuits, 372
  communications services, 24, 372–82
  modems, 381–2
  narrowband, 185
  signals, 185
  transmission, 185, 263, 383
  voiceband, 185
  wideband, 185
Davey, J.R., 182, 186, 217, 238
Deadband, 53
Deal, J., 277
Declination, 73–75, 80
Demand assignment, 269
Dicks, J., 278
Digital:
  hierarchies, 216–19
  microwave radio, 25
  modulation, (*See* Modulation)
  modulator, 219–20
  signals, 216–17
  termination service (DTS), 23
Dill, G.D., 277
Diplexers, 17
Direct Broadcast Satellite Corporation (DBSC), 12
Direct broadcast satellites, 12–13
Dixon, J.T., 199, 238
Dodds, D.E., 238
Down conversion, 314–15
Downlinks (*See* Links)
DSCS-2, 14
DSCS-3, 14
Duplexers, 283
Duttweiler, D.L., 385

E

Early Bird, 2
Earth, 77–78
  nonspherical, 77–79
  oblate, 48
  rotational velocity, 93
  triaxial, 51, 53
Earth stations, 19–21, 183, 241, 308–47
  antennas, 68, 72, 308, 320–37
  beams, 68, 72
  geometry, 60, 70
  multiplex equipment, 247–9
  primary power, 310, 341
  receiver, 308, 314–20
  subsystems, 308–10
  terrestrial interface, 310, 340–1
  test equipment, 310, 342–6
  tracking systems, 310, 337–340
  transmission equipment, 253–257, 308, 310–314
  TVROs, 317–18
Echo:
  cancelers, 370–1, 382
  control, 367–71, 381–2
  suppressors, 369–70
Echo I, 1
Eclipses, 62–68, 123
  geometry, 62–68
ECS, 10
ECS 2, 52
Edelson, Burton I., 27, 277
Efficiency,
  aperture, 330–2
  blockage, 330, 332
  cross-polarization, 330, 332
  data communications, 377–9
  ohmic and mismatch, 330
  spillover, 330, 332
  surface, 330
Ekman, D., 56
Electronics systems, 16
Elevation, 73
Embratel, 11
End-to-end satellite links, 15
Energy-to-noise ratio, 223–4
Equinoxes, 37, 66–67
Equivalent isotropic radiated power (eirp), 19
Error (*See also* Bit error rate)
  forward error control (FEC), 380–1
  rate, 227–31, 380–1
European Space Agency (ESA), 13
Eurosatellite, 13
Eutelsat, 10
Expansion ratio, 92–93
Explanatory Supplement to the Astronomical Ephemeris, 75, 78–79, 81
Explorer I, 1

Index

F

Fashano, M., 238
Federal Communications Commission (FCC), 28
Feher, K., 217, 238, 239, 277
Feller, W., 139, 143
Fiber optic:
  networks, 26
  transmission, 25
Filters, 285–8, 293, 371
Fitzpatrick, P.M., 56
Fixed-service communications satellites, 6–8, 10–11
Fleetsatcom, 14
Forcina, G., 278
Ford Aerospace & Communications Corporation, 5, 7, 10, 12
Ford Aerospace Satellite Services Corporation, 7
Fordsat, 7
Frame efficiency, 266
France, 13
Franks, L.E., 238
Free space loss, 148
Frequency, 18
  automatic control (AFC), 259
  bandwidths, 279, 281
    C-band, 18, 281
    K-band, 281
    K_u-band, 18
    X-band, 18, 281
  downlink, 281
  modulation (*See* Modulation)
  uplink, 281
Frick, R.H., 48, 56

G

Gagliardi, R.M., 181
Galaxy, 6
Galaxy 1, 52
Galaxy 2, 52
Galaxy 3, 52
Galko, P., 217, 238
Garber, T.B., 48, 56
Gardner, F.M., 239
Germano, G.V., 374, 385
Germany, West, 13
G.E. Space Division, 14
Glasgal, R., 186, 238
Glave, F.E., 217, 238, 240
Gold, R., 273, 277

Goode, B., 277
Gough, G.R., 306
Gravity gradient effect, 54, 113
Gronemeyer, S., 217, 238
Groumpas, P.P., 359, 365
Ground traces, 75
GSTAR, 7
G/T, 19, 343–6
GTE, 7
  Lenkurt *Demodulator*
  Spacenet, 7

H

Handbook of Chemistry and Physics, 56
Handbook of Mathematical Tables, 38
Harrington, E.A., 385
Hatch, R.W., 370, 385
Haviland, R., 108, 126, 143
Heiter, G.L., 306
Historical background, 1–8
Hodson, K., 277
Hohmann, 40, 83, 99
Holbrook, B.D., 199, 238
Horwood, D.F., 238
Hour angles, 73–76, 79
House, C.M., 108, 126, 143
Huang, J., 217, 238
Hubbard Broadcasting Company, 12
Hughes Aircraft Company, 4, 5, 6, 7, 8, 10, 11, 14
Hughes Communications Company, 6, 7
Hughes-K_u-band, 7
Hybrid coil, 367–8

I

Impairments, 242–4 (*See also* Transmission)
  earth segment, 19
  grade, 359–60
  satellite, 18, 282, 289, 300
Inclination, 77, 98
Indian Space Research Organization (ISRO), 10
Industry growth, 27
Ingram, D., 205, 216, 217, 238
Insat, 10
Insat 1B, 52
Institute of Electrical and Electronic Engineers (IEEE):
  standards, 165
  Transactions on Communications, 27

Intelsat, 2–5, 234, 253, 277
Intelsat I, 2, 4
Intelsat II, 4
Intelsat III, 4
Intelsat IV, 4
Intelsat IV-A, 5
Intelsat V, 5, 234
Intelsat V-A, 5
Intelsat V F-4, 52
Intelsat V F-6, 52
Intelsat V F-7, 52
Intelsat VI, 5
Intercept point, 300–1
Interference, 60, 348–65
Interim Upper Stage (IUS), 103, 105
Intermodulation, 153, 302–5, 319–20
International Business Machines (IBM), 372
International Telecommunication Union, (ITU), 28, 239, 350, 359
  CCIR, 357, 359
    antenna sidelobe formulas, 355
    loading factors, 200
    Recommendations and Reports, 27, 233, 277, 350, 361, 365
    standards, 203, 250
  CCITT, 234, 385
  Regional Administrative Radio Conference (RARC), 360
Inverse-square law, 147
Ion engines (*see* Propulsion)
Ippolito, L.J., 173, 181
Ishiguro, T., 186, 204, 239

J

Jasik, J., 147, 238, 322, 324, 346
Jayant, S.N., 205, 239
Jefferies, A., 277
Julian days, 74

K

Kalil, F., 51, 56
Kamel, A., 47, 51, 56
Kaneko, H., 186, 204, 239
Kanion, T., 386
Kaplan, M., 42, 56, 112, 143
Kaul, R.D., 181
Kepler's laws, 32
Keying, 220–6
  binary phase-shift (BPSK), 221–6
  frequency-shift, 220

  minimum-shift (MSK), 222
  on/off (OOK), 220
  phase-shift (PSK), 220–8, 242, 251–7, 255–7, 260–7, 275
  quaternary PSK (QPSK), 223–6
  rate, 220
Kirchhoff's laws, 164
Kraus, J.D., 181, 322, 330, 347
Kwan, R.K., 277

L

Latitude, 77
  geographic, 77
Launch:
  sequence, 102
  sites, 93, 98
  vehicles, 82–107 (*See also* names of individual vehicles)
    gravity turn, 95, 100
    parking orbit, 95
    payload capabilities, 101
    velocity, 95
Levy, G.F., 327
Lijima, Y., 239
Lin, S.H., 181
Lindsey, W., 239
Links, 9–11, 25
  baseband (earth terminal—user), 9, 21, 193–4
  downlinks, 16, 151–6, 285
  end-to-end satellite communications, 15
  no baseband processing (NBP), 257–60
  radio-frequency (RF), 9, 11, 16, 21, 145–81, 250, 289
    bandwidth, 244
    limits, 150
    power, 244
    television, 233–8
  uplinks, 16, 151–6, 285, 292
Local distribution radio, 26
Longitude, 44–45, 73
Lopriori, M., 239
Loss:
  echo return, 368
  return, 368
  via net (VNL), 368
L-Sat, 13
Lucky, R.W., 186, 239
Lundquist, 195, 217, 220
Lundsford, J., 277
Lyons, R.G., 278

Index

M

Mahle, C., 306
Mamey, R., 385
Maral, G., 140, 144, 181
Marecs, 14
Marisat, 14
Marsh, H., 270, 277
Martikan, F., 51, 56
Martin, J., 239
Mass:
 estimating, 134–9
 ratio, 85
Matra Espace, 11
McBride, A., 217, 238
McClure, R.B., 156, 277, 303, 306
McGlynn, D.R., 186, 239
Memory, first-in-first-out (FIFO) random access, 384
MESH, 10
Meyer, H., 306
Microwave receiver, 171–3
Military satellites, 14
Milstein, L., 278
Minkoff, J.B., 306
Mitsubishi, 10
Miya, K., 28, 144, 239, 314, 324, 336, 344, 346, 347
Mobile satellites, 14
Modulation, 21, 182–240
 adaptive delta (ADM), 212–13
 adaptive differential PCM (ADPCM), 212–13, 251–2, 255
 amplitude (AM), 188–93
 frequency (FM), 194–203, 233, 257–60, 275
 index of, 190
 intermodulation, 294–6
  nearly instantaneous companding, 212–13
 pulse-amplitude (PAM), 205–11
 pulse-code (PCM), 205–12, 251–2, 255
 ratio, 296–8
 single-sideband (SSB), 246–8, 255
Molniya, 55–56
Moon, 1, 45–48, 68
Morais, D.H., 217, 239
Morgan, L.W., 28
Moulton, F.R., 48, 56
Mueller, D., 56
Mueller, I.I., 56

mu law, 210
Multipath, 300
Multiple access, 21–22, 241–77
 code-division (CDMA), 22, 242, 270–4
 frequency-division (FDMA), 22, 242–3, 246–60, 255, 257–60, 267
 system capacity, 244
 time-division (TDMA), 22, 242–3, 260–9
Multiplexing, 21, 182–240, 367
 frequency-division (FDM), 21, 193–5, 242, 247–51, 255, 275
 time-division (TDM), 21, 214, 216–17, 242, 251–2, 260–7, 275
Mumford, W.W., 164, 181, 345, 347
Muratani, T., 277
Musson, J.T.B., 306

N

National Aeronautics & Space Administration (NASA), 2
National Aeronautics & Space Development Agency (NASDA), 10, 13
National Bureau of Standards, 239
National Telecommunications & Information Administration (NTIA), 28
Nearly instantaneous companding (NIC), 212
Networks, 23, 25, 163
 balancing, 367–8
 interconnectivity, 245
 noise temperature, 163
 satellite, 26
 telephone, 367–8
Newton's laws, 30, 90–91
Noise, 19, 146–7, 155, 214
 density, 222
 figure, 164, 292
 frequencies, 201
 impulsive, 300
 power ratio (NPR), 288, 342–3
 quantizing, 208–14
 quantum, 146
 temperature, 19, 146–8, 162–6, 222, 292–3
  gain-to-system, 19
 thermal, 222, 227–9, 292, 349
NTSC, 363–4
Nuclear power, 119–20
Nuspl, P.P., 278, 385
Nyquist, H., 146, 239

O

Oblate earth, 48
O'Neal, J.B., 205, 239
Onfry, M., 385
Orbital transfer vehicle (OTV), 99
Orbital velocity, 94
Orbits, 2, 29–81, 84
  altitude, 97
  arc, 357–8
  circular, 43, 84, 95
  eccentricity, 382–3
  equatorial, 80
  geostationary, 37, 39, 51, 57–81, 96, 357, 382
    geometry, 57–81
    operating (GEO), 99
    transfer (GTO), 99
  inclination, 80, 97
  low earth (LEO), 40
  maneuvers, 96
  mechanics, 29–30, 39
  medium-altitude, 2
  Molniya, 55
  operational, 40, 83, 96
  parameters, 35–36, 38
  parking, 40, 83, 95
  perturbations, 45–46, 51, 54
  synchronous, 2, 382
  transfer, 39–45, 83
  utilization, 358
  variation, 382–5
Oscillators, 291
Outages, 68
Output back-off, 156–7
Owings, James L., 377, 386

P

Palapa A, 10
Palapa B, 11
Panter, P.F., 153, 157, 181, 195, 239
Papoulis, A., 239
Pares, J., 144, 181
Pasupathy, S., 217
Path loss, 17
Payload (*See* Spacecraft)
Pelton, J., 28
Perigee kick motors, 128
Perillan, L., 278
Perlman, 169
Perumtel, 10, 11
Phase-shift keying (*See* Keying)

Phiel, J., 278
Pickholtz, R., 270, 278
Pierce, J.R., 181
Platforms (*See* Spacecraft)
Polarization, 333–5
  circular, 333–4
  counterrotating, 333
  cross-, 333–5
  dual, 333
Pontano, B., 278
Porcelli, Giacomo, 101, 108
Powered flight, 90–91
Power systems, 341
Price, R., 343, 347
Primary power (*See* Spacecraft)
Pritchard, W.L., 28, 307
Propellant, 83, 85
  utilization, 100
Propulsion, 82–107, 117 (*See also* Spacecraft)
  burn, 100–101
  chemical, 85–86
  ion engines, 86–87
  specific impulse, 87–89, 93
  thrust, 87–89, 92–93, 99–103
Protection ratio, 358–65
Protocols, 372–80
  advanced data communications control (ADCCP), 375
  block-by-block transmission, 372, 374–5, 378–9
  continuous block transmission with restart-after-error detection, 375, 378–9, 381
  continuous block transmission with selective block repeat, 375–81
  high-level data link control (HDLC), 375–7
Psophometric weighting, 200–201
PTT (France), 11
Puente, J.G., 246, 249, 253, 278
Pulse:
  -amplitude modulation (*See* Modulation)
  -code modulation (*See* Modulation)
  stuffing, 214

Q

Quantizing noise, 207–11

R

Radioisotope thermo-electric generator (RTG), 120
Radio stars, 343–5

Index

Rain:
  attenuation, 173-81
  path, 173-9
  rate, 173-8
RCA Americom, 6, 7, 12
RCA Astro-Electronics, 6, 7, 8, 12, 118
RCA-K_u-band, 7
RCA-Satcom IV, 52
RCA-Satcom V, 52
RCA Satcom VII, 52
Reaction control system (RCS) (See Spacecraft, propulsion)
Receivers, 162-3, 314-20
  antenna, 147, 163
  distance from transmitter, 147
  noise level, 162
Relay I, 2
Reliability, 139, 142-3
Repeaters:
  broadband microwave, 2
  hard-limiting, 281
  multichannel, 279-80
  passive, 1
  quasilinear, 281-4
  real-time, 2
  regenerative, 281-2, 284-5
  satellite, 17
RF link. (See Links, satellite)
Right ascension, 73-75
Rocket engines, 83, 128-9
  bipropellant, 129-31
  gimballed, 100
  liquid, 101, 127, 129
  monopropellant, 129-30
  solid, 89, 105, 127-8
  solid-liquid, 101
  thrust, 87, 99-103
Roden, M.S., 239
Rosenbaum, A.S., 217, 238, 240
Rossiter, P., 385
Rowbotham, T.R., 278
Ruppel, A.E., 370, 385
Ruze, J., 333, 347

S

Saad, T., 347
Sablatash, M., 386
Salz, J., 239
Sandler, G., 139, 144
Satcom, 6
Satellite Business Systems (SBS), 6
Satellite delay compensator, 379-80

Satellites (See also Attitude control, Spacecraft)
  body-stabilized, 113-16, 119
  coordinates, 61
  eclipses, 62
  failure rates, 139-40
  geometry, 76
  geostationary, 52, 79-80, 98
  homogeneous, 356
  intermodulation, 162
  mass-estimating, 134-9
  spinning-drum stabilized, 113-17
  synchronous, 73, 79-81
  system design, 142
  television links, 233-8
  three-axis stabilized, 113
  topocentric coordinates, 73
  tri-spun, 119
Satellite systems:
  domestic, 3
  regional, 3
Satellite Television Company (STC), 12
S-band, 132
SBS, 6
Scheibe, E.H., 164, 181, 345, 347
Schilling, D.L., 195, 205, 211, 220, 240, 278
Schmidt, W.G., 278
Schwarz, Mischa, 365
Scientific Atlanta, 347
Sciulli, J.A., 386
Score, 1
Seifert, H.S., 92, 108
Sekimoto, T., 278
Sendyk, A.M., 238
Setzer, R., 386
Shanmugan, K.S., 240
Shannon, C.E., 149, 240
Shigaki, S., 239
Shimbo, O., 157, 181, 307
Shuttle (STS), 40, 86, 96, 99, 103, 105
  IUS/SSUS-A/SSUS-D, 103, 104
  Orbiter, 99
Signal (See also Modulation; Multiplexing)
  baseband, 188, 242
  -to-noise ratio, 188-91, 198-201, 211, 214-15, 222, 231-4, 251, 259-60, 271, 358, 363
  processing, 182-90
  television, 233-8
  transmission, 187-8
  video, 233-8
Silver, S., 322, 324, 330, 347

Simon, M., 239
Single-channel-per-carrier, 159–61
Sling effect, 93
Smart, W.M., 56, 62, 78–79, 81
Smith, E.K., 28
Solar power, 119–25
Source:
  coding, 21, 182–3, 214
  digital, 204–5, 214
  signals, 184–93
Space attenuation, 331
Spacecraft, 109–44 (*See also* Attitude control; Satellites)
  apogee kick motor (AKM), 137
  attitude control system (ACS), 137
  cable-harness, 137
  communications equipment (payload), 137
    antennas, 111, 136
    transponders, 111, 136
  mass, 112, 121, 134–9
  primary power, 111, 119–25, 137
    mass, 135–6
  propulsion (*See also* Propulsion), 111, 117, 127–31, 137
  structure, 110, 111, 112, 137
  system reliability, 139
  telemetry, tracking, and command, (TT&C), 111, 137
  thermal control, 111, 125–7, 106–7, 137
Spacenet, 7
Spacenet 1, 52
Space Transportation System (STS) (*see* Shuttle)
SPADE, 253
Specific impulse (*see* Propulsion)
Spar Aerospace Corporation, 8, 11
Speech:
  coding, 204–14 (*See also* Modulation)
  signals, 184
Spherical astronomy, 62
Spilker, J.J., 28, 240
Spread spectrum, 23, 270, 274
Sputnik I, 1
Station-keeping, 53
Stein, Seymour, 365
Sterne, Theodore E., 32
Stiltz, H., 133, 144
Store-and-forward satellites, 1
Storey, J.R., 386
Sun, 46–47, 54, 62–64
  declination, 72
  interference, 68–72
  mean, 67

  outage, 68–69, 71
Suyderhoud, H.C., 385
Sweden, 13
Switches:
  C-switch, 283
SYNCOM, 2
System reliability, 139–43

T

Taub, H., 195, 205, 211, 220, 240
T-carrier, 251
TDF/TV-Sat, 13
TDRS 1, 52, 105
Telecom, 11
Telecom 1A, 52
Telemetry, tracking and command (TT&C), 19, 131–4
Telephone:
  bandwidth digital voice, 205
  speech signal, 184–5, 187–8
Telesat Canada, 8
Television:
  signals, 186–7
  standards, 186–7
  transmission, 231–8
Tele-X, 13
Telstar, 6
Telstar 1, 1–2
Telstar 3A, 52
Telstar 3C, 52
Temperatures:
  antenna, 166
  noise, 146–8, 162–6
  sky, 167–8, 170–2
  sun, 168–9
  system, 148, 163, 171–3, 343–6
Terrestrial interface, 340–1
Tescher, A.G., 186, 204, 240
Test equipment, 342–6
Thor Delta, 99
Thor Delta/PAM, 102, 106
Thrusters, 114, 119, 131
Tibbits, R., 56
Time, 30, 72, 73–77
  apparent solar, 67
  clock, 67, 71, 72
  ephemeris, 30
  equation of, 67
  Greenwich, 71
    mean sidereal time (GMST), 73–74
    mean time (GMT), 72
    sidereal time (GST), 73

Index

mean, 67, 71, 73
scales, 30
sidereal, 71, 73
standard, 67–68
universal (UT), 30, 72, 73–74
Titan III:
 /DI-T Centaur, 106
 /D1-T Centaur/AKM, 102, 106
 E-Transtage, 98
Toshiba, 13
Tracking, 337
Transmission (*See also* Protocols)
 analog, 187, 275
 capacity, 257
 digital, 203–31, 242, 270, 275, 284, 383
 efficiency, 375–6, 380–1
 FDM/FM, 242, 251
 impairments, 282, 291–300
 multiple channel per carrier (MCPC), 188, 242, 246–52, 274–6
 satellite, 24–25
 single channel per carrier (SCPC), 252–60, 274–6, 303–4
 technologies, 24–25
 television, 231–8
Transmitters, 310–14
 antenna gain, 147
 distance from receiver, 147
 power, 147
Transponders, 17–19, 136, 241–3, 279–307 (*See also* Spacecraft)
 bandwidth, 267, 274, 276
 dual-conversion, 283–4
 single-conversion, 283
 utilization, 233–5
Traveling-wave tube amplifiers (TWTAs) (*See* Amplifiers)
Traveling-wave tubes (TWTs), 156, 159, 290–1, 315
TRW, 2, 14
Tsuji, Y., 277
TVROs (*See* Earth stations)
Twisted pairs, 26

U

Unisat, 13
United Satellites, 13

United States Department of Defense, 14
United States Satellite Broadcasting Company (USSB), 12
Universal antenna formula, 150–1
Uplinks (*See* Links)
Upper stages, 102–3, 106–7 (*See also* individual launch vehicles)

V

Van Trees, H.L., 28, 220, 240
Vector algebra, 31
Velocity, 44, 47
VHF, 132
Video signals, 186–7
Voice, 184–5, 250, 263
 activation, 253–5
 circuits, 371
 -frequency (VF), 194
 signal coding, 204–12

W

Wait, D.F., 343, 346, 347
Wallace, R.G., 181
Wass, J., 307
Weldon, J., 239
Werth, A.M., 246, 249, 253, 278
Westar, 6
Westar 4, 52
Westar 5, 52
Westcott, J., 157, 158, 181, 296, 307
Western Union Telegraph Company, 6
White, J., 56
Whyte, W.A., 359, 365
Wohlberg, D.B., 238

Y

Yang, W., 42, 56
Yeh, L.P., 386

Z

Zero-degree isotherm, 173, 179